AF609222

Wave–Particle Duality

WAVE–PARTICLE DUALITY

Edited by

FRANCO SELLERI

University of Bari
Bari, Italy

SPRINGER SCIENCE+BUSINESS MEDIA, LLC

Library of Congress Cataloging-in-Publication Data

Wave-particle duality / edited by Franco Selleri.
p. cm.
Includes bibliographical references and index.
ISBN 978-0-306-44163-9 ISBN 978-1-4615-3332-0 (eBook)
DOI 10.1007/978-1-4615-3332-0
1. Wave-particle duality. I. Selleri, Franco.
QC476.W38W4 1992
530.1'2--dc20 92-28284
CIP

ISBN 978-0-306-44163-9

Originally published by Plenum Press, New York in 1992

Contributors

JEAN BASS • Institute de Mathématiques Pures et Appliquées, Université Pierre et Marie Curie, F-75230 Paris Cedex 05, France

SERGE CASER • Laboratoire de Physique Théorique et H.E., Université de Paris Sud, F-91405 Orsay, France

JAMES T. CUSHING • Departments of Physics and Philosophy, University of Notre Dame, Notre Dame, Indiana 46556, USA

REGIS DUTHEIL • Foundation Louis de Broglie, F-75006 Paris, France

PETER E. GORDON • Physics Department, University of Massachusetts at Boston, Boston, Massachusetts 02125, USA

PHILIPPE GUÉRET • Institute de Mathématiques Pures et Appliquées, Université Pierre et Marie Curie, F-75230 Paris Cedex 05, France

FRANZ HASSELBACH • Institut für Angewandte Physik, Universität Tübingen, D-7400 Tübingen, Germany

DIPANKER HOME • Department of Physics, Bose Institute, 93/1, Acharya Prafulla Chandra Road, Calcutta 700 009, India

YUJIRO KOH • Department of Physics, Ibaraki University, Bunkyo 2-1-1, Mito 310, Japan

GEORGES LOCHAK • Foundation Louis de Broglie, F-75006 Paris, France

PETER MITTELSTAEDT • Institut für Theoretische Physik, Universität zu Köln, D-5000 Köln 41, Germany

WOLFGANG MÜCKENHEIM • Landshuter Allee 1, D-8903 Bobingen, Germany

THOMAS E. PHIPPS, JR. • 908 South Busey Avenue, Urbana, Illinois 61801, USA

HELMUT RAUCH • Atominstitut der Oesterreichischen Universitäten, A-1020 Wien, Austria

LUIZ CARLOS RYFF • Universidade Federal do Rio de Janeiro, Instituto de Física, Cidade Universitária, 21945 Rio de Janeiro-RJ, Brazil

MICHAEL SCHMIDT • Dipartimento di Fisica, Università di Bari, I-70126 Bari, Italy

FRANCO SELLERI • Dipartimento di Fisica, Università di Bari, I-70126 Bari, Italy

AKIRA TONOMURA • Advanced Research Laboratory, Hitachi, Ltd., Hatoyama, Saitama 350-03, Japan

Preface

This volume tries to continue a tradition of reviews of the contemporary research on the foundations of modern physics begun by the volume on the Einstein–Podolsky–Rosen paradox that appeared a few years ago.[1] Its publication coincides with the hundredth anniversary of de Broglie's birth (1892), a very welcome superposition, given the lasting influence of the Einstein–de Broglie conception of wave–particle duality. The present book, however, contains papers based on a broad spectrum of basic ideas, some even opposite to those that Einstein and de Broglie would have liked.

The order of the contributions in this book is alphabetical by first author's name. It is important here to stress the presence of three reviews of fundamental experimental data, by Hasselbach (electron interferometry), Rauch (neutron interferometry), and Tonomura (Aharonov–Bohm effect). Hasselbach reviews several interesting experiments performed in Tübingen with the electron biprism interferometer. Wave–particle duality is brought out in striking ways, e.g., in the buildup of an interference pattern out of single events. The Sagnac effect for electrons is also discussed. The chapter by Rauch presents interesting results on wave–particle duality for neutrons. Of particular interest are the differences between stochastic and deterministic absorption in the neutron interferometer, and the concrete evidence for the quantum-mechanical 4π-symmetry of spinors. In the short chapter by Tonomura, conclusive evidence for the reality of the Aharonov–Bohm effect is reviewed, collected in experiments based on advanced technologies of electron holography and microlithography.

The historical and philosophical discussion of the events that resulted in the hegemony of the Copenhagen view shows that the process was in a way rational but certainly not uniquely required by criteria internal to science, as Cushing convincingly argues in his chapter. Guéret deals with the feature of "duality within duality" by considering the spherical light wave emitted by an isotropic source as the "envelope" of wave packet of limited longitudinal and transversal size arising from single atomic emissions. The importance of the special theory of relativity for the birth of wave–particle duality is stressed by Lochak and Dutheil. The paper by Home and Selleri shows that the Aharonov–Bohm effect can be interpreted rather naturally as the action of a photonic empty wave on the electrons, thus lending some support to the idea of local realism. The contribution by Koh discusses the scanty evidence that the self-interference of quantum objects depends only on wavelength but not on other intrinsic features, such as mass.

In the early days of quantum theory, Bohr upheld the point of view that an atomic system can be made to exhibit either particle or wave properties, but never the two simultaneously. Mittelstaedt develops this idea by showing that an atomic system can in some cases be considered approximately as a particle and, simultaneously, approximately as a wave, always in such a way that Heisenberg relations are satisfied. Bass discusses the open question of providing a physically satisfactory probabilistic interpretation of quantum theory and suggests the introduction of partially disjoint probability spaces. The chapter by Phipps deals with "covering theories" of the existing quantum theory and proposes to discard the Einstein–Bohr debate on the nature of the wave–particle duality as a possibly wrong question in order to concentrate on a new theory (the embryo of which is presented) that could in the future lead to more detailed predictions, e.g., in the domains of nuclear and particle physics. Gordon discusses in his chapter the possibility that "half of the duality," the wave, is not directly observable. For him the wave–particle duality can be a general world view, in spite of the impossibility of resurrecting Einstein's reality. Some arguments against the existence of de Broglie waves are discussed by Mückenheim.

Caser's chapter explores the option that nonlocal features of some recent EPR experiments are only apparent and that quantum mechanics may not be completely valid in its predictions for nearly ideal EPR experiments. The idea is suggested that the nonlocal aspects of quantum mechanics may originate in a local multiplicity of vacua which obey a set of superselection rules. A set of experiments on two-photon interference performed in Rochester after 1987 is discussed by Selleri and shown to be consistent with a local realistic approach based on variable detection probabilities. Schmidt reports on a set of numerical calculations of particle trajectories in double- and triple-slit experiments, based on the idea of objectively real waves, and studies the possibility of performing experiments that could lead to the direct detection of quantum waves. Ryff deals with *nonlocal* realism and proposes investigating the idea experimentally with apparatus designed to evidence nonlocal features connected with the wave–particle dual properties of light.

The topics included in this volume are testimony to the richness and significance of the current research on the foundations of quantum physics. I am sure that it will be a useful review and reference book.

Franco Selleri

Bari, Italy

1. *Quantum Mechanics versus Local Realism: The Einstein–Podolsky–Rosen Paradox* (F. SELLERI, ed.), Plenum Press, New York (1988).

Contents

CHAPTER 1

Probability, Pseudoprobability, Mean Values

Jean Bass

1. INTRODUCTION

I shall at first restate the principles of quantum mechanics. We take an abstract Hilbert space H. According to a definite rule, we associate with every physical quantity a linear operator over H, bounded or not bounded. We choose an element ψ of H such that $\|\psi\| = 1$. We call the *mean value* of A in the state ψ the scalar product

$$Q(A) = \langle A\psi,\psi\rangle \tag{1}$$

If A is hermitian, $Q(A)$ is real.

This definition is justified by the following properties.[1] Let U_s be a continuous abelian group of unitary operators:

$$U_s U_{s'} = U_{s'} U_s = U_{s+s'},\ U_{-s} = U_s^{-1},\ U_0 = \text{identity}$$

Then the function

$$\theta(s) = \langle U_s\psi,\psi\rangle \tag{2}$$

is a positive definite function of s. This means that, for any integer $n > 0$, for any real numbers $s_1, \ldots, s_n$ and any complex numbers $c_1, \ldots, c_n$ the hermitian form

Jean Bass • Institut de Mathématiques Pures et Appliquées, Université Pierre et Marie Curie, F-75230 Paris Cedex 05, France.

Wave–Particle Duality, edited by Franco Selleri. Plenum Press, New York, 1992.

$$\sum_{kl} c_k \bar{c}_l \theta(s_k - s_l) \tag{3}$$

is positive. According to Bochner's theorem, there exists a positive bounded measure P such that

$$\theta(s) = \int_{-\infty}^{\infty} \exp(is\omega) dP(\omega)$$

In the case where the hermitian operator A is bounded (or in certain cases where A is not bounded), it is possible to define $\exp(iAs)$ which generates a unitary group. Then

$$\theta(s) = \langle \exp(iAs)\psi, \psi \rangle \tag{4}$$

By this formula, we associate with A a measure P which can be called a probability measure. Therefore, $\theta(s)$ plays the role of the *characteristic function* of some random variable associated with the measure P. Nevertheless, it must be noted that this interpretation of the positive definite function θ is not the only one possible. In any case it was chosen for quantum mechanics.

The expression (1) of the mean value is a consequence of the expression (4) of the characteristic function. It is then justified. Examples of characteristic functions are well known. Let us choose as Hilbert space H the space of complex valued function ψ, square-integrable over $[-\infty, +\infty]$ and such that

$$\int_{-\infty}^{\infty} |\psi(x)|^2 dx = 1$$

The operator A such that $A\psi(x) = x\psi(x)$ is associated with the position of a particle. Its characteristic function is

$$\langle \exp(isx)\psi, \psi \rangle = \int_{-\infty}^{\infty} \exp(isx)\psi\bar{\psi}\, dx \tag{5}$$

Therefore, $|\psi|^2$ plays the role of a probability density for the position.

The operator A such that $A\psi(x) = (h/i)(d\psi/dx)$ is associated with the velocity of the particle. Its characteristic function is

$$\begin{aligned} &\left\langle \exp\left(is\frac{h}{i}\frac{d}{dx}\right)\psi, \psi \right\rangle = \langle \psi(x + hs), \psi(x) \rangle = \\ &\int_{-\infty}^{\infty} \psi(x + hs)\bar{\psi}(x)\, dx = \int_{-\infty}^{\infty} \psi\left(x + \frac{hs}{2}\right)\bar{\psi}\left(x - \frac{hs}{2}\right) dx \end{aligned} \tag{6}$$

Here it is convenient to introduce the Fourier transform of ψ. If

$$\psi(x) = \int_{-\infty}^{\infty} \exp(ixz)\psi(z)\, dz \tag{7}$$

we can easily verify that the operator $(h/i)(d/dx)$ corresponds to the probability density $(2\pi/h)|\psi(x/h)|^2$.

More generally, the characteristic function associated with a hermitian operator A can be constructed as follows:
If ψ is a given function of x, we introduce the function

$$\chi(x,s) = \exp(isA)\psi(x) \tag{8}$$

It satisfies the following equation:

$$\frac{1}{i}\frac{\partial\chi}{\partial s} = A_\chi \tag{9}$$

χ is a solution of this equation such that $\chi(x,0) = \psi(x)$, and we have

$$\theta(s) = \int_{-\infty}^{\infty} \chi(x,s)\bar{\psi}(s)\,ds \tag{10}$$

Let us suppose in particular that A is the hamiltonian operator H. Equation (9) becomes identical to the Schrödinger equation with $s = ht$. If we decide that, in formula (8), ψ is a solution of the Schrödinger equation, we must introduce explicitly the fact that ψ is a function of time t. χ is a solution of the same equation, such that $\chi(x,0) = \psi(x,t)$. Therefore, the expression of χ is necessarily

$$\chi(x,s) = \psi(x,\tau + hs) \tag{11}$$

and we have

$$\varphi(\lambda) = \int_{-\infty}^{\infty} \bar{\psi}(x,\tau)\psi(x,\tau + hs)\,dx \tag{12}$$

an expression which does not involve the general structure of the operator $\exp(ish)$ independently of the special form of ψ.

2. PAIRS OF OPERATORS

Let A and B be two hermitian operators.[8,9] If $AB = BA$, the product of A and B is well defined and is a hermitian operator. If λ and μ are two real numbers, we can introduce the hermitian operator $\lambda A + \mu B$. The unitary operator $\exp(i(\lambda A + \mu B))$ is defined, and can be written $\exp(i\lambda A)\exp(i\lambda B)$ as well as $\exp(i\lambda B)\exp(i\lambda A)$. The mean value

$$\langle \exp(i(\lambda A + \mu B)\psi,\psi\rangle \tag{13}$$

is the characteristic function of the pair of operators A,B, in the sense of probability theory.

If A and B do not commute, let us write

$$AB - BA = iC \tag{14}$$

C is a hermitian operator. The most important case is $C = hI$, where I is the identity operator and h is a positive number. Then

$$AB - BA = \mathrm{ih\,I} \tag{15}$$

It is verified by $A = x$, $b = (h/i)(d/dx)$. $C = 0$ corresponds to commutative operators.

The study of quadratic means associated with A and B leads to remarkable consequences. The variance of A, B are the mean values

$$Q(A^2) = \langle A^2\psi,\psi\rangle = \|A\psi\|^2,\ Q(B^2) = \langle B^2\psi,\psi\rangle = \|B\psi\|^2 \tag{16}$$

If A and B commute, the covariance of A,B is

$$Q(AB) = \langle AB\psi,\psi\rangle = \langle A\psi,B\psi\rangle \tag{17}$$

Let r be a real or complex parameter. $rA + B$ is or is not hermitian, but it generates a positive norm:

$$\begin{aligned} &\langle (rA + B)\psi,\ (rA + B)\psi\rangle = \\ &r\bar{r}\|A\psi\|^2 + r\langle A\psi,B\psi\rangle + \bar{r}\langle B\psi,A\psi\rangle + \|B\psi\|^2 \geqslant 0 \end{aligned} \tag{18}$$

If $AB = BA$, we choose for r a real number. As a consequence of (18), we have

$$|\langle A\psi,B\psi\rangle| \leqslant \|A\psi\|\cdot\|B\psi\| \tag{19}$$

Or

$$|Q(AB)| \leqslant [Q(A^2)\cdot Q(B^2)]^{1/2}$$

This is the classical Schwarz's inequality.

If $AB - BA = iC$, we choose for r a purely imaginary number $r'i$. Then

$$r'^2\|A\psi\|^2 + r'\langle C\psi,\psi\rangle + \|B\psi\|^2 \geqslant 0$$

Therefore,

$$|\langle C\psi,\psi\rangle| \leqslant 2\|A\psi\|\cdot\|B\psi\| \tag{20}$$

or

$$|Q(C)| \leqslant 2[Q(A^2)Q(B^2)]^{1/2}$$

If $C = hI$,

$$[Q(A^2)Q(B^2)]^{1/2} \geqslant h/2$$

It is the celebrated Heisenberg's uncertainty relation. For position and velocity, this inequality corresponds to the fact that the probability densities of x and $(h/i)\,(d/dx)$ are $|\psi|^2$ and $(2\pi/h)|\hat{\psi}(x/h)^2$, where $\hat{\psi}$ is the Fourier transform of ψ. If for instance $\varphi(x) = (2\pi)^{-1/4}\sigma^{-1/2}\exp(-x^2/4\sigma^2)$, one has

$$|\psi(x)|^2 = \frac{1}{\sigma\sqrt{2\pi}}\exp\left(-\frac{x^2}{2\sigma^2}\right),\ |\hat{\psi}(x)|^2 = \frac{\sigma}{\pi\sqrt{2\pi}}\exp(-2x^2\sigma^2) \tag{21}$$

If we replace A by $A - Q(A)$, and B by $B - Q(B)$, we see that the product of the standard deviations of the probability law associated with A and B is greater than a positive number $h/2$. If the first one tends to zero, the second one tends to infinity. Moreover, if the probability density of A tends to the δ distribution of Dirac, the probability density of B tends to zero at every point. It stretches and flattens out along the axis in such a way that the integral from $-\infty$ to $+\infty$ remains equal to one. For instance, in the above example of a gaussian ψ, σ tends to zero and it is clear that $|\psi(x)|^2$ tends to zero at any point x. In this sense, B becomes entirely indeterminate.

3. WIGNER'S PSEUDODENSITY

The preceding discussion shows clearly that it is not possible to associate to the pair $A = x$, $B = (h/i)(d/dx)$ a joint probability law. Nevertheless, as $\lambda A + \mu B$ is a well-defined hermitian operator, it is possible to compute the mean value of $\exp(i(\lambda A + \mu B))$, namely

$$\theta(\lambda,\mu) = \int_{-\infty}^{+\infty}\left[\exp\left(i\lambda x + \mu h\frac{\partial}{\partial x}\right)\psi(x)\right]\bar{\psi}(x)\,dx \tag{22}$$

We have to compute the function

$$\psi(x) = \exp\left(i\lambda x + \mu h\frac{d}{dx}\right)\psi(x) \tag{23}$$

which derives from the function $\psi(x,s)$ such that

$$\psi(x,s) = \exp\left[s\left(i\lambda x + \mu h\frac{d}{dx}\right)\right]\psi(x) \tag{24}$$

$\psi(x,s)$ reduces to $\psi(x)$ for $s = 0$, and we are interested in $\psi(x,1)$. For this purpose, we introduce the infinitesimal operator of the family $\exp\left[s(i\lambda x + \mu h(d/dx))\right]$, which does not constitute a group: We find that ψ satisfies the differential equation

$$\frac{\partial\psi}{\partial t} = i\lambda x\psi + \mu h\frac{\partial\psi}{\partial x} \tag{25}$$

Its standard resolution gives

$$\psi(x,s) = \exp\left[i\lambda\left(xs + \frac{\mu h}{2}s^2\right)\right]\psi(x + \mu hs) \tag{26}$$

Therefore,

$$\theta(\lambda,\mu) = \int_{-\infty}^{\infty} \exp\left[i\lambda\left(x + \frac{\mu h}{2}\right)\right]\psi(x + \mu h)\bar{\psi}(x)\,dx$$

which is better written as

$$\theta(\lambda,\mu) = \int_{-\infty}^{\infty} \exp(i\lambda x)\psi\left(x + \frac{\mu h}{2}\right)\bar{\psi}\left(x - \frac{\mu h}{2}\right)dx \tag{27}$$

$\theta(\lambda,0)$ is the exact characteristic function of position, and $\theta(0,\mu)$ the exact characteristic function of velocity. But $\theta(\lambda,\mu)$ is not a characteristic function with respect to λ and μ. Indeed, $\theta(\lambda,\mu)$ is the Fourier transform of

$$\xi(x,y) = \frac{1}{2\pi}\int_{-\infty}^{\infty} \exp(-i\mu y)\psi\left(x + \frac{\mu h}{2}\right)\bar{\psi}\left(x - \frac{\mu h}{2}\right)d\mu \tag{28}$$

and this function may have negative values.
If for instance

$$\begin{aligned}\psi(x) &= \frac{1}{\sqrt{2}} \text{ for } |x| < 1\\ &= 0 \quad \text{for } |x| > 1\end{aligned} \tag{29}$$

we have, for $x = 0$:

$$\xi(0,y) = \frac{1}{4\pi}\int_{-2/h}^{2/h} \exp(-i\mu y)\,d\mu = \frac{1}{2\pi}\frac{\sin 2/hy}{y} \tag{30}$$

This function is alternatively positive and negative. As an exception, it may be proved[8] that if ψ is gaussian, namely if $\psi = \exp(-ax^2 + bx + c)$, a real >0, then ξ is a genuine probability density. For instance, if $\psi = \exp(-ax^2)$,

$$\xi(x,y) = \frac{1}{h\pi}\exp\left(-2ax^2 - \frac{y^2}{2ah^2}\right) \tag{31}$$

It is the probability density of a normal law, with standard deviation $1/2\sqrt{a}$ for position, $h\sqrt{a}$ for velocity. The product of these two standard deviations is equal to (and not greater than) $h/2$, according to Heisenberg's inequality.[6]

More generally, any expression as $\exp i(\alpha A + \beta B)\cdot\exp i(\alpha' A + \beta' B)$, with $\alpha + \alpha' = \lambda$, $\beta + \beta' = \mu$, gives an alternative expression of the characteristic function of A,B. For position and velocity, the result is $\theta(\lambda,\mu)\cdot\exp(ih/2)(\alpha'\beta - \alpha\beta')$.

4. PROBABILITY SPACES

It is usual to say that every hermitian operator generates a random variable. But the results of the preceding section show that the set of all these random variables does not constitute a probability space. Only a set of commutative operators is able to generate a probability space. In such a space, the characteristic function of a finite family $\{A_k\}$ of hermitian operators is given by

$$\langle \exp(i\textstyle\sum \lambda_k A_k)\psi,\psi\rangle \tag{32}$$

Let E be the set of all bounded operators over the Hilbert space H. It contains the unitary operators. The space E is the union of a family of vector subspaces E_j such that, if A and B belong to E_j, then $AB = BA$; in particular, if A and B are hermitian operators, AB is a hermitian operator belonging to E_j. If A and B commute, it will be convenient to write $A \smile B$. Then the situation is as shown in Figure 1. If A and B belong to the same space E_1,

$$A \smile B$$

If A and C belong to the same space E_2,

$$A \smile C$$

But B and C do not belong to the same subspace of E. Then we have not $B \smile C$. Example:

$$A = x_1,\ B = x_2,\ C = \frac{h}{i}\frac{\partial}{\partial x_2} \tag{33}$$

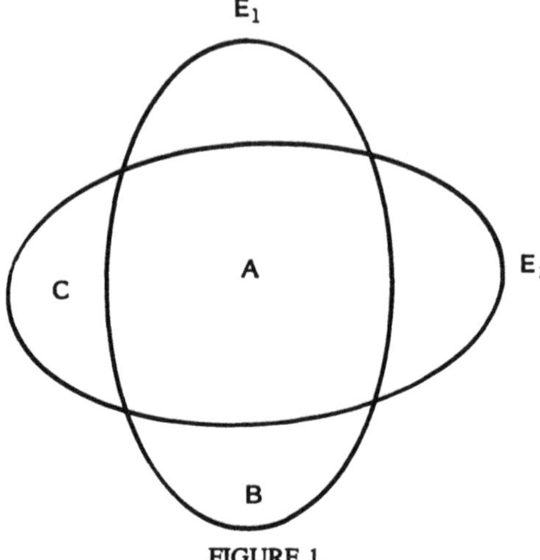

FIGURE 1.

(In this example, the operators A, B, C are not bounded. But the associated unitary groups are bounded, and have the same property.) This discussion is summarized as follows. Quantum mechanics is represented by a vector space E in which is defined a binary relation $A \smile B$, not transitive, for $C \smile A$ and $A \smile B$ does not imply $C \smile B$. The space E is the union of vector spaces E_j, the elements of which are pairwise "comparable," i.e., satisfying the above relation. These subspaces E_j possess a structure of an algebra of bounded operators.

5. QUANTUM MECHANICS AND FLUID DYNAMICS

In the preceding sections, time was given and fixed. If we take in account the evolution in time, we must introduce new concepts. The state ψ describes a trajectory in Hilbert space. This trajectory is a solution of a Schrödinger equation, namely

$$\frac{1}{i}\frac{\partial\psi}{\partial t} - \frac{h}{2}\nabla^2\psi = \frac{V}{h}\psi \tag{34}$$

(We assume that the mass is one.) The potential V is independent of time.

In order to discuss and to solve this equation, several processes can be used. In the first one we write

$$\psi = a\exp\left(\frac{i}{h}\varphi\right), \ \rho = a^2 = \psi\bar{\psi}, \ \varphi = \frac{h}{2i}\log\frac{\psi}{\bar{\psi}} \tag{35}$$

It is well known[7] that (34) gives rise to two real equations, namely:

$$\frac{\partial \rho}{\partial t} + \sum_k \frac{\partial}{\partial x_k}\left(\rho \frac{\partial \varphi}{\partial x_k}\right) = 0 \tag{36}$$

$$\frac{\partial \varphi}{\partial t} + \tfrac{1}{2}\sum_k \left(\frac{\partial \varphi}{\partial x_k}\right) = \frac{h^2}{2}\frac{\Delta a}{a} + V \tag{37}$$

(36) is a conservative equation of the mass, for a flow having density ρ and velocity grad φ. (37) is generally interpreted as a Jacobi equation for a hamiltonian system, in which the potential V is completed by a "quantum potential" $(h^2/2)(\Delta a/a)$. Equation (37) describes the motion of a point under the action of a potential V. The trajectories compatible with this potential are mutually independent, and specified by their initial position. All that is by no means probabilistic. We still interpret $\rho = a^2$ as the probability density of a "particle" (i.e., a moving point) which would have in space a random position belonging to any trajectory.

But (36) is not sufficient to determine that $u_j = \partial\varphi/\partial x_j$ is the velocity of a hydrodynamic flow. In fluid dynamics, the "various trajectories" coexist. They are followed by "fluid particles," not well defined, which have a physical interaction, characterized by internal stresses. These stresses are represented by a symmetrical tensor T_{jk} such that, in accordance with (37),

$$\frac{1}{\rho}\sum_k \frac{\partial T_{jk}}{\partial x_k} = \frac{\partial}{\partial x_j}\left(\frac{h^2}{2}\frac{\nabla_a^2}{a}\right) \tag{38}$$

A solution of (38) is

$$T_{jk} = \frac{h^2}{2}a^2\frac{\partial^2 \log a}{\partial x_j \partial x_k} = \frac{h^2}{4}\left(\frac{\partial^2 \rho}{\partial x_j \partial x_k} - \frac{1}{\rho}\frac{\partial \rho}{\partial x_j}\frac{\partial \rho}{\partial x_k}\right), \ \rho = a^2 \tag{39}$$

(For verification, it is helpful to put $\psi = \exp\chi$.) Other solutions are obtained by adding to (39) the general solution of the homogeneous system

$$\sum_k \frac{\partial T_{jk}}{\partial x_k} = 0$$

namely

$$T_{jk} = \nabla^2 U \cdot \delta_{jk} - \frac{\partial^2 U}{\partial x_j \partial x_k} \tag{40}$$

where U is an arbitrary function.

If we interpret $(h^2/2)(\nabla_a^2/a)$, the "quantum potential," as generating stresses, (37) becomes a genuine equation of fluid dynamics (Bernoulli's equation). The velocity of this fluid derives from a potential, and is given by

$$\rho u_j = \frac{h}{2i}\left(\overline{\psi}\frac{\partial \psi}{\partial x_j} - \psi\frac{\partial \overline{\psi}}{\partial x_j}\right) \tag{41}$$

The question may be asked whether such a flow has a physical existence, and not only a mathematical structure. If so, we are more or less led to abandon the probabilistic interpretation, which we have noticed is incomplete. The function $\rho = |\psi|^2$ becomes a material density, and no longer a probability density. There is no inconvenience to interpret the physical particle (and not the fluid particle) as a fluid possessing an extension in space. Some of its properties are correctly described by the motion of its center of mass, but other more subtle properties take into account the space extension of the fluid. Do the properties of this fluid have an oscillating nature, suggesting the association of a wave notion and a particle notion? It is not obvious, but it seems to be related to the structure of the fluid trajectories. These trajectories are the solutions of the differential equation

$$\frac{dx_j}{dt} = \frac{\partial \psi}{\partial x_j} \tag{42}$$

To get a rough understanding of their nature, we must use a second classical method of integrating the Schrödinger equation. This equation has "stationary solutions" of the form

$$\psi(x,t) = a(x)\exp(i\omega t),\ a = |\psi| \tag{43}$$

The complex function a is a solution of the differential equation

$$\nabla^2 a + 2\left(V - \frac{\omega}{h}\right)a = 0 \tag{44}$$

As $\|\psi\| = 1$, a must satisfy the condition $\int_D a^2(x)\,dx = 1$. Then (44) has nonzero solutions only for the set of eigenvalues of ω. In many cases, this set is countable, and we have for the solution of the Schrödinger equation

$$\psi = \sum_k a_k(x)\exp(i\omega_k t) \tag{45}$$

The a_k are the normed eigenfunctions, multiplied by coefficients such that the series $\sum |a_k(x)|^2$ is convergent.

As a function of t, ψ is a Fourier series, generally nonperiodic, which represents an almost-periodic function. This sort of function is somewhat oscillatory, or "waving." We see that the form of solutions of the Schrödinger equation suggests at once the idea of a wave.

We must now find the solutions of the differential equation

$$\frac{dx_j}{dt} = \frac{1}{2i}\left(\bar{\psi}\frac{\partial\psi}{\partial x_j} - \psi\frac{\partial\bar{\psi}}{\partial x_j}\right) \tag{46}$$

The right-hand member of (46) results from a succession of operations which preserve the structure of almost-periodicity, namely

1. Differentiation of ψ and $\bar{\psi}$ with respect to space
2. Product of two almost-periodic functions
3. Ratio of two almost-periodic functions, except in the case where $\psi\,\bar{\psi}$ vanishes in some points of the domain (density zero at some points)

Therefore, (46) takes the form

$$\frac{dx_j}{dt} = \sum_k b_{jk}(x)\exp(i\alpha_k t) \tag{47}$$

It seems that at present nothing is known about the nature of the solutions of (47). An elementary example where exact integration is possible will show what may happen. Let us consider the equation

$$\frac{dx}{dt} = b(x)\left(\sum_k b_k \exp(i\alpha_k t)\right) \tag{48}$$

with only one scalar variable x. Let us suppose that the function $\int dx/b(x)$ has an inverse function $c(z)$. The solution of (48) takes the form

$$x = c\left(\sum_k \frac{b_k}{i\alpha_k}\exp(i\alpha_k t)\right) \tag{49}$$

If $c(z)$ can be expanded in a power series, we have

$$c(z) = c_0 + c_1 z + c_2 z^2 + \cdots$$

and

$$x = c_0 + c_1\left(\sum \frac{b_k}{i\alpha_k}\exp(i\alpha_k t)\right) - c_2\left(\sum \frac{b_k b_l}{\alpha_k\alpha_l}\exp(i(\alpha_k + \alpha_l)t)\right) + \cdots \tag{50}$$

The spectrum is deduced from the spectrum of the velocity, but is more complicated. It contains linear combinations with integral coefficients, which may be dense over the frequency axis. Now the spectrum $\{\alpha_k\}$ itself is very complicated. It is reasonable to think that the solutions, or some solutions, of (48) are oscillating functions supplied with a spectrum. But we do not know if this spectrum is a ray spectrum or is continuous. In any case it is convenient to investigate the spaces in which we can find these sorts of functions, which we shall call *stationary functions*, and to confront these spaces with the spaces of operators.

6. MARCINKIEWICZ SPACE AND STATIONARY FUNCTIONS

We must now define the stationary functions and investigate their fundamental properties. We shall put aside probability theory and mathematical expectations. We introduce another definition of mean value, suggested by experiment. We are concerned with complex or real valued functions f defined over $[-\infty,+\infty]$. We call an average operator an operator M which maps f into

$$Mf = \lim_{T\to\infty} \frac{1}{2T} \int_{-T}^{T} f(t)\,dt \tag{51}$$

Mf is called the (*temporal*) mean value of f. The set of functions for which Mf exists is a vector space. But it is too large for applications. We introduce now the quadratic mean of f, defined by

$$M|f|^2 = \lim_{T\to\infty} \frac{1}{2T} \int_{-T}^{T} |f(t)|^2\,dt \tag{52}$$

It is easy to see that the set of functions with a quadratic mean is not a vector space. For instance, $f = \exp(i \log|t|)$ has a quadratic mean, $g = 1$ has a quadratic mean, but, as f has no mean value, $f + g = 1 + \exp(i\log|t|)$ has no quadratic mean

It is necessary for functions having a quadratic mean to belong to a vector space. For this purpose, we define a generalized quadratic mean value by

$$\tilde{M}|f|^2 = \lim_{T\to\infty} \sup \frac{1}{2T} \int_{-T}^{T} |f(t)|^2\,dt \tag{53}$$

(Remember that $\lim\sup_{T\to\infty} X_T$ signifies $\lim_{T\to\infty} [\sup_{T'>T} X_{T'}]$.)
The expression

$$\sup_{T>T'} \frac{1}{2T'} \int_{-T'}^{T'} |f(t)|^2\,dt$$

is obviously a positive and decreasing function of T. When $T \to \infty$, it has a finite or infinite limit. It is easy to prove that the set of functions f for which this limit is finite constitutes a vector space. It is a normed space, by $\|f\|^2 = \tilde{M}|f|^2$. It is even a complete space (Banach space). It was introduced by Besicovitch in a particular case, and is called Marcinkiewicz space $\mathfrak{M}^2$. It contains functions having an exact quadratic mean. But likewise it contains functions such as $1 + \exp(i \log|t|)$ without a quadratic mean. After the quadratic mean, we must study the existence of the average of a product. It is clear that the existence of $M|f|^2$, $M|g|^2$ does not imply the existence of $M(f\bar{g})$. An example of this situation is given by $f(t) = (i \log t)$, $g(t) = 1$. Here is a difference between temporal mean values and mathematical expectations. In probability theory, when the random variables X and Y, defined over a well-specified probability space, have a quadratic mean, then XY has a mathematical expectation. If $M(f\bar{g})$ exists, we say that f and g are *comparable* and we write $f \smile g$. The existence of Mf can be written as $f \smile 1$.

Let E' be a set of functions pairwise comparable: $f \smile g$ and in particular $f \smile f$, i.e., $M|f|^2$ exists. It is clear that E' is a vector subspace of Marcinkiewicz space $\mathfrak{M}^2$. It is even a Hilbert space, with the scalar product $Mf\tilde{g}$. The space $\mathfrak{M}^2$ is the union of vector spaces E' of comparable functions which have the same structure as the spaces of operators discussed in Section 4. Figure 1 is valid when the operators are replaced by functions, comparable functions corresponding to commutative operators, and real functions to hermitian operators. Figure 2 gives an example of this situation, when f is a function such that $\lim_{T \to \pm\infty} f(t) = 0$. We see that $Mf = 0$, $M(f \exp i \log|t|) = 0$ but $M \exp i \log|t|$ does not exist.

To a real function f it is possible to associate a characteristic function. When

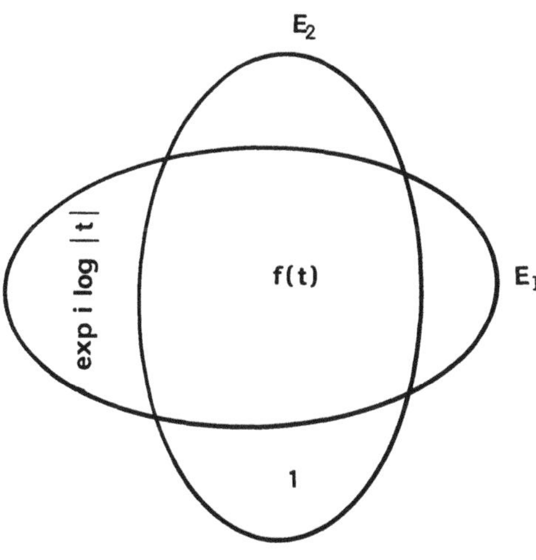

FIGURE 2.

$$M \exp(i\lambda f(t)) \tag{54}$$

exists, this is a positive definite function, the Fourier transform of a measure which has the properties of a probability measure. For instance

$$\text{for } f = 1,\ M \exp(i\lambda) = e^{i\lambda} \qquad (\text{density } \delta(x - 1))$$

$$\text{for } f = \sin t,\ M \exp(i\lambda \sin t) = J_0(\lambda) \quad \left(\text{density } \begin{cases} \dfrac{1}{\pi}\dfrac{1}{\sqrt{1 - x^2}} & \text{for } |x| < 1 \\ 0 & \text{for } |x| > 1 \end{cases}\right)$$

The next step consists of constructing the characteristic function of a pair of real functions f,g. It is the mean value of $\exp[i(\lambda f + \mu g)]$. If $f \smile g$, this mean value exists and is a positive definite function of λ and μ. But if f and g are not comparable, $M[i(\lambda f + \mu g)]$ does not exist.

There is a strong analogy between operators and functions. But there are some differences. For operators, $\langle \exp[i(\lambda A + \mu B)\psi,\psi\rangle$ exists, but is or is not a characteristic function. For functions, $M[\exp i(\lambda f + \mu g)]$ exists or does not. In fact, if it does not exist, it seems possible to give an extended definition of mean value such that various mean values exist, but are not necessarily characteristic functions. In both cases, the notion of characteristic function has no more significance.

Among the functions belonging to $\mathfrak{M}^2$, very important are those for which $f(t) \smile f(t + \tau)$ for any τ. The mean value

$$\gamma(\tau) = M\bar{f}(\tau)f(t + \tau) \tag{55}$$

is called the *correlation* (or autocorrelation) function of f. It is easy to prove that γ is a positive definite function of τ. If $\gamma(\tau)$ is continuous, it is the Fourier transform of a positive bounded measure, called *spectral measure*. This measure is continuous or discontinuous.

The first case corresponds to *almost-periodic functions*, of the form

$$f(t) = \sum c_k \exp(i\omega_k t) \tag{56}$$

The correlation function of f is

$$\gamma(\tau) = \sum |c_k|^2 \exp(i\omega_k \tau) \tag{57}$$

By a convenient extension of their elementary definition, they constitute a complete vector subspace of $\mathfrak{M}^2$.[2–5] In this case, f has a discontinuous (ray) spectrum.

If the spectral measure is absolutely continuous, there exists a spectral density φ such that

$$\gamma(\tau) = \int_{-\infty}^{\infty} \exp(i\omega\tau)\varphi(\omega)\, d\omega \tag{58}$$

Then the function f is called a *pseudorandom function.*(2–4) It has a continuous spectrum. The correlation function is continuous, and $\lim_{\tau\to\infty} \gamma(\tau) = 0$.

Pseudorandom functions have no elementary representation. They are Fourier transforms of vector-spectral measures, and such representations are not concrete. But it is easy to build models of special pseudorandom functions. Here are examples of pseudorandom functions.

We say that a sequence x_n of real numbers such that $0 < x_n < 1$ is *uniformly distributed* if the following condition is satisfied: Let (a,b) be an arbitrary interval included in (0,1). Let N' be the number of terms $x_1, x_2, \ldots, x_N$ belonging to (a,b). We suppose that the ratio N'/N tends to a limit, which is equal to $b - a$, when $N \to \infty$.

In the same way, we define a sequence (x_n, y_n) uniformly distributed in the square $(0,1) \times (0,1)$. We say that the sequence x_n is 2-uniformly distributed if the sequence (x_n, x_{n+1}) is uniformly distributed in the square.

We give a function F, defined and Riemann-integrable over (0,1). We consider the function f such that

$$f(t) = 0 \text{ if } t < 0,\, f(t) = F(x_n) \text{ if } n < t < n + 1 \tag{59}$$

It is possible to prove that this function is pseudorandom (see Refs. 2–4). Its correlation function (defined by $\lim (1/T) \int_0^T$) is equal to

$$(1 - |\tau|)\ (\tau), \text{ where } \ (\tau) = 1 \text{ if } |\tau| \leqslant 1, \ \ (\tau) = 0 \text{ if } |\tau| \geqslant 1$$

Of course, the convolution of this function by an integrable kernel transforms these discontinuous functions into continuous pseudorandom functions.

Concrete examples of 2-uniformly distributed sequences are known. Let

$$P(t) = a_0 t^{\nu} + a_1 t^{\nu-1} + \cdots + a_{\nu}$$

be a real polynomial of degree $\nu \geqslant 2$, such that a_0 is an irrational number (for instance, $P(t) = \sqrt{2}t^2, P(t) = \pi t^2$). Then the sequence $P(n)$ modulo 1 (the decimal part of $P(n)$) is 2-uniformly distributed.

The characteristic function of $f(t)$ is defined by

$$\theta(s) = \lim_{T\to\infty} \frac{1}{T} \int_0^T \exp(isf(t)\, dt = \lim_{N\to\infty} \frac{1}{N} \sum_{n=0}^{N} \exp(isF(x_n)) \tag{61}$$

According to a fundamental theorem of H. Weyl, this limit is equal to

$$\int_0^1 \exp(isF(x)\,dx \tag{62}$$

For this special class of pseudorandom functions, we find a new correspondence between probability theory and temporal mean values. Indeed, (62) is the characteristic function of a random variable $F(x)$, defined over the probability space (0,1) with Lebesgue measure.

The main fact of this section is that almost-periodic functions, as well as pseudorandom functions, are essentially oscillating, without any limit at infinity. We thus are led to conclusions which are no more than hypotheses, and would deserve mathematical complements and physical interpretation.

7. CONCLUSION

The path we have followed begins with a discussion of the standard probabilistic interpretation of quantum mechanics. In a given state ψ, a correspondence is established between operators in Hilbert space H and random variables, through an adequate definition of mean value. Characteristic functions, defined as mean values, enable us to define probability laws, by squaring elements of H. But all these elements are not fully consistent with a customary structure of probability. They suggest the introduction of a family of probability spaces partially disjoint.

The study of the evolution of ψ illustrates the role played by oscillating functions like almost-periodic functions (ray spectrum) and pseudorandom functions (continuous spectrum). These functions belong to a function space (Marcinkiewicz space $\mathfrak{M}^2$ in which mean values are defined neither as mathematical expectations as in probability theory nor by scalar products $\langle A\psi,\psi\rangle$, but as temporal mean values. The structure of Marcinkiewicz space is very similar to the structure of the union of probability spaces used in quantum mechanics. The evolution of ψ suggests the association of a physical particle with a hydrodynamic flow possessing internal stresses. The streamlines of the flow are very likely described by functions belonging to the space $\mathfrak{M}^2$. The question is asked whether this fluid is able to give a physical representation of the particle. It is worth noting that it has the combined properties of a pure particle, an extension in space, and local irregularity, suggesting an oscillatory structure and the word "wave."

REFERENCES

1. E. ARNOUS, *Lois de probabilité en mécanique ondulatoire*, thèse, Paris (1946).
2. J. BASS, *Cours de mathématiques*, Vol. III, Masson, Paris (1971).
3. J. BASS, *J. Math. Anal. Appl.* **47**(2,p. 354; 3, p. 458), 1974.

4. J. BASS, *Fonctions de corrélation, fonctions pseudo-aléatoires et applications*, Masson, Paris (1984).
5. J. P. BERTRANDIAS, *Espaces de fonctions bornées et continues en moyenne asymptotique d'ordre p*, thèse, Paris (1964).
6. L. DE BROGLIE, *Les incertitudes de Heisenberg et l'interprétation probabiliste de la mécanique quantique*, Gauthier–Villars, Paris (1982).
7. E. MADELUNG, Z. Phys. **40**, 322 (1926).
8. C. PIQUET, *C. R. Acad. Sci. Ser. A.* **279**, 107 (1974).
9. K. URBANIK, *Joint Probability Distributions of Observables in Quantum Mechanics*, *Stud. Math.* **21** (1967).

CHAPTER 2

Local Vacua

Serge Caser

1. FROM BELL'S THEOREM TO WHERE?

Ever since Bell's article of 1964,[1] which greatly clarified the debate initiated 30 years earlier by Einstein and co-workers (EPR),[2] nonlocality has been considered—by some physicists at least—as a critical issue in the interpretation of quantum mechanics (QM). On the other hand, the so-called local realistic theories (LRTs) which were proposed as an alternative to QM generally appear as ad hoc constructions with a limited scope, although they certainly are very useful in showing that experiments most often do not prove what they were set up for. The strength of QM, of course, lies in its simplicity and predictive power. But most experiments with some relevance to nonlocality are very difficult to perform, and the only one which appears to be conclusive[3] used photons—particles with very peculiar properties, indeed. As to the measurement process itself, on which the quantum theory is based, it justifiably appears to many as an independent prescription (since all efforts to derive it from the theory actually presuppose what they intend to prove).

In the face of such uncertainty, the physicist seemingly has a choice between three options:

1. QM is exact in all its predictions, and nonlocality, unintuitive as it may be, is probably here to stay (and you better get used to it).
2. Nonlocality is only apparent (and QM may not be true in all its predictions).
3. Why all the fuss about "nonlocality"? It cannot be used to send faster-than-light signals, and so it is irrelevant to physics (and maybe to philosophy as well).

Serge Caser • Laboratoire de Physique Théorique et H.E., Université de Paris Sud, F-91405 Orsay, France.

Wave–Particle Duality, edited by Franco Selleri. Plenum Press, New York, 1992.

Note that options 1 and 3 can easily be reconciled (and often are in practice!), so the real choice is between 1 + 3 and 2.

The aim of this chapter is to explore option 2. One word about our philosophy (or prejudices): when we say that QM may not be true in all its predictions, we expect it to correctly predict spin correlations whenever these correlations are not in conflict with locality; otherwise, we expect the discrepancy between QM and experiment to be large. Why that? Because spin (or isospin) measurements obey beautifully simple rules, and we believe that nature is not vicious enough to *approximate* those rules, with no obvious physical reason. (Maybe Einstein would agree on this point.) Needless to say, when these simple rules are not in conflict with locality, it will remain to show, in the spirit of LRTs, that they are necessary.

Spin correlations and Bell's theorem are the subject of Section 2, where a number of elementary results are given. These will allow us to introduce the concept of local vacua, which will be used in Section 3 in studying a "which way" experiment. Section 4 will be devoted to some heuristic considerations about local vacua, and Section 5 will offer some new insights into Dirac's covariant vacuum—a classical electromagnetic "ether," and a possible candidate for a local vacuum. Conclusions are given in Section 6.

2. SOME BASIC FACTS ABOUT NONLOCALITY

There is considerable confusion in the literature about nonlocality, and many "proofs" of the local character of QM are simply due to their author's personal definition (or lack of definition) of this concept. We know of only one clear definition of this term in correlation experiments, and we will stick to it, namely: Bell's definition (see below). Also, most results in this field can be obtained in the simple model of two spin-$\frac{1}{2}$ particles with total spin zero (Bohm and Aharonov's version of the EPR experiment).[4] This model will now be examined.

Consider the setup of Figure 1, where a spin-zero particle decays into two spin-$\frac{1}{2}$ particles. The two fermion spins are measured along directions $\vec{a}$ and $\vec{b}$ (by Stern–Gerlach instruments, say). If $P_{QM}^{\sigma\sigma'}(\theta)$ is the quantum mechanical joint probability for the detection of particle 1 with a spin component σ along $\vec{a}$ and of particle 2 with a spin component σ' along $\vec{b}$ (where θ is the angle between $\vec{a}$ and $\vec{b}$), Bell's theorem[1] states that it is impossible to write it as

$$P^{\sigma\sigma'}(\theta) = \int_{\Lambda} d\lambda \rho(\lambda) p_1^{\sigma}(\lambda, \vec{a}) p_2^{\sigma'}(\lambda, \vec{b}) \tag{1}$$

where p_i ($i = 1, 2$) are real functions satisfying $0 \leq p_i \leq 1$ (probabilities), and Λ is any set of fields with a normalized distribution function $\rho(\lambda)$. By the way, the factorized form in the right-hand side of (1) is Bell's (and our) definition of locality.

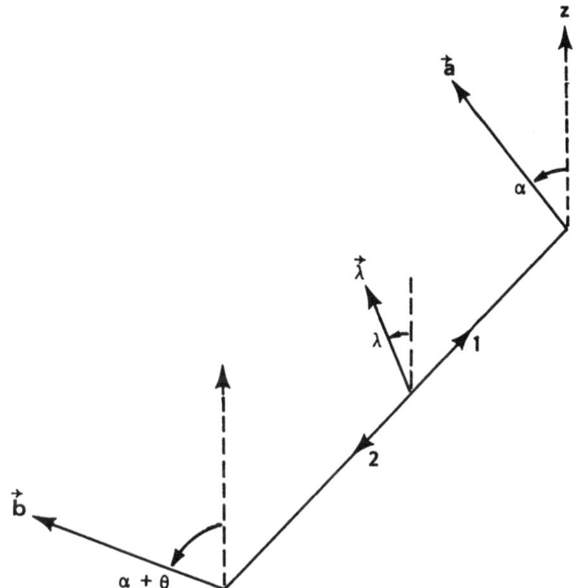

FIGURE 1. Spin-correlation measurement by two analyzers $\vec{a}$ and $\vec{b}$.

(Note that ρ does not depend on $\vec{a}$ and $\vec{b}$, nor p_1 on $\vec{b}$, nor p_2 on $\vec{a}$.*) Therefore, according to Bell's theorem, QM is nonlocal.

Or is it? Take the case $\sigma = \sigma' = +1$. QM tells us that

$$P_{\mathrm{QM}}^{++}(\theta) = \tfrac{1}{2}\sin^2\frac{\theta}{2} \tag{2}$$

(see below). Now, the right-hand side of (1) cannot equal this function (Bell's theorem), but it *can* equal, say, $(1/2)C\sin^2(\theta/2)$, with C a constant less than 1. This only means that some events do not give rise to a coincidence count—in total agreement with the introduction of detection probabilities in (1)! When the experimenter compares coincidence rates for different values of θ, the C constant will drop out (i.e., the rate will be normalized to its value at $\theta = 0$) and the effective rate will equal (2). So the question is: can one set up an experiment where all the particles are detected? Unfortunately, the only clean experimental results

*In a less restrictive definition of locality, $\rho(\lambda)$ can be influenced by some *retarded* action of the analyzers, as in Ref. 7; or the probabilities p_1, p_2 may depend on the results of previous detections through some kind of *memory* effect [S. Caser, Orsay preprint LPTHE 84/51 (1984)]. Such local (in the sense of field theory) models violate Bell's inequality. They will not be considered here.

presently available are those of Aspect and co-workers for photons,[3] and the detection rate there is very poor. The best one can do in comparing (1) with experiment is to first derive an inequality which involves no single particle counts (Clauser and Horne's homogeneous version of Bell's inequality).[5] To do this, however, an extra (and *untestable*) assumption is needed. It is a simple matter to see why. But let us first go into the details of our model for $\sigma = \sigma' = +1$. According to QM, the state of the system is

$$|\psi\rangle = \frac{1}{\sqrt{2}}(|\uparrow\rangle^1|\downarrow\rangle^2 - |\downarrow\rangle^1|\uparrow\rangle^2) \tag{3}$$

where $|\uparrow\rangle^i$ and $|\downarrow\rangle^i$ are eigenstates of $\sigma_z^{(i)}$ for particle $i = 1, 2$. $P_{\text{QM}}^{++}(\theta)$ is the probability of finding the system in the state

$$|a_+b_+\rangle = e^{-i(\alpha/2)\sigma_y^{(1)}}|\uparrow\rangle^1 e^{-i[(\alpha+\theta)/2]\sigma_y^{(2)}}|\uparrow\rangle^2 \tag{4}$$

where α is the angle between $\vec{a}$ and the z direction:

$$P_{\text{QM}}^{++}(\theta) = |\langle a_+b_+|\psi\rangle|^2 \tag{5}$$

Using the identity $e^{i(\alpha/2)\sigma_y} = \cos(\alpha/2) + i\sigma_y \sin(\alpha/2)$ and $\sigma_y = \sigma_+ - \sigma_-$, one easily gets (2). [A similar calculation gives $P_{\text{QM}}^{--}(\theta) = P_{\text{QM}}^{++}(\theta)$, $P_{\text{QM}}^{+-}(\theta) = P_{\text{QM}}^{-+}(\theta) = (1/2)\cos^2(\theta/2)$.] Note that this probability is independent of α, as it should be, because of rotational invariance. We could have used, instead of $|\psi\rangle$, the state

$$|\psi\rangle_\lambda = \frac{1}{\sqrt{2}}(|\uparrow\rangle_\lambda^1|\downarrow\rangle_\lambda^2 - |\downarrow\rangle_\lambda^1|\uparrow\rangle_\lambda^2) \tag{6}$$

where all the kets $|\ \rangle_\lambda^i$ are deduced from the $|\ \rangle^i$ of (3) by application of the rotation operator $= e^{-i(\lambda/2)\sigma_y^{(i)}}$. Obviously, since $S_y|\psi\rangle = 0$,

$$|\psi\rangle_\lambda = e^{-i\lambda S_y}|\psi\rangle = |\psi\rangle \tag{7}$$

$(2S_y = \sigma_y^{(1)} + \sigma_y^{(2)})$. One can also write this as

$$|\psi\rangle = \frac{1}{2\pi}\int_0^{2\pi} d\lambda |\psi\rangle_\lambda \tag{8}$$

What happens if instead of (6) one chooses

$$|F\rangle_\lambda = |\uparrow\rangle_\lambda^1|\downarrow\rangle_\lambda^2\,? \tag{9}$$

(This is equivalent to Furry's hypothesis.[6]) A straightforward calculation yields

$$P^{++}_{F,\lambda}(\theta) = |\langle a_+ b_+|F\rangle_\lambda|^2 = \cos^2\frac{\alpha - \lambda}{2}\sin^2\frac{\alpha - \lambda + \theta}{2} \tag{10}$$

and by averaging over λ (the only way to recover rotational invariance, since we lost it by using $|F\rangle_\lambda$):

$$P^{++}_{F}(\theta) = \frac{1}{2\pi}\int_0^{2\pi} d\lambda P^{++}_{F,\lambda}(\theta) = \tfrac{1}{8} + \tfrac{1}{4}\sin^2\frac{\theta}{2} \tag{11}$$

At this stage, several remarks are in order. First, if one considers (11) together with (10), one concludes that Furry's hypothesis leads to an expression in the form (1), with $\rho(\lambda) = 1/2\pi$. The model is clearly an LRT, but it fails to reproduce QM: the extra factor 1/2 multiplying $(1/2)\sin^2(\theta/2)$ in (11) is only a renormalizing factor and is not important (see remark above), but the constant term 1/8 is! since, according to QM, no ++ events should be present at $\theta = 0$. This is an important point: Furry's model has $p_1^+(\lambda,\vec{a}) = \cos^2[(\alpha - \lambda)/2]$ and $p_1^-(\lambda,\vec{a}) = \sin^2[(\alpha - \lambda)/2]$, hence $p_1^+ + p_1^- = 1$ (similarly for p_2). But the quantum correlations are such that for every λ in (1) only one outcome is possible at each analyzer: if, for instance, the result of the measurement of spin 1 along $\vec{a}$ is − with a nonzero probability $p_1^-(\lambda,\vec{a})$, it must, for that particular λ, be + with *zero* probability (or else the particle must go undetected), and vice versa. Otherwise, when $\vec{a}$ and $\vec{b}$ are set parallel ($\theta = 0$), double + (and double −) events will be present. (Remember that p_1 does not depend on $\vec{b}$, nor p_2 on $\vec{a}$.) In other words, if a model like (1) is to reproduce QM, *it must be deterministic* as to the + or − outcome (but it may, of course, allow undetected events): + and − events are mutually exclusive.

A comparison of (10) and (9) with (5) and (6) also shows that the quantum correlation is given by the half sum of two Furry terms (no averaging over λ necessary) plus an *interference term*: it is this term which is responsible for the strict correlation at $\theta = 0$. Or else, the quantum result may also be found by selecting the component $\lambda = \alpha$ in the Furry ensemble $|F\rangle_\lambda$ (or the component $\lambda = \alpha + \theta$ in the ensemble $|F'\rangle_\lambda = |\downarrow\rangle^1_\lambda|\uparrow\rangle^2_\lambda$) and by multiplying this component by $1/\sqrt{2}$: this operation is equivalent to the reduction of the wave function, a *non-local* process! [Compare with (6).]

One last remark: for photons [no 1/2 in the arguments of (10)] Furry's probabilities are those of Malus'law.

How does one make (1) equal to the QM correlation, up to a multiplicative constant *C*? We know that *C* must be less than 1 (Bell's theorem). For instance, a choice like

$$P^{++}_{\lambda}(\theta) = \frac{1 + \sqrt{2}\cos(\alpha - \lambda)}{2}\,\frac{1 - \sqrt{2}\cos(\alpha - \lambda + \theta)}{2} \tag{12}$$

together with $\rho(\lambda) = 1/2\pi$ [this differs from Furry's term (10) by the presence of the $\sqrt{2}$ factors] exactly yields the QM correlation (2) ($C = 1$). This choice, however, is ruled out, since the corresponding "probabilities" p_1 and p_2 are not between 0 and 1. But consider the following choice[7,8] (see Figure 1):

$$\begin{aligned}
p_1^+(\lambda,\vec{a}) &= |\cos(\lambda - \alpha)| \qquad \text{for } -\frac{\pi}{2} \leqslant \lambda - \alpha \leqslant \frac{\pi}{2} \\
&= 0 \text{ otherwise} \\
p_1^-(\lambda,\vec{a}) &= |\cos(\lambda - \alpha)| \qquad \text{for } \frac{\pi}{2} \leqslant \lambda - \alpha \leqslant \frac{3\pi}{2} \\
&= 0 \text{ otherwise} \\
p_2^+(\lambda,\vec{b}) &= 2/\pi \qquad \text{for } \frac{\pi}{2} \leqslant \lambda - \alpha - \theta \leqslant \frac{3\pi}{2} \\
&= 0 \text{ otherwise} \\
p_2^-(\lambda,\vec{b}) &= 2/\pi - p_2^+(\lambda,\vec{b})
\end{aligned} \tag{13}$$

together with $\rho(\lambda) = 1/2\pi$ (all functions are 2π-periodic). For this model, (1) gives

$$P^{++}(\theta) = \left(\frac{2}{\pi}\right)^2 \cdot \tfrac{1}{2} \sin^2 \frac{\theta}{2} \tag{14}$$

and this is equal to the QM correlation (2) multiplied by $C = (2/\pi)^2$. (C is less than 1, as it should be. Similar relations hold for the other $P^{\sigma\sigma'}$.) Note that the choice $2/\pi$ for p_2 is not arbitrary: after averaging over λ, the four mean probabilities

$$p_i^\sigma = \int d\lambda \rho(\lambda) p_i^\sigma(\lambda, \vec{z}) \tag{15}$$

($i = 1, 2$, $\sigma = \pm$, $\vec{z}$ arbitrary) should be equal (an experimental fact). One finds $p_i^\sigma = 1/\pi$, which is $2/\pi$ times the observed probability 1/2 [a result consistent with the $(2/\pi)^2$ factor in (14)].

This model, therefore, agrees with QM, *provided the ratio of the number of coincidences to the number of events does not exceed* $(2/\pi)^2$. In photon correlation experiments, the coincidence rate is much lower: this is why up to now these experiments failed to "prove" nonlocality. As mentioned earlier, to close this loophole, Clauser and Horne[5] made use of an extra assumption, which they called "no enhancement." This means that

$$p_i^\sigma(\lambda, \vec{a}) \leqslant p_i(\lambda,\infty) \tag{16}$$

for every λ, where $p_i(\lambda, \infty)$ is the probability of a count of detector i in the absence of the polarizer. If one looks at (13), one sees that this inequality is violated.

Indeed, $p_i(\lambda, \infty) = p_i^+ + p_i^-$ [as given by (15)] $= 2/\pi$. But for $\lambda = \alpha$, for instance, the p_i^+ of (13) is equal to 1. Let us simply remark here that the no-enhancement assumption is by no means a physical prerequisite (because the λ's are not accessible to experiment). (see also Ref. 9.)

A striking feature of the above model is the dissymmetry between p_1 and p_2 [see (13)]. In fact, if one thinks of photons as classical fields, one would expect a relation of the kind

$$p_2^\sigma(\lambda, \vec{b}) = p_1^\sigma(\lambda + \pi, \vec{b}) \tag{17}$$

There is a theorem which says that, whatever the number of λ's (in the plane of the polarizers or not), no such symmetry between the analyzers is possible.[8] How close to the QM curve can one get in a symmetric model? The answer is given by a second theorem:[10] arbitrarily close, but in the QM limit no particles are detected (which is why the first theorem is valid!). The limit, therefore, is, physically speaking, a singular one. Why is it so? Nobody knows. But this simple fact explains why people working with symmetric LRTs had such a hard time finding an approximate fit to QM. According to our philosophy, since the QM correlation *is* compatible with locality, it should be exact. In other words, the LRT should be dissymmetric. Is this dissymmetry really a problem?* After all, two functions which are different can be considered as two realizations of the same function for two values of an extra discrete variable τ ($\tau = 0,1$). This leads us to introduce detection probabilities $p_i^\sigma(\lambda_i, \vec{a}, \tau_i)$ with correlations of the type

$$\begin{aligned} \lambda_2 &= \lambda_1 + \pi \\ \tau_2 &= 1 - \tau_1 \end{aligned} \tag{18}$$

($\tau_i = 0, 1$). The model defined by (13) belongs to this category. (See also Ref. 11.)

What is the meaning of the τ variable? Obviously, nature is not "classical" enough to reproduce the QM correlation without appealing to some discrete "quantum" number. Classical fields? Possibly, but probably a whole collection of them (for a given type of particle). We will therefore postulate that a particle is made of (or accompanied by) a collection of fields, labeled by a discrete variable τ, which we call "local vacua." Some properties of these will now be investigated.

3. HOW TO GET RID OF THE WAVE FUNCTION COLLAPSE

In Section 2 we saw that, to reproduce QM, LRTs must be deterministic, in the sense that the result of a spin measurement is determined by mutually exclusive

*We thank Professor N. D. Mermin for convincing us that it is not.

probabilities (or else the particle must go undetected). This not too well-known fact shows that QM, in its essence, is not so dramatically different from classical physics—at least as far as correlations are concerned. On the other hand, we saw that a quantum element (the discrete variable τ) is a necessary ingredient of classical model-making. We believe, therefore, that if a theory must replace QM, it will not be some trivial extension of classical field theory. It could, in fact, be something *between* this and the orthodox QM—but local.

Let us consider the action of a Stern–Gerlach magnet (with its magnetic field gradient along z) on a spin-1/2 particle. The instrument can be found in three different (macroscopic) states: its ground state $|0\rangle$, a state $|u\rangle$ after recording the passage of the particle with its spin up, and a state $|d\rangle$ after recording the spin-down particle. These states form an orthonormal basis for the instrument in that particular measurement. (Actually, each state should be a subspace, but this is not important for the present discussion.) When the particle enters the instrument with its spin up, it is not necessarily recorded: after its passage, the instrument is in the state

$$|10\rangle = \alpha_1|0\rangle + \beta_1|u\rangle \tag{19}$$

with $|\alpha_1|^2 + |\beta_1|^2 = 1$, and the state of the system particle + instrument undergoes during the measurement the transition

$$|\psi(t)\rangle = |\varphi(t)\rangle|\uparrow\rangle|0\rangle \rightarrow |\varphi_u(t)\rangle|\uparrow\rangle|10\rangle \tag{20}$$

where $|\varphi(t)\rangle|\uparrow\rangle$ is the particle wave function. Similarly, when the particle enters the instrument with its spin down, the total system will undergo the transition

$$|\psi(t)\rangle = |\varphi(t)\rangle|\downarrow\rangle|0\rangle \rightarrow |\varphi_d(t)\rangle|\downarrow\rangle|01\rangle \tag{21}$$

and the instrument will be left in the state

$$|01\rangle = \alpha_2|0\rangle + \beta_2|d\rangle \tag{22}$$

with $|\alpha_2|^2 + |\beta_2|^2 = 1$. Note that if the instrument works symmetrically, $|\alpha_1| = |\alpha_2|$, $|\beta_1| = |\beta_2|$. We will suppose it does. If now the particle enters the magnet in a superposition $|\varphi(t)\rangle(|\uparrow\rangle + |\downarrow\rangle)/\sqrt{2}$ (spin perpendicular to the magnetic field gradient and to the line of flight), the total system will evolve according to

$$|\psi(t)\rangle = |\varphi(t)\rangle\frac{1}{\sqrt{2}}(|\uparrow\rangle + |\downarrow\rangle)|0\rangle \rightarrow \frac{1}{\sqrt{2}}(|\varphi_u(t)\rangle|\uparrow\rangle|10\rangle + |\varphi_d(t)\rangle|\downarrow\rangle|01\rangle) \tag{23}$$

Let $\alpha_1 = \epsilon e^{i\delta_1}$, $\alpha_2 = \epsilon e^{i\delta_2}$ ($|\alpha_1| = |\alpha_2| \equiv \epsilon$), $\beta_1 = \sqrt{1 - \epsilon^2}\, e^{i\theta_1}$, $\beta_2 = \sqrt{1 - \epsilon^2}\, e^{i\theta_2}$ ($|\beta_1| = |\beta_2| = \sqrt{1 - \epsilon^2}$). The end result of (23) becomes

$$|\psi(t)\rangle = \frac{\epsilon}{\sqrt{2}}(e^{i\delta_1}|\varphi_u(t)\rangle|\uparrow\rangle + e^{i\delta_2}|\varphi_d(t)\rangle|\downarrow\rangle)|0\rangle + \sqrt{\frac{1-\epsilon^2}{2}}e^{i\theta_1}|\varphi_u(t)\rangle|\uparrow\rangle|u\rangle + \sqrt{\frac{1-\epsilon^2}{2}}e^{i\theta_2}|\varphi_d(t)\rangle|\downarrow\rangle|d\rangle \tag{24}$$

There are two extreme cases:

(1) Complete measurement: $\epsilon = 0$. This gives

$$|\psi(t)\rangle = \frac{1}{\sqrt{2}}(e^{i\theta_1}|\varphi_u(t)\rangle|\uparrow\rangle|u\rangle + e^{i\theta_2}|\varphi_d(t)\rangle|\downarrow\rangle|d\rangle) \tag{25}$$

At this point, one argues that $|u\rangle$ and $|d\rangle$ being macroscopic states of the (classical) instrument, they cannot be superposed: therefore, one concludes that the final state of the system is either the first term on the right-hand side of (25), or the last one—but not both. This is a *superselection* rule. [More generally, the rule applies to the three states in (24).]

(2) No measurement: $\epsilon = 1$. In this case

$$|\psi(t)\rangle = \frac{1}{\sqrt{2}}(e^{i\delta_1}|\varphi_u(t)\rangle|\uparrow\rangle + e^{i\delta_2}|\varphi_d(t)\rangle|\downarrow\rangle)|0\rangle \tag{26}$$

In the interference domain, $\varphi_u(\vec{x}, t) = e^{i\delta}\varphi_d(\vec{x}, t)$: whenever $\delta_1 + \delta = \delta_2$ one recovers the initial state in (23) (no spin rotation).

In general, (24) shows that the probability for no measurement is ϵ^2, and the probabilities for an up-spin or a down-spin measurement both equal $(1 - \epsilon^2)/2$.

What is the physical criterion for measurement? Since $\epsilon = |\alpha_1| = |\alpha_2| = |\langle 0|10\rangle| = |\langle 0|01\rangle|$, complete measurement will take place whenever the initial and the final states of the instrument are orthogonal ($\epsilon = 0$). If, for instance, the instrument initially is in a coherent state (as for the magnetic field of a Stern–Gerlach), and the interaction with the particle spin adds a few photons to it (or turns it into a neighboring coherent state), one cannot expect a complete measurement to take place, since these states are not orthogonal. On the contrary, measurement will take place if the instrument is in a pure photon-number state; that is, if the state of the instrument (or whatever plays that role in the measurement process) is *quantum* rather than *classical*.

Obviously, apart from considerations about the state of the instrument or the aforementioned superselection rule, there is no reason for appealing to the experimenter's mind (or his cat's) to build up a consistent viewpoint. One can even go one step further: look at (23), for instance. Instead of supposing that the states $|0\rangle$, $|10\rangle$, and $|01\rangle$ belong to the *instrument*, why not suppose that they are in some way attached to the *particle* itself, or to something that accompanies it? The wave aspects of the particle are already contained in $|\varphi(t)\rangle$, but we saw in the previous section that a discrete quantum number τ appeared necessary to account for correlation measurements in LRTs: this could be the manifestation of these extra states we are looking for—"local vacua." We therefore write the state of the particle as

$$|\psi(t)\rangle = |\varphi(t)\rangle|s\rangle|0\rangle_\tau \tag{27}$$

where $\langle\vec{x}|\varphi(t)\rangle = \varphi(\vec{x}, t)$ is the usual ("space") wave function of QM, $|s\rangle$ is the spin (isospin . . .) ket, and $|0\rangle_\tau$ is the ket of the associated local vacuum. The local vacua are supposed to form a complete orthonormal basis.

$${}_\tau\langle 0|0\rangle_{\tau'} = \delta_{\tau,\tau'} \tag{28}$$

and one has the *superselection rule*: the local vacuum cannot be observed in a superposition of the $|0\rangle_\tau$ states. These characteristics of the local vacua are a straightforward consequence of the instrument's behavior, so that the above discussion can be reinterpreted step by step according to the new rules. In particular, there is a close correspondence between the instrument and the particle's local vacuum: whenever the instrument is in a coherent (classical) state, the interaction between the instrument and the particle does not change the particle's local vacuum: interference is preserved [see (23), with $|0\rangle$, $|10\rangle$, and $|01\rangle$ all replaced by the same local vacuum $|0\rangle_\tau$]. When this is not the case, the (quantum) state of the instrument is changed by the measurement process, and so is the local vacuum: superselection at work between the two local vacua thus created at random (one up: $|10\rangle \equiv |0\rangle_\tau$, the other down: $|01\rangle \equiv |0\rangle_{\tau'}$, $\tau' \neq \tau$) destroys the interference by preventing their recombination at the end of the process of measurement. In this interpretation, *no reduction of the wave function ever occurs*—no mystifying rule, only superselection. As to the particle itself, we expect it to accompany one of the local vacua (up or down). (Photons could behave differently—see below.) What we have achieved by the introduction of these local vacua, therefore, is a complete decoupling between the particle and the instrument (or observer), as in *classical physics*. Note the strong interplay of classical and quantum concepts at every stage of the investigation: local vacua are described by quantum states, but a superselection rule applies, as for the classical macroscopic states of an instrument, or else, the quantum states of a system which differ by their electric charge. (This rule makes them quite different from, say, classical states of the electromagnetic field.) Also note the intuitive side of the local vacuum–instrument interaction: whenever the local vacuum "bounces off" a classical instrument, its internal state (as described by τ) does not change. This is similar to a ball bouncing off a wall elastically: neither the wall nor the ball undergoes an internal state change. On the contrary, if the collision is inelastic, one can expect an irreversible change to take place in the wall's internal state, and (according to our rule of the game) in that of the ball as well: no doubt a permanent recording of the collision—in other words, a measurement.

4. THE PHYSICAL NATURE OF THE LOCAL VACUUM

One does not always realize that de Broglie's relation $\lambda = h/mv$ has its origin in the relativistic invariance, to order v/c, of the phase of the wave function $\exp(-ip\cdot x/\hbar)$. If one supposes, as de Broglie did, that to a particle of mass m is attached a natural clock of frequency $\omega_0 = mc^2/\hbar$, and that a similar clock, synchronized with the particle's clock in the particle's rest frame, can be found at every neighboring point in space, then de Broglie's wavelength λ in a frame where the particle velocity is v is equal to the distance between two consecutive planes of equal phases of the set of moving clocks (the latter appearing out of synchronization to the observer at rest because of Lorentz's transform, but, to this order, not slowed down by time dilation). Periodic synchronization of the clocks of the atomic electron then leads to quantization of its orbit (the "resonance condition" of wave mechanics). Simple as they are, these arguments probably contain in germ a more profound (dynamical) view of space than is presently accepted. Dirac quantized the electromagnetic field by turning its degrees of freedom into elementary quantum oscillators. If a particle's clocks, as explained above, are elementary oscillations of space,[12] then the particle's wave function would appear as some kind of fundamental field through which it acts on its neighborhood. (In this context, Einstein's postulate of the constancy of the velocity of light can be thought of as a resonance condition between these oscillators.) This is to say that a particle creates its own *local vacuum*, whose extent is given by that of its wave function. Quantum effects would then originate in the oscillatory character of this local vacuum and its evolution under the influence of the limiting action of measuring instruments. For instance, a photon entering a polarizer should adapt its local vacuum to the polarizer's vacuum, an operation that could lead to the collapse of the photon's vacuum and, if the photon is to be identified with its local vacuum, to its subsequent nondetection, as in Section 2.

In (27), the wave character of the local vacuum is described by the "space" wave function $|\varphi(t)\rangle$, while the rest of it is contained in $|0\rangle_T$, which is responsible for the superselection rule. One can of course unite both concepts in a single vector

$$|\varphi(t)\rangle_T \equiv |\varphi(t)\rangle|0\rangle_T \qquad (29)$$

The interesting question now is: why should local vacua obey a superselection rule, as in Section 3? The only thing one can say is that this rule is, for the above experiments at least, the first genuine manifestation of the quantum world—since classical fields do mix together! This does not mean, however, that classical physics cannot be used in our understanding of local vacua. A little known theory of Dirac will be helpful.

5. DIRAC'S COVARIANT VACUUM AS AN EXAMPLE

Many physicists today think that the idea of a universal ether is not only useless, but incompatible with the special theory of relativity. To see that the ether concept does not deserve the first quality, it is enough to change its name: the old ether has become the present-day quantum vacuum, whose manifestations are numerous and predictable. As to the second point, every student in field theory knows what a covariant vacuum is. A classical version of this vacuum is the zero-temperature blackbody spectrum, with spectral density $\rho(\omega) \sim \omega^3$: it is covariant, and it does not cause friction on a charged particle moving with constant velocity.[13] (Friction is proportional to its acceleration, as in Newton's law.) In fact, the ω^3 law is the only one compatible with covariance,[14] besides the empty vacuum ($\rho = 0$), and the renormalization process in quantum field theory replaces the first one with the second. The ether, therefore, is at the center of modern physics. (By the way, Einstein himself never ceased to believe in it.[15])

A fluid with no viscosity is a superfluid. But why couldn't it *flow*? This is the subject of Dirac's beautiful theory of classical electrodynamics.[16,17] Dirac simply observed that a possible gauge for the electromagnetic vector potential A_μ is

$$A_\mu A^\mu = k^2 \tag{30}$$

with k a universal constant (Dirac's gauge). This makes it possible to regard $A_\mu(x)$ as a local velocity (up to a multiplicative constant), and Dirac concludes[17]: "Thus with the new theory of electrodynamics we are rather forced to have an ether." Why is it a new theory? Because Dirac derived his gauge from a *particular Lagrangian*, and this in some instances leads to deviations from the usual theory. [Dirac's ether is not only a superfluid, but a superconductor ($j_\mu \sim A_\mu$), a feature that made him believe in a better renormalization program once the theory is quantized.] What we are interested in here, however, is not Dirac's theory but Dirac's gauge (30), a perfectly legal choice in classical electrodynamics, as we shall now see.

Consider the motion of a particle of charge $q = \pm e$ and mass m in a given field $F_{\mu\nu} = \partial_\mu A_\nu^{(0)} - \partial_\nu A_\mu^{(0)}$. The Hamilton–Jacobi equation

$$\left(A_\mu^{(0)} + \frac{c}{q}\partial_\mu S\right)\left(A^{\mu(0)} + \frac{c}{q}\partial^\mu S\right) = k^2 \tag{31}$$

with $k = mc^2/e$, is identical with Dirac's gauge condition (30), provided one sets

$$A_\mu = A_\mu^{(0)} + \frac{c}{q}\partial_\mu S \tag{32}$$

This means that the vector potential A_μ in Dirac's gauge is connected with $A_\mu^{(0)}$ by a gauge transformation, whose gauge function is the Hamilton–Jacobi characteristic function S. The potential A_μ is not unique: The Hamilton–Jacobi equation in that potential now reads

$$\left(A_\mu + \frac{c}{q}\partial_\mu S'\right)\left(A^\mu + \frac{c}{q}\partial^\mu S'\right) = k^2 \tag{33}$$

This produces a new Dirac potential ($A'_\mu A'^\mu = k^2$):

$$\begin{aligned} A'_\mu &= A_\mu + \frac{c}{q}\partial_\mu S' \\ &= A_\mu^{(0)} + \frac{c}{q}\partial_\mu (S + S') \end{aligned} \tag{34}$$

Let us now concentrate on (33), which gives the motion of the charge in the potential A_μ ($A_\mu A^\mu = k^2$). Its solution S' yields the particle 4-momentum:

$$p_\mu = -\partial_\mu S' \tag{35}$$

and the particle 4-velocity:

$$\begin{aligned} v_\mu &= \left(p_\mu - \frac{q}{c}A_\mu\right)/m = -\frac{q}{mc}A'_\mu \\ v_\mu v^\mu &= c^2 \end{aligned} \tag{36}$$

We now take as the ether 4-velocity:

$$U_\mu = -\frac{q}{mc}A_\mu \tag{37}$$

Hence, from (36),

$$v_\mu = U_\mu - \frac{1}{m}\partial_\mu S' \tag{38}$$

These relations signify that any pair of solutions A_μ, S' of (33) [with A_μ deduced from the given $A_\mu^{(0)}$ by the relation (32)] represents a possible motion of the ether and of the charge, respectively, in the given field $F_{\mu\nu}$. Note that the velocity of the ether in (37) depends on the particle's charge and mass (through the ratio q/m, since this is a classical theory). We will come back to this later.

It is a simple matter to derive the expression of the Lorentz force $\vec{F} = q(\vec{E} + \vec{v} \times \vec{B}/c)$ acting on the charge, from the above equations. What is not so obvious, however, is that the same result can be obtained in the spirit of *Mach's principle*[18] by drawing an analogy between the above motion of the ether and that resulting from the motion of a reference frame: it turns out that the total sum of the inertial forces thus generated is equal to the Lorentz force on the charge. We will only sketch the proof. The calculation is more transparent in the nonrelativistic approximation for the motion of the ether and of the charge, where the expression of the inertial forces (centrifugal + Coriolis) is well known. In that approximation, one gets from (37)

$$\vec{u} = -\frac{q}{mc}\vec{A}$$
$$mc^2 + \tfrac{1}{2}m\vec{u}^2 = -qA^0 \tag{39}$$

where $\vec{u}$ is the ether 3-velocity ($|\vec{u}| \ll c$, or $e|\vec{A}| \ll mc^2$) and $A^0 = \sqrt{\vec{A}^2 + k^2}$.

Mach's principle (in its local version) says that a motion of the ether with the velocity field $\vec{u}$ at some point P in space has the same effect on a particle at P as the inertial forces that would result from a motion of the reference frame, relative to some inertial frame, with the local velocity field $-\vec{u}$.

If $\vec{\omega}$ is the instantaneous rotation vector of the moving frame, one has

$$\vec{\omega} = -\tfrac{1}{2}\nabla \times \vec{u} \tag{40}$$

Introducing the vector radius $\vec{r}$ of the charge with its origin at the instantaneous center of rotation ($\vec{\omega} \times \vec{r} = -\vec{u}_t$, $\vec{u}_t$ being the transverse component of $\vec{u}$), the fundamental equation for the motion of the charge is obtained by equating to zero its acceleration in the inertial frame:

$$\frac{d^2\vec{r}}{dt^2} = 0 \tag{41}$$

(No forces are present in this frame, since by hypothesis *all forces come from inertia.*) This should be written in terms of the velocity of the charge $\vec{v} \equiv D\vec{r}/Dt$ and its acceleration $D^2\vec{r}/Dt^2$ in the rotating (laboratory) frame. The total inertial force acting on the charge

$$\vec{F} = m\frac{D^2\vec{r}}{Dt^2} \tag{42}$$

can then be shown, after some calculation, to be equal to the Lorentz force.[19]

This result draws a curious analogy between gravitation (the equivalence principle, at least) and electromagnetism. We already mentioned the presence of the q/m factor in the expression for the ether velocity (37): this, of course, is what makes the two theories intrinsically different (no genuine equivalence principle for electromagnetism), and the above considerations a pure analogy. Remember, however, that a local vacuum is attached to every particle (with given q and m). If the local vacuum were a quantized version of the above A_μ in Dirac's gauge (in a ground state configuration), the fact that the ether velocity (37) is different from one particle to the next would not be surprising: every charged particle has its own ether. (This is a weaker version of our superselection rule.) Dirac's ether, therefore, as a local manifestation of a classical covariant vacuum, and by its crucial dependence on some of the particle's quantum numbers, seems to support the idea of local vacua.

6. CONCLUSION AND OUTLOOK

A careful examination of all the available experimental evidence certainly conveys the impression that the nonlocal character of QM is far from being established. On the theoretical side, however, there is an urgent need for more physics. The predictions of QM clearly are statistical in character. For instance, the time description of detection is poor, if not totally absent.* How does a polarizer work? What is the criterion for a measurement to take place? (Irreversibility is not a useful criterion, since QM is reversible.) QM treats photons and massive particles on an equal footing, but spin correlations at a distance could be different in these two cases. As we saw, since the photon number is not conserved, photon spin correlations are compatible with a local theory plus enhancement (a phenomenon perfectly compatible with a nonexistent polarizer theory!). Note that there is no conflict at this point with energy or momentum conservation, the latter being satisfied in the mean. But energy–momentum conservation can be checked for individual high-energy photons: in this case, however, polarizers are inefficient, and no direct test of the spin correlation can be made. The question is: is this really a coincidence? After all, QM teaches us that the uncertainty relation will always turn out true, whatever the observer's ingenuity, because it (supposedly) is a built-in characteristic of nature. Why couldn't some new principle like this, linking the particle character to the polarization character, be at work here? On the other hand, a charged particle like the electron is a well-localized object, and one does not expect electrons to disappear in the electromagnetic field of a Stern–

*The role of time in connection with Bell's inequality was stressed in particular by S. PASCAZIO, *Phys. Lett. A* **118**, 47 (1986); in *Microphysical Reality and Quantum Formalism* (A. VAN DER MERWE, F. SELLERI and G. TAROZZI, eds.), Kluwer Academic, Dordrecht (1988).

Gerlach instrument. But in this case, the uncertainty principle makes the spin measurement impossible.[20] Another coincidence? As for neutrons or molecules, it could turn out that for such localized neutral objects Furry's hypothesis is valid, with a corresponding breakdown of the quantum correlation. Or else, neutrons could behave as photons (see above). For heavier objects, of course, the discrepancy between Bell's limit and QM gets smaller, as does the probability for their disappearing in the course of measurement. It is no little irony that the nonlocality "proof" with photons could rest on the (wrong) hypothesis that two photons emerge from every atomic cascade and give rise to detection—a hypothesis suggested by a *classical mental image*.

All these considerations make it impossible to ignore the role of the vacuum. It can readily be seen, for instance, that the normal ordering prescription of quantum field theory (which sets the energy of the vacuum equal to zero) is responsible for photon antibunching, a phenomenon "typical" of the quantum nature of light.[21] The argument can be reversed: take away that normal ordering, and light may appear classical. (This does not mean that the usual procedure is wrong, but only that proofs of the existence of an absolute gap between the quantum and classical worlds do not prove anything!) On the other hand, laser light, a "classical" object, has a well-defined phase, but its photon number is not determined. One could be tempted to conclude that no such thing as a phase exists for a single photon. This conclusion is invalid, because a photodetector does not take notice of vacuum effects.[22] The vacuum, therefore, whether directly or indirectly, appears as a necessary ingredient in the interpretation of a wide range of quantum phenomena—and, in general, a classical vacuum is good enough. This idea stimulated a number of physicists.[23] It seems, however, that if nonlocality is one of the characteristics of QM to be explained this way, an even more sophisticated vacuum should be conceived. In any case, since the problem of inertia cannot be fully understood in the old scheme, this line of research is fully justified by itself. The concept of local vacua, we hope, could be the starting point of such an investigation.*

REFERENCES

1. J. S. BELL, *Physics* **1**, 195 (1964).
2. A. EINSTEIN, B. PODOLSKY, and N. ROSEN, *Phys. Rev.* **47**, 777 (1935).
3. A. ASPECT, P. GRANGIER, and G. ROGER, *Phys. Rev. Lett.* **47**, 460 (1981); A. ASPECT, J. DALIBARD, and G. ROGER, *Phys. Rev. Lett.* **49**, 1804 (1982).
4. D. BOHM and Y. Aharonov, *Phys. Rev.* **108**, 1070 (1957).
5. J. F. CLAUSER and M. A. HORNE, *Phys. Rev.* D**10**, 526 (1974).

*Local vacua bear some resemblance to empty de Broglie waves. On the latter see, e.g., F. SELLERI *Found. Phys.* **12**, 1087 (1982); *Quantum Paradoxes and Physical Reality* (A. VAN DER MERWE, ed.), Kluwer Academic, Dordrecht (1990).

6. W. H. FURRY, *Phys. Rev.* **49**, 393, 476 (1936).
7. S. CASER, *Phys. Lett. A* **92**, 13 (1982).
8. S. CASER, *Phys. Lett. A* **102**, 152 (1984).
9. T. W. MARSHALL, E. SANTOS, and F. SELLERI, *Phys. Lett. A* **98**, 5 (1983); T. W. MARSHALL, *Phys. Lett. A* **99**, 163 (1983), **100**, 225 (1984); A. GARUCCIO and F. SELLERI, *Phys. Lett. A* **107**, 164 (1985); F. SELLERI, *Phys. Lett. A* **108**, 197 (1985).
10. S. CASER *Phys. Lett. A* **121**, 331 (1987).
11. M. FERRERO, T. W. MARSHALL, and E. SANTOS, in: *Quantum Mechanics versus Local Realism: The Einstein–Podolsky–Rosen Paradox* (F. SELLERI, ed.), Plenum Press, New York (1988).
12. See D. BOHM, *Quantum Theory*, Prentice–Hall, Englewood Cliffs, N.J. (1951).
13. A. EINSTEIN and L. HOPF, *Ann. Phys.* **33**, 1105 (1910).
14. T. H. BOYER, *Phys. Rev.* **182**, 1374 (1969).
15. See, e.g., A. EINSTEIN, Conférence à l'Université de Leyde (5 mai 1920), Gauthier–Villars, Paris (1972).
16. P. A. M. DIRAC, *Proc. R. Soc. London Ser. A* **209**, 291 (1951).
17. P. A. M. DIRAC, *Nature* **168**, 906 (1951).
18. A. EINSTEIN, *The Meaning of Relativity*, 3d ed., Princeton University Press, Princeton, N.J. (1950).
19. S. CASER, *Found. Phys. Lett.* **4**, 179 (1991).
20. N. F. MOTT and H. S. W. MASSEY, *The Theory of Atomic Collisions*, Clarendon Press, Oxford (1965).
21. D. F. WALLS, *NATURE* **280**, 451 (1979).
22. H. PAUL, *REV. MOD. PHYS.* **58**, 209 (1986).
23. To cite a few: T. W. MARSHALL, *Proc. Cambridge Philos. Soc.* **61**, 537 (1965); T. H. BOYER, *Phys. Rev. D* **11**, 790, 809 (1975), and in *Foundations of Radiation Theory and Quantum Electrodynamics* (A. O. BARUT, ed.), Plenum Press, New York (1980); A. O. BARUT, *Phys. Scr.* **21**, 18 (1988), and in above-cited book; P. W. MILONNI, *Phys. Scr.* **21**, 102 (1988).

CHAPTER 3

Causal Quantum Theory

Why a Nonstarter?*

James T. Cushing

1. INTRODUCTION

The standard view of quantum mechanics, almost universally accepted by practicing physicists and most often by philosophers of science concerned with such issues, is what may be (somewhat elusively) termed the "Copenhagen" interpretation. This interpretation requires complementarity (e.g., wave–particle duality), inherent indeterminism at the most fundamental level of quantum phenomena, and the impossibility of an event-by-event causal representation in a continuous space-time background. However, it is important to stress that quantum mechanics as a theory has two conceptually distinct but practically related components (as does any modern theory in physics): a formalism and an interpretation. Very loosely, the *formalism* refers to the equations and calculational rules that prove empirically adequate (i.e., "getting the numbers right") and the *interpretation* refers to the accompanying representation the theory gives us about the physical universe (i.e., the picture story that goes with the equations or what our theory "really" tells us about the world). Since a (successful) formalism does not uniquely determine its interpretation, there may be two radically different interpretations (and ontologies) corresponding equally well to *one* empirically adequate formalism. This can be taken as an instantiation of the Duhem–Quine thesis of underdetermination of

*This paper is a much expanded version of a talk, "Copenhagen Hegemony: *Need* It Be So?," delivered at the Symposium on the Foundations of Modern Physics 1990 held in Joensuu, Finland, on August 13–17, 1990, the proceedings of which are being published by World Scientific Publishing Co.

James T. Cushing • Departments of Physics and Philosophy, University of Notre Dame, Notre Dame, Indiana 46556, USA.

Wave–Particle Duality, edited by Franco Selleri. Plenum Press, New York, 1992.

theory by an empirical base. Even if one wants to restrict (and, arguably, that would be a mistake) the Duhem–Quine thesis to different *formalisms* each handling equally well a given body of empirical information, there nevertheless remains the interesting and important point of opposing ontologies equally well supported by a common empirical base.

While the rather abstract logical observation that no set of data uniquely determines a theory may be relatively uncontroversial, the *practical* determination of a physical theory by empirical considerations and by logic is often taken to be quite another matter. Einstein himself offered the following opinion about this in an address he delivered in 1918 before the Physical Society of Berlin on the occasion of Max Planck's 60th birthday.

> There is no logical path to these laws; only intuition, resting on sympathetic understanding of experience, can reach them. In this methodological uncertainty, one might suppose that there were any number of possible systems of theoretical physics all equally well justified; and this opinion is no doubt correct, theoretically. But the development of physics has shown that at any given moment, out of all conceivable constructions, a single one has always proved itself decidedly superior to all the rest. Nobody who has really gone deeply into the matter will deny that in practice the world of phenomena uniquely determines the theoretical system, in spite of the fact that there is no logical bridge between phenomena and their theoretical principles. . . .(1)

Now Einstein to the contrary notwithstanding, there *are* people who have looked carefully at the development of certain major episodes in the history of physics and who have concluded that factors other than just "the world of phenomena" have been essential for specific theory choice and that, but for contingency, the final choice might have turned out other than it did.(2,3) This is *not* to claim that just *any* theory can be made to work, but rather to emphasize that logic and physical phenomena *alone* are not sufficient to select uniquely *one* theory(4) (although, in many cases, they *alone* are in practice enough to rule out or reject some theories as viable candidates). This is not to deny Einstein's claim that "in practice" one theory *is* finally chosen as "decidedly superior to all the rest." The question, then, is *how* in fact is such a choice made? In this chapter we point out a known equivalence between two opposing interpretations of quantum mechanics and discuss the factors that were involved in accepting the so-called "Copenhagen" interpretation over its rival.

Although it is not widely appreciated (even if it is somewhat more widely, but darkly, "known"—at the level of hearsay), there do exist interpretations alternative to the standard Copenhagen one. The so-called causal interpretation that we focus on here is not only equally as well confirmed as the Copenhagen one (since it is based on *exactly the same* formalism), but it preserves event-by-event causality in space-time (albeit with the same type of nonlocality present in the Copenhagen interpretation).

We begin with brief summaries of the formalism of (nonrelativistic) quantum mechanics, of the Copenhagen interpretation and of one causal interpretation and show why these two interpretations are necessarily equally well supported empirically. Each interpretation is then applied to the example of actual neutron interferometry experiments, which are, essentially, modern-day versions of the double-slit "thought" experiment.(5) Such considerations are necessary to block any claim that a causal interpretation is incoherent so that the (clever) founders of quantum mechanics would (surely) have spotted that flaw and hence not bothered pursuing such an interpretation. We then turn to the central question raised in this chapter. Since causal interpretations of quantum mechanics are not refuted by logic and/or by empirical (in)adequacy, *why* have they never (to this day) been seriously considered by any sizable faction of the theoretical physics community? To construct a plausible explanation for this lack of interest in such causal interpretations, one must look (circa 1925–1927) at the roots of the Copenhagen interpretation and of the nascent causal interpretation and at the subsequent historical developments. If one simply begins the discussion today (i.e., with the present situation as it now stands without asking how we arrived there), a charge of ad hocness is too easily (even if invalidly) raised by an opponent wishing to reject out of hand any interpretation alternative to the accepted, "correct" Copenhagen one.

2. FORMALISM AND INTERPRETATION: AN EXAMPLE

Let us expand a bit upon this distinction between the two components of a theory that we mentioned above: namely, a formalism and an interpretation. We do not mean to imply that this division of a theory into the two components of formalism plus interpretation is necessarily unique, complete, or exhaustive. For our purposes in this chapter we need only the recognition that a formalism and an interpretation *are* two distinct, even if related, parts of a theory.

An interpretation is based on a (necessarily) incomplete examination of a formalism, since it is not possible to apply a given formalism to *all* conceivable situations and experiments (either actual or of the "thought" variety) in arriving at an interpretation of that formalism. That is, our "intuition" about the world is based very largely on those relatively few (but hopefully "typical") cases or problems we can solve (often exactly). Thus, in the case of classical mechanics applied to the motion of a planet (mass m) about the sun (mass M), we might in thumbnail sketch represent the appropriate *formalism* in terms of Newton's second law of motion.

$$\mathbf{F} = m\mathbf{a} \tag{1}$$

and of his law of universal gravitation

$$\mathbf{F} = \frac{-GMm}{r^2}\hat{\mathbf{r}} \tag{2}$$

On the basis of many applications (basically, the two-body problem and perturbations thereof), we then develop an *interpretation* of the nature of the world governed by this system of laws. The picture, or folklore belief, that emerged historically was that of a completely deterministic, causal, (in principle) predictable physical universe. This intuition was based on the class of problems (today termed "integrable") that yielded to the analytical tools available. However, the lesson of modern chaos theory, which has emerged in the last couple of decades or so, is that such integrable dynamical systems are quite *atypical* of classical mechanical systems. We now appreciate that a "typical" mechanical system (of which there are many even simple examples) can exhibit chaotic behavior so that we have really *no* predictive power about its long-term future behavior. That is, our intuition about the nature of classical mechanical systems was seriously wrong for about 300 years! The formalism (or equations) of classical mechanics has not changed, but, for many people, the interpretation most definitely has (although one could still accept an ontological determinism). With this as an elementary illustration of the difference between a formalism and its interpretation, let us now turn to the case of quantum mechanics.

3. THE "COPENHAGEN" INTERPRETATION

Entire books[6–8] have been written on the formalism of (nonrelativistic) quantum mechanics and we intend here only to sketch in the briefest (if somewhat vague) form, in terms of a few simple rules, the types of postulates that are usually employed in making quantum-mechanical calculations.*

1. A state vector (e.g., ψ)—a vector, in a Hilbert space $\mathcal{H}$, representing the state of the physical system.
2. A dynamical equation (e.g., the Schrödinger equation),

$$\mathrm{H}\psi = i\hbar\frac{\partial\psi}{\partial t} \tag{3}$$

   giving the time evolution of the state vector ψ under the influence of the Hamiltonian H for the physical system.
3. A correspondence between (hermitian) operators A in $\mathcal{H}$ and physical observables a. These physical observables a can take on only the eigenvalues a_j where

$$A\psi_j = a_j\psi_j \tag{4}$$

*No claim is made that these postulates are complete, independent, or the most general ones possible. They are intended only as an *illustration* of a formal structure when a state vector ψ can be used to represent a specific physical situation.

4. Ensemble averages for a series of observations of a given as $\langle\psi|A\,|\psi\rangle$.
5. A projection postulate (either explicitly or effectively assumed) upon measurement

$$\psi = \sum_k a_k \psi_k \rightarrow \psi_j \tag{5}$$

From this formalism follows the Heisenberg uncertainty, or indeterminacy, principle. This is related to Bohr's complementarity principle. For our purposes here, a special case of complementarity that will serve as an illustration is the well-known wave–particle duality according to which a physical system (e.g., an electron or a photon) behaves *either* as a wave *or* as a particle, depending upon the context or environment. Applications of the formalism of quantum mechanics to (idealized) position–momentum measurements, double-slit arrangements, and the like lead to a picture, or interpretation, in which definite space-time trajectories cannot be maintained, specific possessed values of observables (such as *all* components of spin) are not possible at *all* times, event-by-event causality must be abandoned (to be replaced, perhaps, by "statistical causality," whatever that may be), the process of measurement (the necessary and sufficient conditions for which are not spelled out *in advance* of its occurrence) assumes a central and highly problematic role in nature (i.e., the projection postulate or collapse of the wave function), and the passage to a classical limit (in terms of an underlying physical ontology) defies any coherent description. An examination of the formalism in specific Einstein–Podolsky–Rosen (EPR) or Bohm correlation-type experiments shows the nonseparable nature of the theory and this gives rise to correlations that may imply the existence of nonlocal influences between spatially separated regions (really, at spacelike separations). So, on the Copenhagen interpretation of quantum mechanics, physical processes are, at the most fundamental level, both inherently indeterministic (perhaps acausal?) and nonlocal. The ontology of classical physics is dead.

The Bell[(9)] theorems play an important role in this discussion.[(10)] The Bell inequality (of which there are actually several versions) and experimental results[(11)] undercut the possibility of a local, deterministic theory to account for quantum phenomena. Also, the theory that is already widely accepted, namely quantum mechanics with the Copenhagen interpretation, is *both* indeterministic *and* nonlocal. So, it *is* tempting to leave things alone and stay with the present theory.

4. A LOGICALLY POSSIBLE, EMPIRICALLY VIABLE ALTERNATIVE: CAUSAL INTERPRETATIONS

Now that we have indicated that the *formalism* of quantum mechanics is not identical with, or need not include, the Copenhagen *interpretation* of that formalism, let us outline an alternative, equally as empirically adequate interpretation of that same formalism. Once the existence of such a "causal" interpretation has been pointed out, we then turn to the question of its historical origin and of its fate. Perhaps the most direct way to introduce this is to discuss David Bohm's causal interpretation.[(12)] His basic idea is the following.* Beginning with the (non-relativistic) Schrödinger equation (which is *accepted, not derived*, there)

$$i\hbar\frac{\partial\psi}{\partial t} = \frac{-\hbar^2}{2m}\nabla^2\psi + V\psi \tag{6}$$

one defines two *real* functions R and S as

$$\psi = R\exp(iS/\hbar) \tag{7}$$

Substitution of Eq. (7) into Eq. (6) and separation of the real and imaginary parts of the resulting expression yields

$$\frac{\partial R}{\partial t} = \frac{-1}{2m}[R\nabla^2 S + 2\nabla R\cdot\nabla S] \tag{8}$$

$$\frac{\partial S}{\partial t} = -\left[\frac{(\nabla S)^2}{2m} + V - \frac{\hbar^2}{2m}\frac{\nabla^2 R}{R}\right] \tag{9}$$

The *quantum potential* U is defined as

$$U \equiv -\frac{\hbar^2}{2m}\frac{\nabla^2 R}{R} \tag{10}$$

With the definition $P = R^2 = |\psi|^2$, Eq. (8) can be rewritten as

$$\frac{\partial P}{\partial t} + \nabla\cdot\left(P\frac{\nabla S}{m}\right) = 0 \tag{11}$$

If U were identically zero, then Eqs. (9) and (11) together would represent a continuous "fluid" of particles of momentum

$$\mathbf{p} = \nabla S \tag{12}$$

following well-defined classical trajectories. With this assignment for $\mathbf{p} = m\mathbf{v}$, the $P = |\psi|^2$ of Eq. (11) can be given the interpretation of a probability density for the

*A reader not interested in seeing the origins of the "Newtonian" form of the quantum-mechanical equation of motion can simply skip to the next paragraph.

distribution of particles, since Eq. (11) then becomes the standard continuity equation. However, even when $U \not\equiv 0$, we can use Eq. (12) to write

$$\begin{aligned}\frac{\mathrm{d}\mathbf{p}}{\mathrm{d}t} &= v_j \frac{\partial}{\partial x_j}(\nabla S) + \frac{\partial}{\partial t}(\nabla S) = \frac{1}{m} p_j \frac{\partial}{\partial x_j}(\nabla S) + \frac{\partial}{\partial t}(\nabla S) \\ &= \frac{1}{m} \nabla S \cdot \nabla(\nabla S) + \frac{\partial}{\partial t}(\nabla S) \\ &= \nabla \cdot \left[\frac{1}{2m}(\nabla S)^2 + \frac{\partial S}{\partial t} \right] \end{aligned} \tag{13}$$

The last form of this expression, plus Eq. (9), imply that

$$\frac{\mathrm{d}\mathbf{p}}{\mathrm{d}t} = -\nabla(V + U) \tag{14}$$

or that

$$\frac{\mathrm{d}\mathbf{p}}{\mathrm{d}t} = \mathbf{F} \tag{15}$$

where $\mathbf{F}$ is the gradient of the potential energy, $V + U$. This potential energy now includes the familiar "classical" potential energy V as well as the "quantum" potential energy U. The quantum potential of Eq. (10) introduces highly nonclassical, nonlocal effects.

The reader, of course, need not be particularly concerned about or interested in the mathematical manipulations displayed in Eqs. (6) through (15). What is relevant, though, for our purposes here is that the dynamics of quantum mechanics can be put into the "Newtonian" form $\mathbf{F} = m\mathbf{a}$ [Eq. (15)] and given a causal interpretation in which microscopic particles, such as electrons, follow well-defined trajectories in space-time. However, because of the influence of the quantum potential, these trajectories are very sensitive to the initial conditions $(\mathbf{r}_0, \mathbf{v}_0)$ of the particles.* Let us put the following gloss on this presentation of Bohm's (1952) "causal" interpretation. The "wave function" ψ represents the effect of the environment on the microsystem under consideration (here, a particle of mass m). This ψ is a solution to the Schrödinger equation (by analogy, perhaps a "generalization" of Poisson's equation, which determines the potential V in classical mechanics) and it yields the quantum potential U via Eq. (10). This is the *fundamental* (ontological or epistemological) meaning of ψ. The causal interpreta-

*Although Bohm's original papers were written in 1952, well before the advent and popularity of modern chaos theory, his general approach and several of his insights are forerunners of, and certainly consonant with, this current field of activity.

tion and the standard Copenhagen one are based on the same formalism and are indistinguishable in their predictions if the following three assumptions are made[13]:

1. The field ψ satisfies the Schrödinger equation [Eq. (6)].
2. The particle velocity is restricted to $v = (1/m)\nabla S$ [Eq. (11)].
3. The precise location of a particle is not predicted or controlled, but has a statistical (ensemble) distribution according to the probability density $P(\mathbf{x},t) = |\psi(\mathbf{x},t)|^2$.

These are logically independent assumptions. In particular, notice that ψ plays *very* different roles conceptually in (1) and (3). In our gloss of the formalism presented above, (1) and (2) would be taken as representing the quantum dynamics of a microsystem (influenced by a quantum potential through the wave function ψ). In response to why it should in addition be the case that the probability density P should *happen* to have the value $|\psi|^2$, Bohm[13] has given an argument to show that any initial P such that $P \not\equiv |\psi|^2$ would be "driven" to $P = |\psi|^2$ by random interactions and by the quantum dynamics [(1) and (2) above], much as an arbitrary initial distribution in (classical) statistical mechanics is driven to an equilibrium (Maxwell–Boltzmann) one through random interactions.*

It is not our purpose here to show in great detail the empirical indistinguishability of this and of the standard interpretation of quantum mechanics. Such details can be found elsewhere.[12,14] We have indicted how a radically different interpretation can be based on the standard formalism of quantum mechanics. It is also worth pointing out that there is no measurement problem in this causal interpretation and no collapse of the wave function,[12] although all of the standard results, such as the Heisenberg uncertainty relations, still obtain. That is, there is no ontological rift between the classical and quantum worlds or domains. Furthermore, it is easy to state precisely when a system behaves classically: when the quantum potential [Eq. (10)] is negligible. Since Eq. (10) involves the wave function ψ, this becomes a property that the wave function must satisfy to be in the classical domain (a much more coherent criterion than some unrealizable and conceptually ill-defined "limit" such as $\hbar \rightarrow 0$). Bohm's causal interpretation gives us a more (nearly) understandable (picturable) view of microphenomena than does the Copenhagen interpretation, which merely leaves us with nothing comprehensible to say about the detailed physical behavior of a system between one preparation and a subsequent measurement. True, at this level of the causal interpretation, or "theory," we have no understanding of the physical origin of the highly nonlocal quantum potential U that is responsible for those nonseparable features that are the hallmark of specifically quantum phenomena. Still, we

*Of course, form a purely logical point of view, one could simply demand (3) by fiat, as is essentially done for the Copenhagen interpretation.

are better off with regard to understanding than with the Copenhagen interpretation. There is a reasonable analogy, perhaps, with classical Newtonian gravitational theory with its (instantaneous) action at a distance. *That* property remained a mystery, even though a causal story could still be told about, say, planetary motion. A successor theory, Einstein's general theory of relativity, replaced action-at-a-distance with gravitational waves propagating through a space-time geometry.

Another aspect of the causal quantum mechanics program has been various attempts to provide a physical underpinning for that interpretation. This project has proven to be quite difficult and it is important both to distinguish these efforts from Bohm's logical exercise of his interpretation (outlined above) and to appreciate that Bohm's substructure in terms of his implicate order[(14)] is quite different from Vigier's covariant ether approach.[(15,16)] We do not have space here to discuss these programs. At present, neither is complete and without its problems.

5. A MODERN "DOUBLE-SLIT" EXPERIMENT

Let us illustrate how the Copenhagen and causal interpretations of quantum mechanics handle the results of modern neutron interferometry experiments performed by Helmut Rauch and his co-workers in recent years in Vienna. This is the modern-day version of the familiar double-slit experiment. Figure 1 illustrates the experiment. The planes of a *single* crystal are used to Bragg-reflect/transmit an incident, polarized, monochromatic (really, monoenergetic) beam of neutrons into two coherent subbeams (of equal intensity). The (four) crystal planes (all actually part of *one* single crystal that has been machined to form this interferometer) perform the same function as the double-slit arrangement for the older optical-type experiments. The small vertical arrows in Figure 1 represent the spin of the neutron. The incident beam of neutrons is completely polarized, which means here that the neutron spin points up along the z axis. The magnetic spin-flip

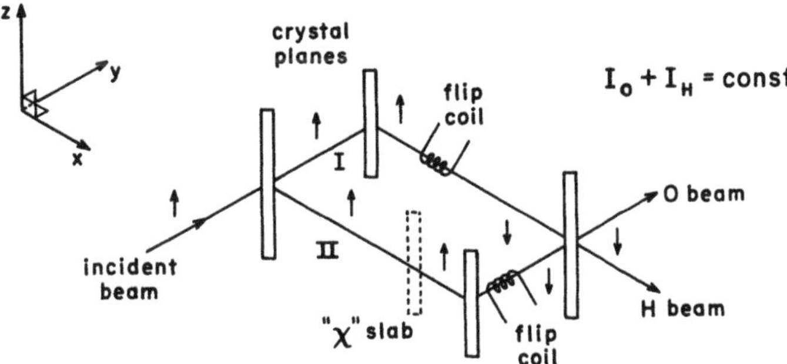

FIGURE 1. A neutron interferometry experiments.

coils in beams I and II flip (with essentially 100% efficiency) the spin of the transmitted neutrons (to spin down along the z axis). The intensity of the incident beam is sufficiently low that just *one* neutron at a time is in the interferometer. (That is, a neutron enters and leaves the interferometer before the next neutron is incident upon the interferometer.) Consequently, each neutron can only interfere with *itself*—a truly quantum phenomenon. The radio-frequency spin-flip coils operate by exchanging a *single* photon of energy $\hbar\omega_r$ with the neutron (where ω_r is the resonant frequency at which the coil is driven). Each coil is driven at slightly different resonant frequencies, ω_{r_1} and ω_{r_2}. Hence, the two beams, when recombined at the crystal plane on the far right of Figure 1, have slightly different energies (or "frequencies") and so will exhibit a "beat" phenomenon in the intensity of this recombined beam (labeled the O beam). A straightforward application of the *formalism* of quantum mechanics (independent of any particular interpretation)[17] leads to the prediction that the intensity of the recombined O beam should have the time variation*

$$I_o(t) \sim 1 + \cos[\alpha - (\Delta\omega)t] \tag{16}$$

where

$$\Delta\omega \equiv \omega_{r_1} - \omega_{r_2} \tag{17}$$

The actual experimental results[17] show this interference or beat effect (see Figure 2). That is, even though the "neutron" (via beam I or beam II) has exchanged a *single* photon of definite energy with *one* of the flip coils, no "collapse" of the wave function has taken place, since interference effects are exhibited in the recombined O beam.†

On the Copenhagen interpretation, we are effectively stranded with the formalism and its predictions, leaving as a mystery just how the neutron (wave? particle?) interacts with *one* flipper and yet produces interference. The exchange of a photon with one flipper would seem to indicate the neutron behaves as a

*To simplify the form of Eq. (16), we have written as the *constant* phase factor α what is actually the sum of a nuclear phase shift χ (produced by the "χ" slab of Figure 1) and a phase difference Δ between the two radio-frequency (rf) generators driving the two flip coils. However, all that is important for our discussion is that $I_0(t)$ of Eq. (16) has a time dependence [i.e., the $(\Delta\omega)t$ term in the argument of the cosine].

†A point for development at length elsewhere is that one might expect, on the basis of an "intuition" based on the Copenhagen interpretation, that the photon exchange with one of the coils could (in principle, at least) constitute a *measurement* process, in which case there could be no interference in the final O beam. This is indicative of the problematic status of the projection postulate, which is to be applied whenever a measurement occurs. Unfortunately, "Copenhagen" does not specify (in advance) necessary and sufficient (physical) conditions for an (arbitrary) interaction to constitute a measurement. When a measurement has occurred (i.e., no interference, say), then one applies, *after the fact*, the projection postulate. This is, at least, an incompleteness in that interpretation.

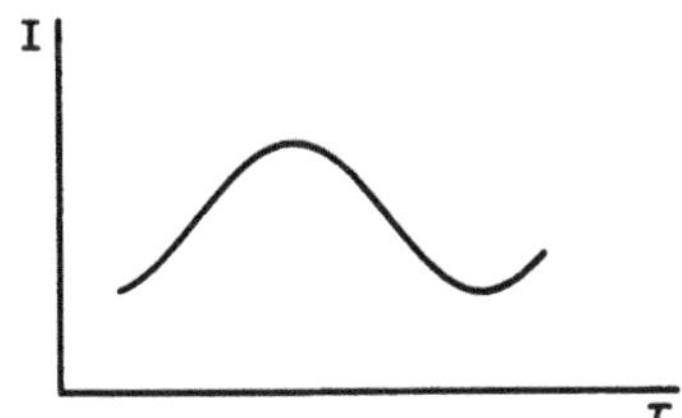

FIGURE 2. The experimental interference curve.

particle, yet subsequently it behaves as a *wave* (part traveling along path I and part along II within the interferometer) to produce interference in the emerging O beam. Copenhagen intuition (perhaps only a folklore gloss, though) typically leads one to expect wave *or* particle, depending upon the environment or experimental arrangement. On the other hand, the causal interpretation (as outlined earlier in this chapter) assigns *both* a wave *and* a particle to the neutron. Detailed calculations[18,19] using the quantum potential show how energy and angular momentum are transferred by the quantum potential between the environment and a *localized* neutron (a *particle*) that passes through just one coil. Whether or not such experiments and analyses of them take the discussion of rival interpretations beyond the level of mere preference and empty heated debate may still depend upon one's prior predilections pro or con "Copenhagen." It remains relevant, though, that the *experimentalist* largely responsible for this neutron interferometry work remarks:

> All of the observed interference phenomena can easily be explained in terms of wave mechanics, but it has to be considered that well-defined particle properties can be attributed to the neutron, too. . . . So far it can hardly be imagined how the neutrons propagate as localized particles through the interferometer.[20]

Similarly, the *theorist* Jean-Pierre Vigier sums up the tension between these two interpretations or world views as:

> We are confronted by a stark alternative. Either (i) we renounce the independent existence of the neutron and with it any possibility of describing what happens in the neutron interferometry experiments. There exists then no possibility of explaining quantum phenomena, not even in terms of a wave/particle duality which only leads to ambiguity. Individual quantum phenomena are in principle and irreducibly indeterminist in character and there can be no form of physical determinism appropriate in the quantum domain or (ii) we adopt the quantum potential approach as the only known consistent manner in which the quantum world can be conceived and explained in terms of a physically determinist reality. Then, even if the quantum potential approach is not taken as the finally satisfactory description of quantum mechanical reality it at least shows in a clear way the features that such a description must entail.[21]

Our purpose here has not been to resolve this tension definitively one way or the other. Rather, it has been to outline a case that at least one viable interpretation alternative to the widely accepted Copenhagen view exists. Even today this issue remains actively debated so that a causal interpretation need not be rejected on grounds of logic or empirical adequacy.

6. THE VALUE OF AN ALTERNATIVE INTERPRETATION

However, before we turn to the historical details of the emergence of quantum mechanics, let us consider some of the motivations for and values of studying an interpretation alternative to the standard Copenhagen one.

We might feel that there is little point in such an exercise since "Copenhagen" works and is consistent. But, even if this latter claim is accepted, the question of *understandability* remains. Does the Copenhagen interpretation give us a description of the world that we can understand in any meaningful sense of that term? That question is certainly open to debate. The quest for a more (nearly) understandable world view can be a motivating factor in seeking another interpretation of a quantum formalism.* These and other relevant factors have been concisely and elegantly stated by David Bohm in his classic 1952 paper:

> The usual interpretation of the quantum theory is self-consistent, but it involves an assumption that cannot be tested experimentally, *viz.*, that the most complete possible specification of an individual system is in terms of a wave function that determines only probable results of actual measurement processes. The only way of investigating the truth of this assumption is by trying to find some other interpretation of the quantum theory in terms of at present "hidden" variables, which in principle determine the precise behavior of an individual system, but which are in practice averaged over in measurements of the types that can now be carried out. In this paper and in a subsequent paper, an interpretation of the quantum theory in terms of just such "hidden" variables is suggested. It is shown that as long as the mathematical theory retains its present general form, this suggested interpretation leads to precisely the same results for all physical processes as does the usual interpretation. Nevertheless, the suggested interpretation provides a broader conceptual framework than the usual interpretation, because it makes possible a precise and continuous description of all processes, even at the quantum level.(25)
>
> . . .
>
> As a matter of fact, whenever we have previously had recourse to statistical theories, we have always ultimately found that the laws governing the individual members of a statistical ensemble could be expressed in terms of just such hidden variables.(26)
>
> . . .
>
> The usual interpretation [i.e., in its finality and completeness] . . . presents us with a considerable danger of falling into a trap, consisting of a self-closing chain of circular hypotheses which are in principle unverifiable if true.(27)

*This is not the first time philosophers of science have discussed alternative causal interpretations of quantum mechanics.(22–24)

7. OPPOSING COMMITMENTS, OPPOSING SCHOOLS*

If neither empirical (in)adequacy nor logical (in)consistency provides a sufficient explanation, perhaps we might profitably consider other factors in such a choice between theories. Criteria such as fertility, beauty, and coherence, while often important, can have a Whiggish aspect to them if they are defined in terms of the successful, victorious, or accepted theory and then applied to a competing theory. We shall discuss such criteria when applied to the Copenhagen versus the causal interpretation. However, let us now turn to the influence of contingent historical events in the development of quantum theory. Could the philosophical outlooks and backgrounds of the creators of the "Copenhagen hegemony" have been important factors?

While it is not a new insight that there were two essentially independent routes to quantum mechanics, one leading to wave mechanics and the other to matrix mechanics,[28] it is nevertheless important for our purposes here to see what were the basic philosophical or methodological commitments of these two opposing "schools." This difference is often characterized loosely as determinism versus indeterminism. Although we indicate below that such a characterization is both too simplistic and also misplaced in its focus, we want first to emphasize that the concept of indeterminism as an essential feature of nature did not for the first time become a seriously considered option just in the early part of the 20th century with the advent of quantum physics. While it does not appear defensible to take seriously at face value a strong Forman-type thesis[29–31] according to which (external) social factors determine the very content of science, it is still the case that science is undeniably a historical entity in which contingency does play a major role. That is, the cultural/philosophical milieu within which science develops and fashions its theories does provide a background of ideas and concepts that can influence (but neither uniquely nor solely) the direction of science. This is not intended as a denial of the crucial role played by internal factors (such as logical deduction and consistency, empirical adequacy, the scope of a theory, etc.), but it is meant to indicate that such external factors are not irrelevant to the scientific enterprise.[4] Even though science often creates new concepts or significantly modifies current ones, still, those concepts on offer at any given time can exert an influence on the debate over theory selection.

7.1. *Cultural Milieux*

By the late 19th century there were significant philosophical precedents for the concept of indeterminism in nature, as opposed to the straightforward determinism often associated with classical physics.[32,33] For example, Charles-Bernard Renouvier (1815–1903) questioned the causality principle for physical processes, challenged the (Kantian) doctrine that acceptance of causality was a

*There is today a huge secondary literature on the early history of quantum mechanics and here we can only touch on it.

precondition for human understanding, and held that an object and its representation cannot be divorced even in principle. Émile Boutroux (1845–1921) stressed contingency and inherent (i.e., not just representing *our* ignorance) chance in nature's actions. Their thinking opposed belief in a completely rational universe and proposed instead an element of irrationality in a nature having contingent laws. In a given situation, there would be equally possible alternatives to what does, *in fact*, occur. Henri Poincaré (1854–1912) read and was influenced by Renouvier, while Louis de Broglie and other founders of the quantum theory studied the writings of Poincaré. It has been quite well documented that Søren Kierkegaard's (1813–1855) philosophy made an impression on Niels Bohr through the teachings of Harald Høffding (1843–1931). Not only had Bohr attended Høffding's lectures as a student, but he also read his works and corresponded with him later in life. Bohr explicitly acknowledged the influence of Høffding's philosophy on his own formulation of complementarity.(34) One of Høffding's tenets was that in life decisive events proceed through sudden "jerks" or discontinuities. The point here of these brief comments is not that these philosophical currents alone determined the course of quantum theory in the early part of this century, but rather that these concepts were available to, and in the minds of, the creators of quantum theory and, we shall argue, did exert *an* influence on the choice of the final, "accepted" form of quantum theory. And, of course, logical positivism was a part of this backdrop with its emphasis on the central role of the empirical determination or definition of a term.

Consonant with this theme is Brush's study(35) of the philosophical background to quantum mechanics. Brush sees the rise of indeterminism (or the fall of determinism) as having its roots in the 19th century and growing gradually. For him, the primary opposition of concepts in the evolution of quantum mechanics was realism versus instrumentalism ("positivism"). In his reconstruction, the acceptance or consideration of indeterminism had its roots in thermodynamics. In the larger arena of Western intellectual history, Brush contrasts romanticism with realism and claims there has been an oscillation between these two poles and that this has gone through several cycles in modern times. To look ahead a bit in our argument, we can see how these different philosophical commitments came to be focused, through an examination of specific problems in quantum physics, into the following opposing positions.(36,37)

Einstein (yes)		Bohr (no)
	light quanta	
	space-time	
	differential equations	
	continuity and causality	

Our goal here is to see what were the *prior* philosophical or metaphysical commitments that the key figures in the formulation of quantum theory brought with them to their study of problems in physics. We do not claim that such commitments *alone* determine the theory formulated and finally accepted. Logic, data, and fertility are central factors, but, we argue, neither are they *alone* fully determinate of the theory. *Both* sets of factors must be taken into account. Of course, *if* the final decision between the two theories were a wholly rational one, *independent of* these prior metaphysical commitments (and this is by no means obvious in the case of quantum mechanics), then the origins of and reasons for those commitments, while perhaps interesting in themselves, would actually be irrelevant for the ultimate theory choice and for its "logical" justification. (This is just the usual distinction between the logic of discovery and the logic of justification.)

Since the Einstein–de Broglie–Schrödinger route will be considered first, let us look at the roots of Einstein's philosophical commitments.

7.2. *The Wave-Mechanics Route*

Degen[38] argues that Einstein's general position on foundational questions in physics—relativity, his stance on quantum theory, and his long-standing commitment to a unified field theory—can best be understood as a search for the God of Spinoza. Einstein believed, according to Degen, that ". . . this God manifests himself in the rational structure of the external physical world, which the physicist tries to capture in a causal space-time theory." Einstein's basic *Weltanschauung* was that of a rational, causal world that could be comprehended in terms of an objective reality. He also acknowledged explicitly the philosophical influence on him of Poincaré and Mach.[39,40] It is, of course, the intersection of general predilections like these with the puzzles presented by physical phenomena that results in a definite theory or research program.

To illustrate this, we now sketch how the opposing positions—wave- versus matrix-mechanics—were arrived at and what the arguments were for each view. The one traces its roots back to the nature of electromagnetic phenomena, the other to a study of (the discrete) spectral lines. So, not only does it turn out that the general philosophical outlooks of these two groups of key players were quite different, but also it was with very different classes of physical phenomena that each group began. The discontinuity versus continuity dichotomy is contingently rooted in philosophical commitments and in physical phenomena.

In 1909 Einstein used the Planck black body radiation law

$$\rho(\nu, T) = \frac{8\pi\nu^2}{c^3}(h\nu)\left\{\exp\left(\frac{h\nu}{kT}\right) - 1\right\}^{-1} \tag{18}$$

and Planck's energy quantization condition

$$\epsilon = h\nu \tag{19}$$

to compute[41] the mean square energy fluctuation for blackbody radiation in thermal equilibrium with another system[42]

$$\langle \epsilon^2 \rangle = (V d\nu)\left\{ h\nu\rho + \left(\frac{c^3}{8\pi\nu^2}\right)\rho^2 \right\} \tag{20}$$

Einstein pointed out that the first term would be expected if radiation were composed of independent particles (his "photons") and the second term if radiation consisted of waves that could interfere with each other. At a conference in Salzburg in 1909, he stated[43]:

> It is my opinion, therefore, that the next phase of the development of theoretical physics will bring us a theory of light that can be interpreted as a kind of fusion of the wave and emission theories.[44]

In retrospect, we tend to see this as an early flirtation with the concept of wave–particle duality.

In a paper on the quantum theory of radiation published in 1917, Einstein[45] showed that when molecules emit and absorb radiation under the influence of an external radiation field, then momentum and energy must each be conserved, so that such radiation was termed "needle radiation" by Debye.[46] It is well known that Compton's experimental work[47] on the scattering of radiation by electrons gave support to the hypothesis of free electromagnetic quanta.[48] In 1922 Einstein and Ehrenfest analyzed the implications of the Stern–Gerlach results and concluded that either energy conservation had to go or only radiation-emitting systems can be quantized.[49] This was an early indication of the conceptual difficulties encountered when one attempts to construct pictures of atomic events. This would sharpen into a conflict between a representation in a continuous space-time background ("visualizability" of microevents) versus strict (event-by-event) energy conservation (and causality). In this same 1917 paper, Einstein stated that the recoil direction of the molecule, which has emitted radiation, is ". . . only determined by 'chance,' *according to the present state of the theory*" (italics added) and that "the weakness of the theory lies . . . in the fact . . . that it leaves the duration and direction of the elementary process to 'chance.' "[50] Here, as later, Einstein took this to be a shortcoming of the theory, a provisional fault to be overcome (hopefully) in the future.

The next key figure in this "continuity" school is, of course, Louis de Broglie. Not only was there little interest in quantum physics in France after World War I[51] (and France and Germany remained scientifically insulated from each other then), but even earlier in the century, theoretical physics there had fallen very much behind the times. Léon Brillouin (1889–1969) recalls:

> The situation in France was serious. There was no regular course of applied mathematics when I attended the Sorbonne—nothing at all. . . . There were really few people working in theoretical physics in France at that time. Theoretical physics was really at a low level when I was a student.[52]

There was certainly no existing tradition of research in quantum theory in France when Louis de Broglie was a student. What theoretical research there was was dominated by classical physics. It was Louis's brother, Maurice, who was actively involved in experimental work on the photoelectric effect and who introduced the younger de Broglie to quantum physics. Abragam[(53)] has pointed out that Louis de Broglie was poor in foreign languages and rarely went abroad. In 1923 de Broglie initiated the theory of wave mechanics in attempting to understand the dual nature of Einstein's photon. In his early youth, Louis de Broglie had been impressed by the well-known formal (mathematical) analogy between wave optics and classical particle mechanics.[(54)] Building on this analogy and on some of his own previous work[(55–57)] on Einstein's light quanta, de Broglie built a model of a particle that follows the trajectory of its associated phase wave. There was a great affinity of views between Einstein and de Broglie. In 1925 Einstein stated his opinion that de Broglie's ideas "involve more than merely an analogy."[(58)] This is not surprising since Einstein had previously, in the context of general relativity, attempted to treat "particles" as the singularities in an underlying field.

This notice by Einstein drew attention to de Broglie's work. Heitler recalls:

> Upon this remark by Einstein, de Broglie's paper was widely studied in Germany. I was a student at the time preparing for my Ph.D. in Munich under Sommerfeld, and de Broglie's paper was discussed there too. Everyone had objections (they were not very difficult to find) and no one took the idea seriously. I believe this applied to most theoretical physicists—except Schrödinger.[(59)]

Indeed, Schrödinger recalled that "My theory was stimulated by de Broglie's thesis and by short but infinitely far-seeing remarks by Einstein."[(60,61)] In a paper on Einstein's gas theory, Schrödinger[(62)] concluded that photons can be seen as the energy levels of the "aether" oscillators, that cavity radiation need not ". . . correspond to the extreme light-quantum representation" and that

> This means nothing else but taking seriously the de Broglie–Einstein wave theory of moving particles, according to which the particles are nothing more than a kind of "wave crest"* on a background of waves.[(63)]

In that same year, Schrödinger[(64,65)] exploited Hamilton's analogy between mechanics and optics to obtain his wave equation. Quantization was implemented by boundary conditions imposed on a continuous wave function. It is interesting to note in this connection that Schrödinger had early on acquired a mastery of eigenvalue problems in the physics of continuous media and that he was familiar with Courant and Hilbert's book on mathematical methods in physics (in which such eigenvalue problems are treated in detail).[(66)] Also, since Schrödinger

*Actually, de Broglie's earlier theory had both a wave and a particle for each microentity, rather than just a wave with the particle represented by a crest on it.

worked out the *relativistic* case for hydrogen first (no spin, of course) and obtained an answer other than the Balmer formula and then put this work aside for some months before returning to the nonrelativistic case,[(67)] it would appear as though his motivation and results owed nothing to Heisenberg's formulation of matrix mechanics or to Pauli's solution of the hydrogen atom with it. This is consistent with Schrödinger's own comment in a note in his paper on the equivalence between wave- and matrix-mechanics:

> I did not at all suspect any relation to Heisenberg's theory at the beginning. I naturally knew about his theory, but was discouraged, if not repelled, by what appeared to me as a very difficult method of transcendental algebra, and by the want of perspicuity.[(68,69)]

Our last thread in this part of our story is de Broglie's proposal in 1927 of a principle of the double solution in which the basic entity is a wave, but having a singularity whose trajectory would correspond to a particle path.* Because of mathematical difficulties with that proposal, at the 1927 Solvay Congress he presented a simplified pilot-wave model. We return to this subject later.

What we have sketched, only in outline above, is one (historical) route to *a* formulation of quantum mechanics—namely Schrödinger's wave mechanics. It should be evident that this small group (Einstein, de Broglie, and Schrödinger) of the creators of that theory shared a commitment to a continuous wave as the basic entity subject to a causal description. Visualizability and self-consistency had become accepted hallmarks of classical physical theories. As Hendry puts it, ". . . better to have an intelligible classical wave theory with flaws than a totally unintelligible 'dual' theory."[(70)] Furthermore, the creation of wave mechanics owed essentially nothing to any interaction with the other group to whom we turn—those who shaped matrix mechanics. Again, we shall see that this second group was also fairly small, independent, and quite closed. Now the position taken by the wave-mechanics school was the more "natural" one relative to the then-accepted concepts of classical physics (i.e., it represented a less radical departure). The other program, based fundamentally upon discontinuity (and a lack, even in principle, of event-by-event causality), however, finally carried the day in spite of its much more radical nature. So, it is all the more important that we examine the arguments that were seen as convincing in this choice.

7.3. *The Matrix-Mechanics Route*

The small number of central players (Bohr, Heisenberg, Pauli, Jordan, and Born) involved in the program that led to matrix mechanics suggests that we ask whether this, too, was a closed group. This becomes all the more plausible when we realize that Pauli and Heisenberg were both Ph.D. students with Sommerfeld at

*If de Broglie had succeeded with this theory, he might have satisfied Schrödinger's wish for a theory based on waves only, rather than on waves and particles.

Munich, each in succession were then Bohr's assistants at Göttingen and later worked with Bohr at Copenhagen. Each of these young students was greatly impressed by Bohr's 1922 Göttingen lecture.[(71)] Jordan was also a student at Göttingen at this time. We have already indicated (philosophical) factors in Bohr's own background that inclined him toward, or at the very least made him receptive to, a discontinuous structure in nature at the most fundamental level and, eventually, to a doctrine of complementarity between opposites. This element of discontinuous transitions is a central feature in his 1913 "semiclassical" model[(72)] for the hydrogen atom. This was certainly the current language for discussing atomic phenomena in Sommerfeld's school. So, *prima facie*, there is a case that Pauli and Heisenberg as young students were impressionable and naturally accepted this central tenet of atomic theory. Incidentally, throughout his life, the ever-critical Pauli remained deferential toward his old teacher, Sommerfeld.[(73)] However, this *inclination* toward credence is scarcely sufficient to account for the prevailing strength of this conviction on the discontinuity versus continuity issue. Beller[(74)] sees *discontinuities* as the key issue in this formulation of quantum mechanics. Hendry[(75)] also stresses that *causality*, as such, was not the central question initially in the development of this program.

Largely due to the *failure* of certain classical approaches, the main players took up various *philosophical* positions on what was and was not *possible* in principle. That is, these were not logical or in-principle refutations, but strong, practical *beliefs* that became dogma.* Thus, Bohr's own Ph.D. dissertation argued that the failure of the classical electron theory of metals was attributable to a fundamental insufficiency of the classical principles themselves.[(76)] Pauli, in his work on general relativity and related field-theory generalizations, convinced himself (again, because of a *failure*) that a continuum field theory, with the particles as singularities, was not possible.[(77)] In his famous 1921 *Theory of Relativity*, Pauli already was of the opinion that ". . . there is no point in discussing . . . quantities [that] cannot, in principle, be observed experimentally."[(78)] And, in the same vein, we find:

> Finally, a conceptual doubt should be mentioned. The continuum theories make direct use of the ordinary concept of electric field strength, even for the fields in the interior of the electron. This field strength is however defined as the force acting on a test particle, and since there are no test particles smaller than an electron or a hydrogen nucleus, the field strength at a given point in the interior of such a particle would seem to be unobservable, by definition, and thus be fictitious and without physical meaning.[(79)]

This certainly has a strong operationalist air about it. By 1923, in a letter to Eddington, Pauli required operational definitions of anything used in physics and

*Bell's theorem provides an instructive contrast to this situation. That is, prior to, say, 1964, most physicists *believed* that a "hidden-variables" completion of quantum mechanics was impossible. However, after Bell's work it was *proven* that a local, deterministic theory, agreeing in all of its predictions with quantum mechanics, is impossible.

the replacement of continuous concepts by discrete ones.[80] After all, observations are essentially localized and instantaneous, with measurements being discrete and, he felt, this structure should be carried over into the foundation of any theory that accounts for such observations and measurements. He believed this would require major conceptual revisions. Both Pauli and Heisenberg had been involved in Bohr's program of attempting to apply the old quantum theory, with its orbitals, to molecular systems and the utter failure of this approach convinced them that electron orbitals were meaningless.[81,82] Also, a major logical difficulty of the old quantum theory with its visualizable orbitals was that, due to the quantization requirement for orbital angular momentum, the orbitals had to be oriented in certain special ways in any field, *however weak*.[83] If a field in one direction were reduced slowly to zero and then gradually increased to some small value in another, it is not at all clear how the orbit could properly align itself (rather than simply precessing about the initial axis). The Stern–Gerlach experiments (1921–22) supported quantum predictions (notice the name—*space quantization*—originally used for this effect!), rather than a *continuous* (classical) magnetic moment.[84] Pauli's success with the Zeeman effect (1924) in terms of a classically nondescribable two-valuedness in the quantum-theoretical properties of the electron further strengthened his belief in nonvisualizability.[85] He was convinced as well that the exclusion principle could not follow from classical mechanics or from the old quantum rules.[86] Heisenberg began to be converted to Bohr's and Pauli's views on the failure of mechanics.[87] The failure of the Bohr–Kramers–Slater theory (in which energy conservation had been given up) in 1925 indicated to Bohr that ". . . a complete renunciation of the usual space-time methods of visualization of the physical phenomena. . ." would be necessary for further progress.[88] Heisenberg's matrix mechanics provided this.

Dirac was an exception in this group in that he had no particular interest in philosophical questions.[89] He admitted finding de Broglie's ideas beautiful, but could not take those waves seriously because he was so much embedded in the (old) Bohr theory with its orbits (which he took literally).[90] Dirac cared only about the equations and what could be calculated with them. Later, he took from Bohr the rejection of mental pictures in space-time. Perhaps like many (most?) physicists, Dirac was willing to leave the philosophical considerations to someone else. Dirac also made the interesting observation that, in his opinion, part of the reason for the great impact that general relativity made in 1919 after the end of World War I was the need to forget the old and to focus on something wonderful and new.[91] This desire for a radical conceptual revolution was prevalent in the general cultural milieu of the time[92] and also, as we show below, in Pauli's and Heisenberg's expectations about quantum theory.

It was the "collision" between matrix mechanics and wave mechanics that provided the impetus for the formulation of a consistent interpretation of quantum mechanics. Although it is not uncommon for scientists to believe that there is just one (unique) law or theory, Kalckar[93] recounts that Bohr even as a child believed in the uniqueness (necessity) of natural laws. Such a belief would justify one in looking for, or attempting to formulate, *the* correct version of quantum mechanics. Heisenberg's faith in the finality of quantum mechanics was essential for his struggle to fashion the "Copenhagen" interpretation via his uncertainty relations. Beller puts this very forcefully:

> In his recollections, Heisenberg repeatedly stressed that his belief in the completeness of this mathematical scheme of quantum mechanics led him to assume that nature works only in such a way as not to violate the quantum mechanical formalism. Heisenberg's recollections regarding his belief are supported by correspondence at the time. This belief in the completeness of the mathematical scheme was essential—without it, Heisenberg, in the case of a discrepancy between "nature" and "formalism," would seek to improve the formalism rather than to reinterpret nature. It is in this context . . . that Einstein's dictum ("it is the theory which decides what we can observe") "suddenly" approached the status of a guiding principle.(94)

Born believed that microscopic coordinates were unmeasurable and, therefore, irrelevant.(95) Through his analysis of scattering processes with Schrödinger's formalism, Born came to the opinion that even *perfect* initial information still led to uncertainty in the result and this implied, for him, a lack of causality.(96) Apropos of these issues, Heilbron(97) asks what was the warrant for the complementarity principle as being complete and the final word in forbidding even the in-principle possibility of a description of microphenomena that is both causal and pictured in a continuous space-time. His response is that (*thus far*) experience has shown the validity of complementary pairs of descriptions and that belief in the ultimate *necessity* of complementarity rests on the *subjective* epistemological criterion of the need for classical concepts and on the indivisibility of atomic phenomena (i.e., Bohr's act of faith). Rosen summarized Bohr's position as "[physical] reality is whatever quantum mechanics is capable of describing."(98) In Heilbron's representation, the Copenhagen interpretation defined itself as true and strengthened its hold on physics, rewriting history so that Einstein, de Broglie, and Schrödinger largely fade from view, thus leaving "Copenhagen" as the only intelligible version of quantum mechanics.(99)

Interestingly enough, it was some of the more mathematically inclined (and at times less philosophically committed) contributors who maintained an openness on the question of the (physical) interpretation of the formalism. According to Hendry:

> [Dirac] took great care to keep formalism and interpretation distinct and emphasized that the probabilistic interpretation did not follow from the formalism, as Born had suggested, but must rather follow from a separate association of theoretical and physical terms that included probabilistic assumptions.(100)

Similarly, Jammer tells us that Hilbert recognized a certain freedom of choice in the interpretation of a formalism.(101) These were, in fact, intertwined in the *historical* development of matrix mechanics. This contingency made it easy to accept a particular interpretation as an essential part of the formalism. Heisenberg's position on the relation between *the* formalism and *the* interpretation fits such a scenario. A bit later we expand on this point.

Let us interject here that, once the Copenhagen interpretation had established its hegemony in Europe, this hegemony was essentially automatically extended to the United States. In her study of quantum physics in America prior to 1935, Sopka[(102)] shows that the American connection of young physicists (e.g., Urey, Lindsay, Dennison, Slater) was mainly through Copenhagen, Göttingen, and Cambridge (where the orthodoxy had already spread). Many American physicists were first alerted to the "new quantum theory" by Max Born when he visited MIT and other U.S. institutions in the winter of 1925–1926.[(103)] Of 32 American visitors to Europe, all went to "Copenhagen"-doctrine centers (and none, for example, to France).[(104)] America was still in a learning, catch-up phase prior to, say, 1930, and the pragmatic American approach to quantum mechanics led to an acceptance here, by and large without philosophical qualms. In the United States, notable achievements in quantum theory were in applications of a largely established formalism.[(105)]

One can perceive, perhaps, a parallel (and not just in America) between the rapid spread of orthodox quantum mechanics (once the formalism had been fixed and the Copenhagen interpretation forged) and Newtonianism. There was an emphasis, by most of the rapidly growing band of practitioners, on computation as opposed to thinking about foundational questions. It is simplest and most efficient for disciples to follow the path of the master and use the scheme to calculate, carrying along without much reflection a certain amount of philosophical "baggage."

8. FORGING THE "COPENHAGEN" INTERPRETATION

With this as background for the historical development that led to two opposing interpretations of the formalism of quantum mechanics, we can now summarize rather briefly how the Copenhagen interpretation came to be formulated under the challenge that Schrödinger's wave mechanics presented to the Göttingen–Copenhagen matrix-mechanics program and how this latter interpretation established its hegemony. Since we have argued that matters of logic and of empirical adequacy alone are insufficient for rejecting out of hand a causal interpretation, it can be little surprise that we must turn to other factors to provide an adequate explanation.

In fact, we return to the early days of the quantum formalism and the Solvay Congress of 1927. In 1923 Louis de Broglie put forward his concept of wave–particle duality as extended to electrons and he further developed these duality arguments in his doctoral dissertation.[(106)] Although, as we have seen, Einstein was impressed with de Broglie's thesis, and Schrödinger's own work on wave mechanics was influenced by de Broglie's insight, de Broglie nevertheless had the reputation of being an unorthodox theoretician[(107)] and that did not condition the scientific community at large to consider his subsequent speculations seriously.

Also, the French scientific community had isolated the Germans for reasons of nationalism after World War I and this was not conducive to a flow of ideas between the Germans and the French.[(108)] But, of course, this can scarcely have been an absolutely determining factor since, after all, Schrödinger was Austrian and *was* receptive to de Broglie's ideas. We have argued that a crucial factor in the relevant historical developments was a split in philosophical outlook along generational lines: the "older" essentially classical, worldview of people like Einstein, Schrödinger, and de Broglie versus a radically different, eventually indeterministic conception of physical processes engendered by the younger generation including Heisenberg, Pauli, Jordan, and Dirac. This tension between what we might term the "Nordic gang" (e.g., Bohr, Heisenberg, Pauli, Jordan, Born, with Dirac being an "honorary" member in this classification) and the likes of Einstein, Schrödinger, and de Broglie is a key element in appreciating how "Copenhagen" gained the ascendancy. Nor were the differences between these two groups *merely* philosophical. As we discuss in a bit more detail below, there was a very practical "social" or "professional" dimension to it as well.

We have cited some relevant personal, or even psychological, views of prominent members of the Copenhagen school. The prevalence of an empiricist-operationalist philosophical tendency among Heisenberg, Pauli, and Bohr can be traced in part (somewhat ironically, given Einstein's later views) back to Einstein's 1905 relativity papers. This operationalist approach seems to have made a great impression and to have exerted a profound influence upon young German physicists. Several such factors help us in understanding the vehemence of the Copenhagen school's reaction against Schrödinger's wave mechanics. Matrix mechanics had been formulated by Heisenberg, and developed by other members of the Copenhagen school, as an essentially abstract mathematical formalism with no physical interpretation. Heisenberg's views on the nature of a successful mathematical formalism of a physical theory and of its relation to an interpretation were the following.[(109)] He believed that a successful formalism, such as classical mechanics, was of a piece or whole and that it could not be modified in any essential way without destroying the entire structure. Thus, when such a formalism encounters difficulties (as classical mechanics did with quantum phenomena), it is not possible to modify that formalism successfully. Rather, a radically new formalism must be found to accommodate these new features of the physical world. (In his later years Heisenberg came to see "Kuhnian" revolutions in science as a natural outcome of this sharp break between formalisms.) Not only did Heisenberg see a successful formalism as unique and of a whole, but he also held a remarkable view of the relation between a formalism and its interpretation.

In an interview in Munich on February 22, 1963, for the *Archive for the History of Quantum Physics*,[(110)] Thomas Kuhn discussed with Werner Heisenberg Heisenberg's own reaction (in 1926) to Erwin Schrödinger's development of wave mechanics. Heisenberg believed that one simply had to examine the formalism (here, of matrix mechanics) to find its proper interpretation. This view that a formalism gives (uniquely) its own (proper) interpretation made the appearance of

Schrödinger's (apparently) very different formalism (theory) quite disturbing for Heisenberg. This possibility, coupled with Bohr's "defection" to the enemy camp when he wrote his paper on the statistical interpretation of (Schrödinger's) wave function, produced in Heisenberg (so he told Kuhn in this interview) the fear that the physics community might take a wrong fork in the road and be led into error. Before any equivalence between the two formalisms had been established, the possibility of a wrong choice was a major concern (according to Heisenberg here) so that it remained essential to find *the* correct interpretation of matrix mechanics as quickly as possible. This became a major undertaking in Copenhagen with Bohr, Heisenberg, and Pauli. Later in this interview, Kuhn pressed Heisenberg about what might have happened if the equivalence (between matrix and wave mechanics) had taken longer to establish and if the theoretical physics community had gone the ("wrong") Schrödinger route—might not physics have developed quite differently? While Heisenberg allowed that it might have *for a while*, he was certain that things would have come back to where they in fact have ended (to the truth? to matrix mechanics?). The point is that here Heisenberg affirms a belief in a *unique*, correct theory (one truth).

While Heisenberg's responses to Kuhn in this interview remain on the lofty plane of objective truth and concern with physics not being misled into a wrong turn, there were other, much more practical and mundane, factors that lent urgency to the dispute/choice between matrix mechanics (Copenhagen) and wave mechanics. As Beller[(111)] and Cassidy[(112)] have made clear, the formalism of matrix mechanics had *not* had many successful applications (and, in fact, appeared to be bogged down in a mathematical morass) before Schrödinger's wave mechanics allowed theorists to make a stunningly wide variety of well-supported calculations. Wave mechanics, not matrix mechanics, was the formalism employed by most theorists! This danger of losing the war on the calculational front threatened further consequences. Not only did Heisenberg have personal ambitions for advancement, but several chairs in theoretical physics were opening up in Germany. There was a conscious realization by members of the Copenhagen school that control of the future direction of theoretical physics was at stake. This group (our "Nordic gang") had the talent, organization, and drive to carry the day in establishing the hegemony of the Copenhagen view.[(113)] Heisenberg's uncertainty relation paper was a major step in accomplishing this. They worked in concert, while their opponents (Einstein, Schrödinger, de Broglie) pulled each in his own direction. The influence of the Bohr Institute in Copenhagen was enormous on an entire generation of leading theoretical physicists who passed through it (e.g., most of those who played dominant roles in establishing theoretical physics in the United States). It is also interesting to note that *no* French physicist ever worked at the Bohr Institute (at least not for any extended period of time).*[(114)]

*This is not to imply that there was a conspiracy by Bohr and his followers to keep the French from visiting Copenhagen. In large measure the French excluded themselves (even if unconsciously) by

A crucial encounter occurred at the 1927 Solvay Congress. In 1927 Louis de Broglie proposed a "principle of the double solution," according to which he suggested a synthesis of the wave and particle nature of matter.[(115)] At the Fifth Solvay Congress in 1927, he presented some of these ideas in a form he termed the pilot-wave theory.[(116)] Here a physical particle is pictured as being guided by the pilot wave. In discussion at that Congress, Wolfgang Pauli[(117)] criticized de Broglie's theory on the basis of the example of the inelastic scattering of a plane wave by a rigid rotor. Although de Broglie felt he understood the general outlines of a suitable response to Pauli's[(118,119)] objection, he in fact did a poor job in attempting to rebut Pauli at the 1927 Congress so that his response appeared *ad hoc* and was not convincing then.[(120)] In addition to Pauli's negative reaction to de Broglie's paper, neither Einstein nor Schrödinger gave positive support to de Broglie's ideas: Einstein because he did not like the nonlocal (or nonseparable) nature of the theory and Schrödinger because he wanted a theory based only on waves (not on waves *and* particles). And, de Broglie's reputation of being an unorthodox theoretician did not help the situation. In addition, the people who were producing results with the matrix-mechanics formalism for problems involving spin (e.g., Heisenberg and Born who also spoke at the 1927 Solvay Congress) strongly favored the indeterministic or noncausal picture.[(121)] Bohr was for a long time against the concept of the photon,[(122)] so that de Broglie's ideas had never spread rapidly in the Copenhagen school. The Institute at Copenhagen was a very closed community and those invited there were identified as the "respectable" theorists. De Broglie was never a member of this group. By 1930 when he wrote a *very* standard quantum-mechanics book, de Broglie had himself changed his mind about the pilot-wave theory: "It is not possible to regard the theory of the pilot-wave as satisfactory."[(123)] In that same book he rehearsed other arguments against the pilot-wave theory, both general conceptual ones[(124)] and a specific thought experiment involving the reflection of light from an imperfect mirror.[(125,126)] Von Neumann's 1932 impossibility "proof"[(127)] for hidden variables theories further confirmed de Broglie's position against his own previous theory.[(128)]

9. AN ALTERNATIVE HISTORICAL SCENARIO?*

There matters essentially stood until 1952 when David Bohm published two papers on a causal interpretation of quantum mechanics. (We have discussed this in Section 4.) Initially, de Broglie was against Bohm's ideas (which were similar to his own pilot-wave theory of 1927) and he raised the same objections against Bohm's theory that had been raised against his own.[(130)] Interestingly

their choice of the areas of theoretical physics in which they worked. Quantum physics simply was not a field of great activity in France at this time.

*A detailed summary and analysis of current work in the causal quantum theory program will be presented in a forthcoming case study. Discussion of some recent work on alternative interpretations of quantum mechanics can be found in Selleri (1990).[(129)]

enough, when Bohm sent Pauli a copy of the paper in which Bohm showed Pauli's objections to the causal interpretation to be specious, Pauli never responded.[131] Furthermore, Pauli's views on the nature of science and its relation to his conception of God (basically, a cosmic bookkeeper who enforced statistical causality) made it inconceivable to him that anything like a return to a "classical" world with causality and picturable, continuous processes in space-time was either possible or anything less than a disgusting loss of nerve and a return to darkness.[132] Bohm did, as we have seen, produce a causal version of quantum mechanics—one capable of a *realistic* interpretation with a largely classical (micro) ontology. Bohm's work of the early 1950s reconverted de Broglie to his former ideas.[133] For de Broglie the issue at stake was not (classical) determinism, but rather the possibility of a precise space-time representation for a clear picture of microprocesses.[134] In this, his expectations were similar to Einstein's. de Broglie felt[135] that classical Hamilton–Jacobi theory provided an embryonic theory of the union of waves and particles, all in a manner consistent with a realist conception of matter. With the concept of the quantum potential,[136] one could provide a *model* for fundamental processes. This strong commitment by de Broglie to a realistic interpretation of the quantum formalism is consistent with his own high estimate of the work of Émile Meyerson (1859–1933), as stated by de Broglie in a preface to Meyerson's *Essais* (1936).[137] Meyerson, an influential French philosopher of science, attempted to dispel the positivist bias and held that the goal of science is an *ontological* one.[138]

Subsequent to Bohm's papers, Edward Nelson[139,140] showed that a single particle subject to Brownian motion, with a diffusion coefficient ($\hbar/2m$) and no friction, and responding to imposed forces in accord with Newton's second law, $\mathbf{F} = m\mathbf{a}$, obeys (*exactly*) the Schrödinger equation. Although there is randomness, a radical departure from classical physics is unnecessary so that the resulting theory is probabilistic in a *classical* way.[141] William Lehr and James Park[142] generalized Nelson's work to the relativistic case. This alternative program has thus shown a great deal of fertility for generalization within its own resources, not just as ad hoc moves. That is, *if*, say in 1927, the fate of the causal interpretation had taken a very different turn and been accepted (over the "Copenhagen" one), it would have had the resources to cope with the generalizations essential for a broad-based empirical adequacy. We could today have arrived at a *very different* worldview of microphenomena. If someone were then to present the (merely) empirically equally as adequate Copenhagen version, with all of its own counterintuitive and mind-boggling aspects, who would listen!

That is, a highly "reconstructed" but entirely plausible bit of history could run as follows (all around 1925–1927). Heisenberg's matrix mechanics and Schrödinger's wave mechanics are formulated and shown to be mathematically equivalent. Hence, the Dirac transformation theory and an operator formalism are available as a *convenience* for further development of the formalism to provide algorithms for calculation. Study of a classical particle subject to Brownian motion (about which Einstein surely knew something!) leads to a "classical" understanding of the already discovered "Schrödinger" equation, which is then given a "de Broglie–Bohm" realistic interpretation. A "Nelson" model[139,140] underpins this interpretation with a visualizable model of microphenomena. "Bell's" theorem[9] is proven and taken as convincing evidence that there is a type of nonlocality present in quantum phenomena. A "no-signaling" theorem[143,144] for quantum-mechanical correlations is established and this puts to rest Einstein's objections[145] to the nonseparability of quantum mechanics. This important point is the following. If one considers a system S consisting of two subsystems S_1 and S_2 which are spatially separated at some time, then Einstein felt that ". . . the real factual situation of the system S_2 is independent of what is done with the system S_1. . . ."[146] Einstein worried what it would even mean to do science if such were not the case. But, a no-signaling theorem would have shown that relativity could be respected at the practical or observational level and that the nonlocality present in nature was of a "benign" variety. This could reasonably have been enough to overcome his objections to the nonlocal or nonseparable nature of a "de Broglie–Bohm" interpretation of the formalism of quantum mechanics. Exhibiting explicit wave solutions with particle-like singularities[147,148] could also have satisfied Schrödinger who wanted a theory with *waves* as the fundamental physical entities. That is, these developments, which *could* (conceptually and logically) have taken place around 1927, could have overcome the resistance of Einstein and of Schrödinger to supporting a "de Broglie–Bohm" program. As is well known, Madelung[149] in 1926 already had the same *equations* Bohm would employ in 1952, but his *interpretation* was very different from Bohm's and did not carry conviction. Bohm's interpretation would certainly have been possible in 1927. These models and theories could be generalized to include relativity and spin. The program is off and running! Finally, quantum statistics follow naturally in a causal stochastic interpretation[150] and this causal interpretation can be extended to quantum fields.*[14,16,151]

It is essential to appreciate that this "story" is neither ad hoc (in the sense of these causal models having as their sole justification an origin in successful results of a rival program) nor mere fancy, since all of these developments exist in the physics literature. However, "Copenhagen" got to the top of the hill first and, to most practicing scientists, there seems to be no point in dislodging it.

10. INTERNAL VERSUS EXTERNAL EXPLANATIONS

It has not been our intention here to argue in favor of a Bohm type of interpretation of quantum mechanics over the standard Copenhagen one. Rather,

*It is not the intention of this sketch to imply that one would have to go a stochastic route as opposed to Bohm's quantum potential one.

our interest has been to see whether a very different choice (from the actual historical one) *might* reasonably have been made. In the present case, we have seen that neither internal factors (such as logical consistency, empirical adequacy) *alone* nor external ones (say, sociological, psychological) *alone* are sufficient to account for the wide acceptance of the Copenhagen worldview in place of a causal one. That is, while "facts" alone do not uniquely constrain theory construction and selection, neither can one's predilections alone enforce just *any* theory.(152) Nature provides (often tight) constraints, but there still remains latitude in theory choice (here, an interpretation or worldview). The actual course of theory construction and selection is a rich and involved one with many overlapping factors. Science, even in its products or laws, remains historical or contingent in an essential manner. That is, how things might have gone a very different way at certain crucial junctures and why they did not may be as important as the reasons for the "right" choices that science has made. We are not particularly uncomfortable with a lack of inevitability in other areas of history. That point is nicely made in the review of a book which examines whether it was inevitable that the Confederacy should lose the United States Civil War in the 19th century.

> Inevitability is an attribute that historical events take on after the passage of sufficient time. Once the event has happened and enough time has passed for anxieties and doubts about how it was all going to turn out to have faded from memory, the event is seen to have been inevitable. Different outcomes become less and less plausible, and before long what did happen appears to be pretty much what had to happen. To argue about what might have happened or whether and why the presumably inevitable turned out to be thought so strikes many people as a waste of time.(153)

ACKNOWLEDGMENT. The research on which this paper is based has been supported in part by the National Science Foundation under Grant DIR-89-08497.

REFERENCES

1. A. EINSTEIN, in: *Ideas and Opinions*, Dell Publishing, New York (1954), pp. 219–222, on pp. 221–222.
2. A. PICKERING, *Constructing Quarks: A Sociological History of Particle Physics*, University of Chicago Press, Chicago (1984).
3. J. T. CUSHING, in: *Proceedings of the 1984 Biennial Meeting of the Philosophy of Science Association*, Vol. 1, Philosophy of Science Association, East Lansing, Mich. (1984), pp. 211–223.
4. J. T. CUSHING, *Theory Construction and Selection in Modern Physics: The S Matrix*, Cambridge University Press, London (1990).
5. R. P. FEYNMAN, in: *The Character of Physical Law*, MIT Press, Cambridge, Mass. (1965), pp. 127–148.
6. J. VON NEUMANN, *Mathematical Foundations of Quantum Mechanics*, Princeton University Press, Princeton, N.J. (1955).
7. A. MESSIAH, *Quantum Mechanics*, Vol. I, North-Holland, Amsterdam (1965).

8. B. D'ESPAGNAT, *Conceptual Foundations of Quantum Mechanics* 2nd ed., Benjamin, Reading, Mass. (1976).
9. J. S. BELL, *Physics* **1**, 195–200 (1964).
10. J. T. CUSHING and E. MCMULLIN, *Philosophical Consequences of Quantum Theory: Reflections on Bell's Theorem*, University of Notre Dame Press, Notre Dame, Ind. (1989).
11. A. ASPECT, J. DALIBARD, and G. ROGER, *Phys. Rev. Lett.* **49**, 1804–1807 (1982).
12. D. BOHM, *Phys. Rev.* **85**, 166–193 (1952).
13. D. BOHM, *Phys. Rev.* **89**, 458–466 (1953).
14. D. BOHM, B. J. HILEY, AND P. N. KALOYEROU, *Phys. Rep.* **144**, 321–375 (1987).
15. J.-P. VIGIER, *Astron. Nach.* **303**, 55–80 (1982).
16. J.-P. VIGIER, in: *Proceedings of the 3rd International Symposium on the Foundations of Quantum Mechanics in the Light of New Technology* (S. KOBAYASHI, H. EZAWA, Y. MURAYAMA, and S. NOMURA, eds.), The Physical Society of Japan, Tokyo (1990), pp. 140–152.
17. G. BADUREK, H. RAUCH, and D. TUPPINGER, *Phys. Rev. A* **34**, 2600–2608 (1986).
18. J.-P. VIGIER, in: *Quantum Uncertainties* (W. M. HONIG, D. W. KRAFT and E. PANARELLA, eds.), Plenum Press, New York (1987), pp. 1–18.
19. C. DEWDNEY, P. R. HOLLAND, A. KYPRIANIDIS, and J.-P. VIGIER, *Nature* **336**, 536–544 (1988).
20. Ref. 17, p. 2601.
21. J.-P. VIGIER, in: *Symposium on the Foundations of Modern Physics* (P. LAHTI and P. MITTELSTAEDT, eds.), World Scientific Publishing, Singapore (1985), pp. 653–675.
22. D. ALBERT and B. LOEWER, *Nous* **23**, 169–186.
23. R. HEALY, *The Philosophy of Quantum Mechanics: An Interactive Interpretation*, Cambridge University Press, London (1989).
24. Ref. 4, Sect. 9.5.
25. Ref. 12, p. 166.
26. Ref. 12, p. 168.
27. Ref. 12, p. 169.
28. M. KLEIN, in: *The Natural Philosopher*, Vol. 3. (D. E. GERSHENSON and D. A. GREENBERG, eds.), Blaisdell Publishing, New York (1964), pp. 1–49.
29. P. FORMAN, *Hist. Stud. Phys. Sci.* **3**, 1–115 (1971).
30. P. FORMAN, in: *The Reception of Unconventional Science, AAAS Selected Symposium 25* (S. H. MAUSKOPF, ed.), Westview Press, Boulder (1979), pp. 11–50.
31. J. HENDRY, *Hist. Sci.* **18**, 155–180 (1980).
32. M. JAMMER, *The Conceptual Development of Quantum Mechanics*, 2nd ed., Tomash Publishers, New York (1989), pp. 173–186.
33. H. C. SPRINKLE, *Concerning the Philosophical Defensibility of a Limited Indeterminism*, Mennonite Press, Scottsdale, Pa. (1933).
34. Ref. 32, p. 364.
35. S. G. BRUSH, *Soc. Stud. Sci.* **10**, 393–447 (1980).
36. Ref. 35, p. 411.
37. V. V. RAMAN and P. FORMAN, *Hist. Stud. Phys. Sci.* **1**, 291–314 (1969).
38. P. A. DEGEN, *Abstracts of the XVIII International Congress of History of Sciences*, ICHS, Hamburg–München (1989).
39. P. EDWARDS, *The Encyclopedia of Philosophy*, Macmillan Co., New York (1967), Vol. 6, pp. 360–363.
40. A. EINSTEIN, in: *Albert Einstein: Philosopher-Scientist* (P. A. SCHILPP, ed.), Open Court, LaSalle, Ill. (1949), pp. 2–95.
41. A. EINSTEIN, *Phys. Z.* **10**, 185–193 (1909).
42. Ref. 28, p. 8.
43. A. EINSTEIN, *Phys. Z.* **10**, 817–826 (1909).
44. Ref. 28, p. 5.

45. A. EINSTEIN, *Phys. Z.* **18**,121–128 (1917).
46. P. DEBYE, *Phys. Z.* **24**, 161–166 (1923).
47. A. H. COMPTON, *Bull. Nat. Res. Counc.* **4**, 1–56 (1922).
48. A. H. COMPTON, *Phys. Rev.* **21**, 483–502 (1923).
49. Ref. 32, pp. 131–132.
50. B. L. VAN DER WAERDEN (ed.), *Sources of Quantum Mechanics*, North-Holland, Amsterdam (1967), p. 76.
51. J. MEHRA and H. RECHENBERG, *The Historical Development of Quantum Theory*, Vol. 1, Part 2, Springer-Verlag, Berlin (1982), pp. 578–604.
52. F. KUBLI, *Arch. Hist. Exact Sci.* **7**, 55 (1970).
53. A. ABRAGAM, *Biogr. Mem. Fellows R. Soc.* **34**, 27 (1988).
54. Ref. 32, p. 247.
55. L. DE BROGLIE, *C. R.* **177**, 507–510 (1923).
56. L. DE BROGLIE, *C. R.* **177**, 548–550 (1923).
57. L. DE BROGLIE, *C. R.* **177**, 630–632 (1923).
58. Ref. 32, p. 258.
59. W. HEITLER, *Biogr. Mem. Fellows R. Soc.* **7**, pp. 221–228, on p. 222 (1961).
60. E. SCHRÖDINGER, *Ann. Phys.* **79**, 734–756 (1926).
61. Ref. 28, p. 4.
62. E. SCHRÖDINGER, *Phys. Z.* **27**, 95–101 (1926).
63. Ref. 28, p. 43.
64. E. SCHRÖDINGER, *Ann. Phys.* **79**, 361–376 (1926).
65. E. SCHÖDINGER, *Ann. Phys.* **79**, 489–527 (1926).
66. Ref. 32, pp. 257, 264.
67. Ref. 32, p. 258.
68. E. SCHRÖDINGER, *Ann. Phys.* **81**, 109–139 (1926).
69. E. SCHRÖDINGER, *Collected Papers on Wave Mechanics*, Blackie, London (1928), p. 46.
70. J. HENDRY, *The Creation of Quantum Mechanics and the Bohr–Pauli Dialogue*, Reidel, Dordrecht (1984), p. 7.
71. Ref. 32, p. 19.
72. N. BOHR, *Philos. Mag.* **26**, 1–25 (1913).
73. R. E. PEIERLS, *Biogr. Mem. Fellows R. Soc.* **5**, 175–192 (1959).
74. M. BELLER, *The Genesis of Interpretations of Quantum Physics, 1925–1927*, unpublished Ph.D. dissertation, University of Maryland (1983).
75. Ref. 70, p. 132.
76. L. ROSENFELD, in: *Dictionary of Scientific Biography*, Vol. 2 (C. C. Gillispie, ed.), Scribner's, New York (1970), p. 239.
77. Ref. 70, pp. 13–14.
78. W. PAULI, *Encyklopädie der mathematischen Wissenschaften*, Vol. 19, Teubner, Leipzig (1921), p. 4. [Also appears in translation (by G. Field, 1981) as *Theory of Relativity*, Dover Publications, New York.] Page references are to the Dover edition.
79. Ref. 78, p. 206.
80. Ref. 70, p. 22.
81. Ref. 70, pp. 30, 50.
82. D. C. CASSIDY in: *Dictionary of Scientific Biography*, Vol. 17 (C. C. Gillispie, ed.), Scribner's, New York (1977), pp. 394–403; on p. 396.
83. Ref. 73, p. 177.
84. Ref. 70, p. 36.
85. Ref. 70, p. 64.
86. Ref. 73, p. 178.
87. Ref. 70, p. 36.

88. M. Dresden, *H. A. Kramers: Between Tradition and Revolution*, Springer-Verlag, Berlin (1987), p. 212.
89. H. Kragh, *Dirac, A Scientific Biography*, Cambridge University Press, London (1990), pp. 2, 32.
90. P. A. M. Dirac, in: *History of Twentieth Century Physics* (C. Weiner, ed.), Academic Press, New York (1977), pp. 109–146, on p. 118.
91. Ref. 89, p. 5.
92. Ref. 29, pp. 58–63.
93. N. Bohr, *Collected Works*, Vol. 6, North-Holland, Amsterdam (1972–1985), p. xix.
94. M. Beller, *Arch. Hist. Exact Sci.* **33**, pp. 337–349, on p. 340 (1985).
95. Ref. 70, p. 90.
96. Ref. 70, p. 91.
97. J. L. Heilbron, in: *Science in Reflection* (E. Ullmann-Margalit, ed.), Kluwer, Dordrecht (1988), pp. 201–233, on pp. 203–204.
98. Ref. 97, p. 211.
99. Ref. 97, p. 219.
100. Ref. 70, p. 107.
101. Ref. 32, p. 325.
102. K. R. Sopka, *Quantum Physics in America: The Years Through 1935*, Tomash Publishers, New York (1988), pp. 102–104.
103. Ref. 102, p. 162.
104. Ref. 102, pp. 166–167.
105. Ref. 102, pp. 196–201.
106. L. de Broglie, *Ann. Phy.* **3**, 22–128 (1925).
107. Ref. 37, p. 295.
108. Ref. 51, p. 580.
109. C. F. von Weizsäcker, in: *Quantum Theory and the Structures of Time and Space*, Vol. 2 (L. Castell, M. Drieschner and C. F. von Weizsäcker eds.), Carl Hanser Verlag, Munich (1977), pp. 9–19.
110. AHQP, *Archive for the History of Quantum Physics*, interview transcripts on deposit in the Department of History of Science, University of California, Berkeley (1962–1964). We refer here to interview number 7 (of 12).
111. M. Beller, *Isis* **74**, 469–491 (1983).
112. D. C. Cassidy, *Uncertainty: The Life and Science of Werner Heisenberg*, Freeman, New York (1992).
113. Ref. 97, pp. 204–209.
114. P. Robertson, *The Early Years: The Niels Bohr Institute, 1921–1930*, Akadmisk Forlag, Copenhagen (1979).
115. L. de Broglie, *J. Phys. Radium* **8**, 225–241 (1927).
116. *Electrons et Photons, Rapports et Discussions du Ciniquiéme Conseil de Physique*, Gauthier–Villars, Paris (1928), 105–132.
117. W. Pauli, in: *Electrons et Photons* (1928), pp. 280–282.
118. L. de Broglie, *Non-Linear Wave Mechanics: A Causal Interpretation* (translated from the French by A. J. Knodel and J. C. Miller), Elsevier, Amsterdam (1960), p. 183.
119. D. Bohm, and B. J. Hiley, *Found. Phys.* **12**, pp. 1001–1016, on p. 1003 (1982).
120. M. Jammer, *The Philosophy of Quantum Mechanics*, McGraw–Hill, New York (1974), pp. 113–114.
121. L. de Broglie, in: *Wave Mechanics: The First Fifty Years* (W. C. Price, S. S. Chissick, and T. Ravendale, eds.), Butterworths, London (1973), pp. 12–18, on p. 16.
122. J. C. Slater, in: *Wave Mechanics: The First Fifty Years* (W. C. Price, S. S. Chissick and T. Ravendale eds.), Butterworths, London (1973), pp. 19–25, on p. 23.

123. L. DE BROGLIE, *An Introduction to the Study of Wave Mechanics* (translated from the French by H. T. Flint), Methuen, London (1930), p. 7.
124. Ref. 123, pp. 119–121.
125. Ref. 123, pp. 132–133.
126. Ref. 118, pp. 183–184.
127. Ref. 6, pp. 313–328.
128. L. DE BROGLIE, *New Perspectives in Physics* (translated from the French by A. J. Pomerans), Oliver & Boyd, Edinburgh (1962), p. 99.
129. F. SELLERI, *Quantum Paradoxes and Physical Reality*, Kluwer, Dordrecht (1990).
130. Ref. 119, pp. 1003, 1014–1015.
131. D. BOHM, private communication 7/20/89.
132. K. V. LAURIKAINEN, *Beyond the Atom: The Philosophical Thought of Wolfgang Pauli*, Springer-Verlag, Berlin (1988).
133. Ref. 128, p. vi.
134. Ref. 128, pp. vii–viii.
135. Ref. 121, p. 12.
136. L. DE BROGLIE, *Found. Phys.* **1**, pp. 5–15, on p. 12 (1970).
137. E. MEYERSON, *Essais*, Librarie Philosophique J. Vrier, Paris (1936).
138. Ref. 39, Vol. 5, pp. 307–308.
139. E. NELSON, *Phys. Rev.* **150**, 1079–1085 (1966).
140. E. NELSON, *Dynamical Theories of Brownian Motion*, Princeton University Press, Princeton, N. J. (1967).
141. Ref. 139, p. 1084.
142. W. J. LEHR and J. L. PARK, *J. Math. Phys.* **18**, 1235–1240 (1977).
143. G. C. GHIRARDI, A. RIMINI, AND T. WEBER, *Lett. Nuovo Cimento* **27**, 293–298 (1980).
144. A. SHIMONY in: *Foundations of Quantum Mechanics in the Light of New Technology* (S. KAMEFUCHI, H. EZAWA, Y. MURAYAMA, M. NAMIKI, S. NOMURA, Y. OHNUKI and T. YAZIMA eds.), The Physical Society of Japan, Tokyo (1984), pp. 225–230.
145. A. FINE, *The Shaky Game: Einstein, Realism and the Quantum Theory*, University of Chicago Press, Chicago (1986), p. 103.
146. A. EINSTEIN, in: *Albert Einstein: Philosopher-Scientist* (P. A. SCHILPP, ed.), Open Court, LaSalle, Ill. (1949), p. 85.
147. J.-P. VIGIER, *Phys. Lett. A* **135**, 99–105 (1989).
148. A. O. BARUT in: *Symposium on the Foundations of Modern Physics 1990* (P. Lahti and P. Mittelstaedt, eds.), World Scientific Publishing, Singapore (1991), pp. 31–46.
149. E. MADELUNG, *Z. Phys.* **40**, 322–326 (1926).
150. J.-P. VIGIER, *Pramana* **25**, 397–418 (1985).
151. J. S. BELL, in: *Speakable and Unspeakable in Quantum Mechanics*, Cambridge University Press, London (1987), pp. 171–180.
152. Ref. 4, Chapter 10.
153. C. VANN WOODWARD, *The New York Review of Books* **33**(12) (July 17, 1986), 3–6.

CHAPTER 4

Duality of Fluctuations, Fields, and More

Peter E. Gordon

1. INTRODUCTION

Much of the contemporary literature on duality is an attempt to eliminate the concept. That is, it consists of efforts to account for phenomena without introducing the wave–particle duality.

Surdin[1] questions whether the "photon concept is really necessary?" by substituting wave packets. Marshall[2] tries for a duality without particles while Bach's[3] particles simulate wave behavior. Garuccio's[4] enhanced photon detectors produce the effect of duality without the basic property. The list of these ambitious and well-developed strategies could be extended.

Another approach is to dissolve the mystery or dilemma of duality by physically unifying the subject. Chief among these ideas is the coexisting wave and particle of de Broglie and more recently Selleri.[5]

Our approach is to accept duality and to attempt to broaden the concept. Our idea of duality is as a dynamic property whose consequences have not yet been fully explored. In fact, after extending its applicability, we advance duality as a worldview.

In the next three sections, we try to add to the properties of light to which we apply the wave–particle duality. In Section 2 we generalize Einstein's original result for the fluctuations of blackbody radiation to apply to any field. We do this by separating the operators for the fluctuations into wave and particle parts according to criteria drawn from the quantum theory of coherence. We define

Peter E. Gordon • Physics Department, University of Massachusetts at Boston, Boston, Massachusetts 02125, USA.

Wave–Particle Duality, edited by Franco Selleri. Plenum Press, New York, 1992.

orders of wave fluctuation which we relate to *m*th-order coherence. We apply this to the wave–particle fluctuations of nonclassical states and relate the wave fluctuation to squeezed states.

In Section 3 we inquire as to what extent the fields themselves can be separated into wave and particle parts. We reach certain conclusions which we apply to nonclassical states and to the vacuum.

In Section 4 we look at the two radiation processes, spontaneous and stimulated, asking what this duality has to do with the wave–particle duality. We propose a model for stimulated emission in terms of the nonclassical states discussed earlier and compare the results of this model with those of recent experiments. We offer yet another argument relating spontaneous emission and the vacuum fluctuations. For each radiation process we examine the stimulation process separately from the emission process. We conclude this section with a hypothesis regarding radiation.

In Section 5 we analyze a number of experiments proposed and performed with the object of detecting the wave. We try to show in each experiment that the wave is unobservable and propose this as a principle; that half of the duality, the wave, is not directly observable.

In Section 6 we look at the consequences of this unobservability and its relation to measurement, and the relation of wave–particle duality to measurement.

We close by presenting duality as a worldview. Einstein's reality may be difficult to resurrect. Duality offers us an alternative.

2. DUALITY OF FLUCTUATIONS

The first clear indication of the wave–particle nature of light was due to Einstein. He showed that the fluctuations in blackbody radiation, with which he and Planck were concerned at the time, could be divided into two parts: one part is derivable from the wave properties of light, the other is due to its particle properties.

We will begin by deriving Einstein's original separation using the methods of the quantum theory of coherence. We will generalize Einstein's result from blackbody radiation. We will do this by showing that the *operators* for the fluctuations can be separated into wave and particle parts. Our results will then hold for any field.

2.1. Generalized Wave–Particle Fluctuations

One reason for generalizing the separation of the fluctuation is that dualism is not confined to blackbody radiation and so the separation of the fluctuation should not be either. We demonstrate this separation of the fluctuations to second, third, and fourth orders explicitly. By defining orders of "wave" fluctuation we give new definitions of *m*th-order coherence and *m*th-order coherent states. It is interesting to see that the concepts of dualism can be used to reproduce quantitative results of contemporary theory.

The fluctuations in number may be expressed as

$$(\Delta n)^2 = \bar{n} + \bar{n}^2 \tag{1}$$

Einstein[6] showed that the first term would arise from a system consisting of independent particles and the second from a system of waves.

We would like to generalize Einstein's result using the methods of coherent states.[7]

We will separate the operators for the fluctuation into two parts according to the following definitions:

1. The first, or "particle" part, will give the Poisson fluctuation.
2. The second, or "wave" part, will give a zero result when applied to a coherent state.

This is a reasonable definition in view of the association between the coherent state and the classical wave. The definite-valued wave gives zero "wave" fluctuation, i.e., no fluctuation beyond the Poisson. An operator for the fluctuation which gives zero on the coherent state will therefore be called a wave fluctuation.

In general, the wave fluctuation will be nonzero. Its value will depend on the nature of the state. We will see that for blackbody or Gaussian radiation it gives the usual result. Notice that the wave fluctuation is defined with reference to the field, i.e., to the coherent state. The fluctuations are in terms of the number operator. We will discuss fluctuations of the field in the next section.

The simplest case is the second-order fluctuation in n, the case considered by Einstein:

$$\begin{aligned}(\Delta\bar{n})^2 &= (a^+a - \langle n\rangle)^2 \\ &= a^+a + a^+a^+a^{\cdot}a - 2a^+\langle n\rangle + \langle n\rangle^2 \\ &= a^+a + \overline{W}^{(2)}\end{aligned} \tag{2}$$

The first term gives the Poisson fluctuation while the second, which we have labeled $\overline{W}^2$, gives zero for a coherent state,

$$\langle\alpha|\overline{W}^2|\alpha\rangle = 0 \tag{3}$$

while giving of course $\langle n\rangle^2$ for the well-known wave fluctuation of the Bose–Einstein distribution described by the Gaussian density operator

$$P(\alpha) = (\pi\langle n\rangle)^{-1}e^{-|\alpha|2/\langle n\rangle} \tag{4}$$

We note for future reference the third- and fourth-order Poisson fluctuations:

$$\overline{(\Delta n)^3}_{\text{Poisson}} = \overline{n} \tag{5}$$

$$\overline{(\Delta n)^4}_{\text{Poisson}} = 3\overline{n}^2 + \overline{n} \tag{6}$$

For the third-order case we have

$$(\Delta \overline{n})^3 = \overline{n}^3 - 3\overline{n}^2 \langle n\rangle + 2\langle n\rangle^3$$

Using

$$(a^+a)^3 = (a^+)^3a^3 = 3(a^+)^2a^2 + a^+a$$

we calculate

$$\begin{aligned}(\Delta \overline{n})^3 &= a^+a \\ &+ [2\langle n\rangle^3 - 3\langle n\rangle(a^+)^2a^2 + (a^+)^3a^3] \\ &+ [3(a^+)^2a^2 - 3\langle n\rangle^2] \\ &= a^+a + \overline{W}^{(3)}\end{aligned} \tag{7}$$

where we have bracketed terms of the same order to show that $\overline{W}^{(3)}$ goes to zero in the limit of a coherent state.

The wave fluctuation is of course nonzero for general states and may be calculated.

As an example, the third-order wave fluctuation for $|\gamma\rangle$ or blackbody radiation [the distribution Eq. (4)] may be calculated from

$$\begin{aligned}\langle\gamma|W^{(3)}(a^+,a)|\gamma\rangle &= \text{Tr}\{W^{(3)}(a^+,a)\rho\} \\ &= (\pi\langle n\rangle^{-1}W^{(3)}(\alpha^*,\alpha) \\ &\cdot \ \exp -|\alpha|^2/\langle n\rangle d^2\alpha\end{aligned}$$

which holds because $W^{(3)}\ (a^+,a)$ is normal ordered. We then get

$$\langle\gamma|W^{(3)}|\gamma\rangle = 2\langle n\rangle^3 + 3\langle n\rangle^2 \tag{9}$$

In fact, these methods can be used to obtain a quantitative measure of the wave fluctuations of any field.

The fourth-order Poisson fluctuations Eq. (6) differ from the second and third orders so that it is worthwhile to see whether this separation holds in the fourth order.

We make use of an expansion for normal-ordered moments.[8]

$$(a^+a)^\nu = \sum_\mu \left(\sum_\eta^\mu \frac{\eta^\nu(-1)^{\mu-\eta}}{\eta!(\mu-\eta)!} \right)(a^+)^\mu a^\mu \tag{10}$$

[The coefficients (for the normal-ordered moment) are called homogeneous product sums and are equal to the product of the first μ integers taken ν at time, repetitions allowed.] First we expand

$$\begin{aligned}(a^+a - \langle n\rangle)^4 = (a^+a)^4 &- 4(a^+a)^3\langle n\rangle \\ &+ 6(a^+a)^2\langle n\rangle^2 - 3\langle n\rangle^4\end{aligned}$$

Using Eq. (10), we have

$$\begin{aligned}(\Delta\bar{n})^4 &= [(a^+)^4a^4 - 4\langle n\rangle(a^+)^3a^3 + 6\langle n\rangle^2(a^+)^2a^2 - 3\langle n\rangle^4] \\ &+ [6(a^+)^3a^3 - 12\langle n\rangle(a^+)^2a^2 + 6\langle n\rangle^2a^+] + [7(a^+)a^2 - 7\langle n\rangle^2] \\ &+ 3\langle n\rangle^2 + \langle n\rangle = 3\langle n\rangle^2 + \langle n\rangle + \overline{W}^{(4)}\end{aligned} \tag{11}$$

We have again grouped terms of common order to show that $W^{(4)}$ gives zero for a coherent state leaving the fourth-order Poisson fluctuation.

Note that these results are for operators and therefore apply to any field, not just blackbody radiation.

2.2. *Nonclassical States*

Now we would like to apply these methods to nonclassical states, states with a limited number of photons. This will later allow us to consider the dualistic properties of the vacuum.

The concept of wave fluctuation may be used to define *m*th-order coherence[9] and *m*th-order coherent states.

We recently explored properties of second-order coherent states.[10] The states satisfy the second-order coherence condition

$$\langle II|:(a^+a)^2:|II\rangle = \langle II|a^+a|\text{II}\rangle^2 \tag{12}$$

where the dots denote normal ordering.

For a state containing up to two photons in a given mode we have states of the form

$$|II\rangle = \sum_{n=0}^{2} C_n|n\rangle \tag{13}$$

Assuming the normalization condition

$$\sum_n |C_n|^2 = 1 \tag{14}$$

as well as Eq. (12) we obtain (up to a phase factor) a one-parameter family of states in which

$$|C_0| = \left[1 + \frac{\langle n\rangle^2}{2} - \langle n\rangle\right]^{1/2} \tag{15}$$

$$|C_1| = [\langle n\rangle - \langle n\rangle^2]^{1/2} \tag{16}$$

$$|C_2| = \frac{\sqrt{2}}{2}\langle n\rangle \tag{17}$$

with $\langle n\rangle$ denoting the average number of photons in a given mode. We note that Eq. (16) implies that the value of $\langle n\rangle$ is limited to

$$0 \leqslant \langle n\rangle \leqslant 1 \tag{18}$$

The second-order coherent state with maximal $\langle n\rangle$ is

$$|II\rangle = \sqrt{\tfrac{1}{2}}|0\rangle + \sqrt{\tfrac{1}{2}}|2\rangle \tag{19}$$

Calculation shows that although

$$\langle II|\overline{W}^{(2)}|II\rangle = 0 \tag{20}$$

we have

$$\langle II|\overline{W}^{(3)}|II\rangle \neq 0 \tag{21}$$

Wave fluctuations may therefore be used to define orders of coherent states: mth-order coherent state has zero wave fluctuation up to mth-order $\overline{W}^{(m)}$. This definition is equivalent to that for mth-order coherence, e.g., Eq. (12).

The state equation (19) exhibits nonclassical behavior[8]; that is, its $P(\alpha)$ is negative in some regions and cannot be interpreted as a probability or must be interpreted as probability with negative values.[11] In fact, any field which has zero second-order wave fluctuations and nonzero wave fluctuations of a higher order behaves nonclassically.

We will be exploring the dualistic properties of nonclassical states. Recently a type of nonclassical states called "squeezed states" have been introduced into the theoretical and experimental literature.[12] We can type the various states conveniently in terms of the second-order wave fluctuation operator:

$$W^{(2)} = a^+a^+aa - 2a^+a\langle n\rangle + \langle n\rangle^2 \tag{22}$$

Note the presence of the negative term in $W^{(2)}$. It allows for the existence of squeezed states.

Type of behavior	$\langle W^{(2)}\rangle$		Fluctuations
Photon bunching	$\langle W^{(2)}\rangle$	> 0	Above Poisson
Coherent states	$\langle W^{(2)}\rangle$	$= 0$	Poisson
Squeezed states	$\langle W^{(2)}\rangle$	< 0	Below Poisson

3. DUALITY OF FIELDS

Having separated the fluctuations it is natural to attempt to extend this separation to the fields themselves. To what extent can the fields be separated into wave and particle parts?

The duality of fields would, like the duality of fluctuations, show wave and particle parts being displayed simultaneously. We do not, however, have in mind any spatial separation of light into waves and particles. Photons in any case cannot be localized in the same sense that a particle in nonrelativistic quantum mechanics can be located. About the difficulty of placing the wave in space and time we will have more to say later. So, although particles and waves are often described visually, we do not hold out hopes for a visualizable model of duality.

We will treat in turn classical and nonclassical fields because, as we will see, our method of separating classical fields does not apply to nonclassical ones.

3.1. Classical Fields

Classical fields can be described by a positive definite probability function $P(\alpha)$. For these fields the convolution theorem allows us to separate $P(\alpha)$ into two parts:

1. A coherent state with

$$P_w(\alpha) = \delta^2(\alpha - \overline{\alpha}) \tag{23}$$

where

$$\overline{\alpha} = \int \alpha P(\alpha) d^2\alpha \tag{24}$$

is the mean value of the total field.

2. A field $P(\alpha + \overline{\alpha})$ for which the mean value vanishes and is commonly known as an unphased field:

$$\int \alpha P(\alpha + \overline{\alpha}) d^2\alpha = 0 \tag{25}$$

In the terms of the convolution theorem, we have

$$P(\alpha) = \int \delta^{(2)}(\alpha - \overline{\alpha} - \alpha') P(\alpha' + \overline{\alpha}) d^2\alpha' \tag{26}$$

The coherent state, $P(\alpha) = \delta^{(2)}(\alpha - \overline{\alpha})$, is as close as we can get to a classical wave in quantum theory. It is in fact a classical wave plus the (unavoidable) vacuum fluctuations. For these fields the amplitudes add as in classical wave theory; that is, it will interfere as a classical wave.

For unphased fields on the other hand, the intensities add. When combining a number of such fields we can add the number of "particles" in each unphased field to get the number in the total field.

Of course the unphased field is not exactly a set of classical particles any more than the coherent state is exactly a classical wave. Second-order correlations for this state will differ from those for the Gaussian, as an example.

This separation of the field allows us to discuss the fluctuations of the *field* $\overline{(\Delta\alpha)^2}$. All of our results on the duality of fluctuations in the last section were in terms of the number operator. The field was brought in only to define the wave fluctuations $W^{(n)}$.

The fluctuation of the field operator for the coherent part of the field is zero, because the coherent part of the field is a delta function in α.

For the unphased field we have for the mean number of equivalent particles

$$\begin{aligned} \langle n\rangle_{\text{unphased}} &= \int |\alpha|^2 P(\alpha + \overline{\alpha}) d^2\alpha \\ &= \int |\alpha + \overline{\alpha}|^2 P(\alpha + \overline{\alpha}) d^2\alpha - \overline{\alpha}^2 \\ &= |\overline{\alpha^2}| - |\overline{\alpha}^2| = \overline{(\Delta\alpha)^2} \end{aligned} \tag{27}$$

The fluctuations of the field do not divide up neatly into wave and particle parts like the fluctuations for the number operator. But the above result does show dualistic properties:

The means square fluctuation (the variance) of the *amplitude* of the total field is equal to the average *number* of particles in the unphased part of the field.

Let us do a number of "checks" on this conclusion. Consider the number of particles in the other part of the field:

$$\langle n\rangle_{\text{coherent}} = \int |\alpha|^2 \delta^2(\alpha - \overline{\alpha}) d^2\alpha = |\overline{\alpha}|^2 \tag{28}$$

Combining the number of particles in each part of the field

$$\langle n\rangle_{\text{unphased}} + \langle n\rangle_{\text{coherent}} = |\overline{\alpha^2}| - |\overline{\alpha}|^2 + |\overline{\alpha}|^2 = |\overline{\alpha^2}| \tag{29}$$

Since

$$|\overline{\alpha^2}| = \int |\alpha^2| P(\alpha) d^2\alpha = \langle n\rangle_{\text{total field}} \tag{30}$$

we have

$$\langle n\rangle_{\text{unphased}} + \langle n\rangle_{\text{coherent}} = \langle n\rangle_{\text{total field}}$$

Now consider the special case

$$\langle n\rangle_{\text{unphased}} = 0 = (\overline{\Delta\alpha})^2 \tag{31}$$

We have:

1. No fluctuation of the field
2. Total field = $|\alpha\rangle$ (because $\langle n\rangle_{\text{unphased}} = 0$)

Now let us introduce a result in order to draw another conclusion. Hillery,[13] and earlier Aharonov *et al.*,[14] showed that for a classical $P(\alpha)$ any pure state must be a coherent state $|\alpha\rangle$.

In terms of our results we see that in order for a classical state to have particle properties it must be a *mixture*. In fact, we can say that *particles are the mixture properties of states*. That is, insofar as a classical state can be said to consist of particles it must be a mixture.

We will demonstrate the same property in nonclassical states.

A particular example of unphased fields is $P(|\alpha|)$ which does not depend on the phase of α. Such *stationary* fields have density matrices in the number representation which reduce to diagonal form. They are thus equivalent to a *mixture* of number states. Laser light is often described as a convolution of the Gaussian $P(\alpha)$ [Eq. (4), another mixture of number states] and the coherent state.

This then is the extent to which fields with positive-definite $P(\alpha)$ can be separated into wave and particle parts.

3.2. *Nonclassical Fields*

Not all superpositions of the number states can be described by a positive-definite $P(\alpha)$. We met up with such a state in Eq. (19). These nonclassical states have regions where $P(\alpha)$ is negative but which cannot be observed because of the vacuum fluctuations. That is, the negative values are restricted to a range so small that they are always averaged out by the classical observation. We will elaborate on this point after discussing the vacuum in the next section.

Because of this property the diagonal of the density matrix of the field $\langle\alpha|\rho|\alpha\rangle$ is better suited for the description of these nonclassical states. Using it we may attempt to characterize such states dualistically.

To do this let us first write down $\langle\alpha|\rho|\alpha\rangle$ for two states which we have characterized as particlelike:

Gaussian:

$$\langle\alpha|\rho|\alpha\rangle = \frac{1}{1 + \langle n\rangle}\exp\left\{\frac{-|\alpha|^2}{1 + \langle n\rangle}\right\} \tag{32}$$

Number state:

$$\langle\alpha|\rho|\alpha\rangle = \frac{|\alpha|^{2n}}{n!}\exp -|\alpha|^2 \tag{33}$$

Notice that both of these states are functions of $|\alpha|$. Now let us turn to the coherent state

$$P(\alpha) = \delta^2(\alpha - \alpha_0) \tag{34}$$

and

$$\begin{aligned}\langle\alpha|\rho|\alpha\rangle_{\alpha_0} &= \exp -|\alpha - \alpha_0|^2 \\ &= \exp\{-|\alpha|^2 - |\alpha_0|^2 - 2\mathrm{Re}\,\alpha\alpha_0 \cos(\theta - \theta_0)\}\end{aligned} \tag{35}$$

We see a clear dependence on the phase angle θ.

Now let us analyze the nonclassical state Eq. (19) in the same way. The density matrix $\langle\alpha|\rho|\alpha\rangle$ can be calculated as

$$\begin{aligned}\langle\sigma|\rho|\alpha\rangle_{II} &= \frac{1}{2}e^{-|\alpha|^2}\left\{1 + \frac{\alpha^2 + (\alpha^\alpha)^2}{\sqrt{2}} + \frac{\alpha^4}{2}\right\} = \\ &\frac{1}{2}e^{-|\alpha|^2}\left\{1 + \sqrt{2}|\alpha|^2\cos 2\theta + \frac{|\alpha|^4}{2}\right\}\end{aligned} \tag{36}$$

We see that the wave properties of this nonclassical state are separated out explicitly in the middle phase-dependent term of the right-hand member of Eq. (36). As further evidence of this, note that a mixture of $|0\rangle$ and $|2\rangle$ would give the other two "unphased" terms. That is, the density matrices for the vacuum and the two-photon state are the first and third terms of Eq. (36), respectively. The coherence between these states is responsible for the coherent state $|\mathrm{II}\rangle$ and gives the middle term of Eq. (36). This phase dependence is characteristic of stimulated emission.

This is in line with our conclusions about classical fields. The pure state Eq. (19) shows wave properties. It may be of interest to see that the state is pure by noting that its density matrix in the n-representation obeys

$$\begin{bmatrix} \frac{1}{2} & 0 & \frac{1}{2} \\ 0 & 0 & 0 \\ \frac{1}{2} & 0 & \frac{1}{2} \end{bmatrix}^2 = \begin{bmatrix} \frac{1}{2} & 0 & \frac{1}{2} \\ 0 & 0 & 0 \\ \frac{1}{2} & 0 & \frac{1}{2} \end{bmatrix}$$

Wave and phased properties go with the pure state; particles and unphased properties belong to the mixture. This is an indication of a relation between measurement and duality which we will turn to later.

3.3. *The Vacuum*

Our interest in the density matrix formulation and in duality itself extends to the vacuum. Each of our fields Eq. (32) to Eq. (35) reduces to the vacuum according to the following prescriptions:

$$\text{Gaussian} \quad \langle n \rangle \to 0 \tag{38}$$

$$\text{Number state} \quad n \to 0 \tag{39}$$

$$\text{Coherent state} \quad \alpha_0 \to 0 \tag{40}$$

All lead to the same density matrix for the vacuum:

$$\langle \alpha | \rho | \alpha \rangle_{\text{vac}} = \exp -|\alpha|^2 \tag{41}$$

The fact that different limit procedures lead to the same state Eq. (41) shows that the vacuum is continuous with other states of the field. We will consider these limits [Eqs. (38)–(40)] again. According to our "unphased" criterion for particle properties of state

$$\langle \alpha | \rho | \alpha \rangle_{\text{particles}} = f(|\alpha|) \tag{42}$$

we conclude that the vacuum is a particlelike state. This is not a new idea but we are presenting it in the context of our discussion of duality. Born, Heisenberg, and Jordan long ago noted that the particlelike term in the fluctuations (of a thermal field) and the zero-point energy of the harmonic oscillator are closely related. We will explore this relation further in the next section.

The structure of the coherent state itself is of interest in this discussion. It may be considered a classical wave of amplitude α_0 on which is superposed the vacuum fluctuations. When the classical wave—α_0—vanishes, the vacuum fluctuations remain.

Note that the structure of the vacuum itself [Eq. (41)] prevents the observation of the negative values of $P(\alpha)$ for nonclassical states mentioned at the start of Section 3.2. $P(\alpha)$ satisfies the condition

$$\int P(\alpha) \exp -|\alpha - \beta|^2 d^2\alpha \geqslant 0 \tag{43}$$

which must hold for all complex values of β. It corresponds to the positive definite nature of the diagonal of the density matrix

$$\langle\alpha|\rho|\alpha\rangle \geqslant 0 \qquad \text{for all complex } \alpha \tag{44}$$

Equation (43) is not strong enough to keep $P(\alpha)$ from having negative values, but it does place restrictions on the range of these values. Suppose we consider the range of values around the vacuum $\beta = 0$. Equation (43) becomes

$$\int P(\alpha)e^{-|\alpha|^2}d^2\alpha \geqslant 0 \tag{45}$$

We see that the factor in Eq. (45) has the same structure as the vacuum equation (41).

The classical theory only applies to ranges of the field larger than the zero-point fluctuations. The measurement is 'smeared out' over this range. An observation gives us only an average of P(α) over the range.

Equation (45) tells us that such an average will always be positive.

When $P(\alpha)$ tends to vary little over the range, that is, when $P(\alpha)$ is identical with its average over this range, then we are dealing with a classical field.

Finally we can calculate the average number of particles in this particular unphased field, the vacuum. We obtain

$$\langle n\rangle_{\text{vac}} = \frac{\int_0^\infty |\alpha|^2 e^{-|\alpha|^2} d|\alpha|}{\int_0^\infty e^{-|\alpha|^2} d|\alpha|} = \frac{1}{2} \tag{46}$$

4. DUALITY OF THE RADIATION PROCESS

4.1. Two Kinds of Radiation Processes

From the time of Einstein *two* kinds of radiation processes have been postulated, spontaneous emission and stimulated emission. In line with our general program we would like to ask the following question: to what degree can we extend wave–particle duality to the duality of emissions, stimulated and spontaneous? Many physicists question whether there is a duality of emissions at all, whether spontaneous emission really differs from stimulated emission. We will propose an answer to this question at the end of this section.

In order to do this we will investigate the dualistic nature of the radiation process. We will treat the stimulation process and the emission process separately. And we will separate each treatment in turn in terms of spontaneous and stimulated radiation.

We proceed according to the following plan:

- Section 4.2—spontaneous *emission*
- Section 4.3—stimulated *emission*: the Pfleegor—Mandel experiment

- Section 4.4—*stimulation* of stimulated emission: the Blake–Scarl experiment
- Section 4.5—*stimulation* of "spontaneous" emission
- Section 4.6—a hypothesis

4.2. Spontaneous Emission

Quasi-classical models of radiation have long associated stimulated emission with coherence and waves. The author some time ago[15] presented a model of the formation of pulses of coherent radiation through the stimulated emission process. In terms of photons, one says that the stimulated photons have a common phase, resulting in coherence and a classical-like wave. This type of explanation begs the question of how individual photons can have a phase at all.

Spontaneously emitted light, on the other hand, at least at a low density in phase space, adds and behaves like particles.[16] But we must consider that, at a high density in phase space, spontaneously emitted light also behaves like a wave. We can distinguish these two types of emission if we think of a wave in terms of second- or higher-order coherence, or equivalently, as having higher-order wave fluctuations equal to zero. *Only stimulated emission* can produce such light. Spontaneous emission, however intense, produces only first-order coherence. Any second-order experiment performed on light resulting from spontaneous emission will fail to show coherent of that order. The Brown–Twiss experiment was of this type.

Experiments of a second order will show nonzero wave fluctuations $W^{(2)}$ for any spontaneously emitted radiation. On the other hand, stimulated emission will show

$$W^{(n)} = 0$$

for

$$n \leqslant m$$

where m may be a quite high number.

We will define particle behavior as

$$W^{(2)} \neq 0 \tag{47}$$

This has the effect of classifying as particlelike the light used in all of the great interference experiments done before the introduction of lasers. To this objection we can only say that second-order experiments were not available at the time.

This definition states that in any experiment involving second- or higher-order correlation properties between pairs or greater numbers of photons, we can say that spontaneously emitted light behaves like particles.

4.3. *Stimulated Emission*

With stimulated light it is not clear that we should talk about atoms emitting photons at all. Just this difference is brought out by the Pfleegor–Mandel interference between independent lasers.(17) Conditions of the experiment are such that the intensity is low enough so that no more than one photon is likely to be present in the system at any one time.

Interference between the lasers shows that the photons do not come from either laser. Attempts to assume so would lead to contradictions as in interpreting ordinary two-slit interference. We are accustomed to saying that the photon does not come through either slit, but it is not as easy to accept that it does not come from either laser. One difficulty is that there are more theories of the laser (than theories of slits) with energy levels as sources and photons as products.

All this is a direct consequence of the stimulated emission process. Consider the contrast with spontaneous emission. To do this suppose the lasers were emitting $|n\rangle$ states or mixtures of $|n\rangle$ states, such as the Gaussian. $|n\rangle$ states are the product of definite numbers of excited and unexcited atoms. To form these states we must in principle be able to tell which atom had decayed or at least in which laser the atom was in. This is only possible with spontaneous emission. But $|n\rangle$ states do not interfere and so spontaneous emission (at low $\langle n\rangle$) is not compatible with the interference we find.

With stimulated emission we get interference and we cannot tell which laser the photon came from. In this case it hardly makes sense to talk of photons being emitted.

4.4. *Stimulation (of Induced Emission): The Blake–Scarl Experiment*

Stimulated emission is in any case an odd process. What are we to make of the interaction between the stimulating photon and the rest of the system, including the stimulated photon? There is, unlike most interactions, no exchange of energy and momentum. We are accustomed to observing particle properties of light when energy-transferring interactions take place between light and matter. On the other hand, if the "interaction" between the stimulating and the stimulated light is considered, we should expect to see interference—and wave behavior.

The Blake–Scarl experiment(18) presents similar problems for the photon theory of stimulated emission as the Pfleegor–Mandel experiment just discussed. The difference is a focus on the stimulation as opposed the emission process.

A beam of light from a single-mode laser passes through an amplifier tube in which stimulated emission increases the intensity of the laser beam. Blake and Scarl looked for time correlations between pairs of successively detected photons in the output light. The expectation was for correlations in the arrival times for the stimulating and stimulated photons. No change of correlation was found as the ratio of amplified to chaotic light was raised from 0 to 2, raising questions about the photon nature of stimulated emission.

We have proposed the following model of stimulated emission in order to explain the Blake–Scarl experiment. Let us suppose the stimulating photon and the stimulated photon interact—interfere—and form the second-order coherent state $|II\rangle$ described in Section 2. We expect this state because of the phase coherence associated with stimulated emission. The phase dependence of $|II\rangle$ is expressed explicitly in Eq. (36). Moreover, the correlation properties of laser light are given by a series of correlation functions beginning with Eq. (12) which is the defining relation for second-order coherent states. Let us see how the properties of this state can be used to explain the results of Blake and Scarl.

We have discussed the interpretation of the coefficients of the second-order coherent state in terms of the probabilities for spontaneous and stimulated emission.(19)

Defining the ratio of amplified to chaotic light as

$$\frac{I_L}{I_C} = 2\frac{|C_2|^2}{|C_0|^2} \tag{48}$$

We have from the variation of the average occupation number in the second-order coherent state Eq. (18)

$$0 < \frac{I_L}{I_C} < 2 \tag{49}$$

as in the Blake–Scarl experiment.

Now our point is that throughout this variation the correlation properties of this state remain constant. The correlation is independent of the ratio I_L/I_C as observed by Blake and Scarl.

Of course, states involving $|3\rangle$ and higher-order Fock states might be produced by the LGT. But these states, because of their phase coherence, would be higher-order coherent states. They would therefore automatically satisfy the second-order conditions tested by Blake and Scarl. For example, a possible third-order coherent state is

$$|III\rangle \text{e.g} = \sqrt{\tfrac{1}{3}}|0\rangle + \sqrt{\tfrac{1}{2}}|1\rangle + \sqrt{\tfrac{1}{6}}|3\rangle \tag{50}$$

This state is also coherent to second order. The point is that Blake and Scarl detected only second-order coherence properties, correlations between pairs of photons.

According to the criteria of Section 4.2, these observations are consistent with wave behavior. From this point of view, stimulated emission is an interaction of light and light, characterized by the wave behavior of the state $|II\rangle$ formed by the stimulating and the stimulated photon.

4.5. *Stimulation (of Spontaneous Emission)*

It has long been an idea that spontaneous emission is induced by the vacuum fluctuations. Comparison of the Einstein radiation coefficients with the zero-point energy permits us to at least formally make the connection. More recently, lifetime changes for excited atoms in front of mirrors have been explained in terms of reflections of the vacuum fluctuation.

We would like to show a relation between the vacuum fluctuations and spontaneous emission in terms of the present methods.

First we point out several consistencies of this relation.

1. The emitting atom recoils in a random direction when emitting spontaneous radiation. Any stimulating field responsible for this would have to have a random direction.
2. Absence of coherence in the spontaneous radiation is consistent with an unphased character of the stimulating field.

The point is that, supposing spontaneous emission is being stimulated, the vacuum fluctuations satisfy both requirements of the stimulating field.

Now we would like to relate spontaneous emission and the vacuum fluctuations by comparing the classical limits in the number and coherent state representations.

In the number representation the probability for emission is calculated by standard Q.E.D. to be proportional to $[\langle n\rangle + 1]$.

When $\langle n\rangle = 0$ (no photons present) there still remains a probability for emission and this is spontaneous emission.

The semiclassical theory, on the other hand, gives a probability for emission proportional to $\langle n\rangle$. The classical limit, in which the semiclassical result is valid, is

$$\langle n\rangle \gg 1$$

and is equivalent to ignoring the spontaneous emission term in the quantum mechanical probability. From this point of view, spontaneous emission is a quantum effect—it disappears in the classical limit. This conclusion is not shared by all physicists.[20]

Now let us turn to the vacuum fluctuations. The coherent state is described in the P-representation as a delta-function:

$$P(\alpha) = \delta^{(2)}(\alpha - \alpha_0) \tag{51}$$

We have encountered this state in its density matrix description:

$$\langle\alpha|\rho|\alpha\rangle_{\alpha_0} = \exp -|\alpha - \alpha_0|^2 \tag{52}$$

We will argue that Eq. (51) is the classical limit of Eq. (52). This property of the density matrix is closely related to the weak correspondence principle.[21] According to the latter, a classical generator equals the diagonal coherent state matrix elements of the associated quantum generator.

The advantage of the density matrix over the generator for our purposes is that the former becomes the classical distribution only in the classical limit. The density matrix formalism allows us to examine the approach to the classical limit.

In what sense does Eq. (52) approach Eq. (51) in the classical limit? If we take this limit as

$$|\alpha_0| \gg 1$$

Eq. (52) approaches Eq. (51) in that it goes to zero for values of α differing from α_0 by magnitudes slight compared to the magnitude of α_0 itself.

The classical limit is thus equivalent to ignoring the fluctuations around α_0. But these fluctuations around α_0 are just the vacuum fluctuations as we have seen [they remain in Eq. (52) when $\alpha_0 = 0$].

Comparing the physical interpretation of these two classical limits

1. In the number representation spontaneous emission vanishes in the limit $\langle n\rangle >> 1$
2. In the coherent state representation the vacuum fluctuations vanish in the limit

$$|\alpha_0| >> 1$$

We see that the vacuum fluctuations act analogously to spontaneous emission. They are corresponding quantum effects or terms. If we wish the classical limit to be invariant to representation, vacuum fluctuations must play the same role as spontaneous emission. Spontaneous emission and the vacuum fluctuations are part of a larger group of associated quantum effects including the particle term in the wave–particle separation of the fluctuations we started with. The relation between the particle term and the ground state fluctuations was first pointed out by Born, Heisenberg, and Jordan.[22]

4.6. A Hypothesis

Now we are ready to offer a hypothesis. If we accept

1. The particle nature of the vacuum fluctuations (Section 3.3)

2. The role of the vacuum fluctuation in spontaneous emission (Section 4.5)
3. That spontaneous emission results in particles (Section 4.2)

then we can say the following: In spontaneous emission, *particles stimulate the emission of particles*.

In Sections 4.3 and 4.4 we argued that stimulated emission both was stimulated by and resulted in waves. In stimulated emission, *waves stimulate the emission of waves*.

The stimulating light passes on to the stimulated light not only its direction of motion and its phase (both random in the case of spontaneous emission) but also its wave or particle nature.

5. THE UNOBSERVABLE WAVE

5.1. Empty Waves and Duality

Recently a number of experiments have been proposed or performed with the object of showing the existence or nonexistence of the empty wave. In these experiments an attempt is made to detect the wave in a region where no photons are present. The empty wave is so called because if we agree that all of the energy and momentum is carried by the particle, then no energy or momentum is left over for the wave.

In support of the emptiness of the wave we should note that if detectors with threshold energies as low as 1/100 of the photon energy are placed in both arms of a split-beam experiment, light is still detected at only one detector. In this way upper limits can be established on the energy content of the wave.

Our principal point to start with is that attempting to detect the empty wave is equivalent to detecting the existence of duality. The empty wave is the same wave whose existence we accept when photons are present. In cases where photons are rarely present, such as the Pfleegor–Mandel experiment discussed in the last section, we have no trouble explaining the resulting interference as a consequence of the wave. Continuity requires that when we have no photons present, we do not invoke a different wave.

In any case, whether or not the wave is empty, we are accustomed to detecting photons. This is because the detection process, being an interaction of light and matter, necessarily involves light particles.

Detection of the empty wave, on the other hand, because there are no photons, would have to involve the light wave. So a direct detection of the wave would not necessarily involve the interaction of light and matter. This is just as well because the detection process could not involve the exchange of energy or momentum since the empty wave does not have any. All of this reminds us of stimulated emission, which as we have seen does not involve any such exchange between the stimulating agent and the emitted light. Stimulation emission has therefore been proposed as a candidate for the detection process.

We have just discussed stimulated emission as a process involving an interaction between light and light. We proposed that a second-order coherent state is formed.

In this section we will try to show that the correlation properties of this state always prevent the observation of the wave. Assuming the existence of the empty wave, we will show that it is undetectable. That the wave, in itself, should possess properties rendering it unobservable is not entirely new in physics. We have only to think of virtual particles. But just as the uncertainty principle prevents observations violating quantum theory, so (as we will see in Section 6.1) the unobservability of the wave prevents violations of locality from being observable. Moreover, as we will see in Section 6.2 the same property that protects locality guarantees the Hermitian nature of observables.

We have recently[(23)] pointed out difficulties in the detection of the empty wave in various experiments designed to prove or disprove its existence. The reason for this lies in the coherence properties of the two-photon states such as $|II\rangle$ formed by stimulated emission.

The time correlation experiments (so-called type-1 experiments) which have been proposed to detect the wave do not require phase coherence. Nevertheless, the phase coherence of $|II\rangle$ effects its correlation properties.

5.2. *Blake–Scarl*

In the Blake–Scarl experiment discussed in Section 4.4, correlations were expected between the arrival times of the stimulating and stimulated photons. No such correlations were found.

Selleri[(24)] has explained this result in terms of the empty wave. Suppose each primary photon from the first laser is accompanied by an extended empty wave. If the stimulated emission is due to the action of the wave, no time correlation would be expected between the detection of the stimulating photon and the stimulated photon, as found by Blake and Scarl.

In the photon model of stimulated emission, we would certainly expect increasing correlation in the arrival times of pairs of photons as the gain and the amount of stimulated emission increased. All that is necessary for this conclusion is to assume that the primary photon is in the vicinity of the excited atom, practically a statement of the photon model of stimulated emission. It would seem that the observed absence of increased coincidences is evidence against the photon model and in favor of the empty wave as stimulating agent.

But, as we have pointed out, if the state $|II\rangle$ is formed in stimulated emission out of the primary photon and the stimulated photon, we will be unable to observe correlations (beyond accidental correlations). This will be true ***whatever the nature of the stimulating light*** whether due to the primary photon or to an extended empty wave. The coefficients [Eqs. 15–17)] of the state $|II\rangle$ ensure that we would obtain Poisson statistics in any experiment up to the second order, i.e., involving up to two photons. Selleri's model of the amplification of light from the wave point of view produces the same Poisson distribution.

The Blake–Scarl experiment does not provide evidence either for or against the existence of the empty wave.

5.3. Selleri and Others

Another experiment or basic type of experiment has been proposed by Selleri[25] and since elaborated on by many authors. Stimulated emission also plays a role in this experiment but now the stimulating photon and the empty wave associated with it takes different paths after encountering a beam splitter (BS; see Fig. 1). The empty wave is to be detected as the stimulating agent for emissions from the laser gain tube (LGT). This would show up as correlations between detectors P_1 and P_2.

Now let us accept the premise that the empty wave exists and is responsible for stimulating the emissions from the LGT. But suppose the photon to be detected at P_1 (the initial photon which has been reflected at BS) and the photon generated in the LGT by the empty wave associated with the initial photon form the state $|II\rangle$. Now the expected P_1P_2 coincidences will not occur. Again this is because $|II\rangle$ contains no correlations beyond the accidental or Poisson.

Our expectation that the photons in the separate beams form the state $|II\rangle$ is based on the arguments given in Sections 4.4 and 5.2. Such a spatially extended state is familiar to us from EPR.

5.4. Mückenheim et al.

Mückenheim, Lokai, and Burghardt[26] have recently performed an experiment which they claim shows that empty waves do not induce stimulated emission in laser media. In a dispersive medium the phase velocity exceeds the group velocity,

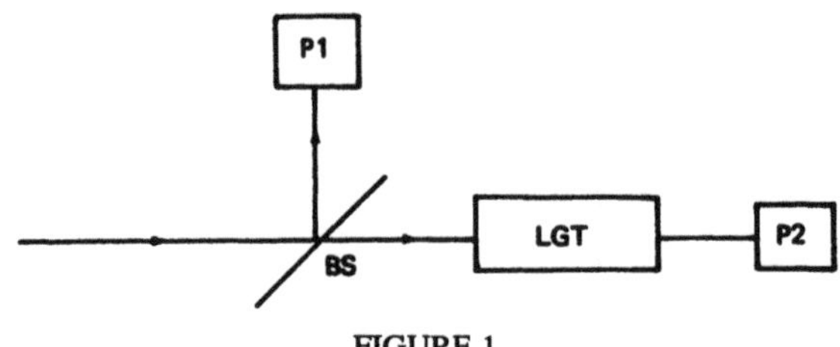

FIGURE 1.

$$C_p = \frac{C_0}{n} \tag{53}$$

$$C_g = C_p\left(1 + \frac{\lambda}{n}\frac{dn}{d\lambda}\right) \tag{54}$$

where C_p is the phase velocity of waves, C_g is the group velocity of the photons, $n(\lambda)$ is the index of refraction, and $dn/d\lambda$ is negative in media with normal dispersion.

Because of this, Mückenheim *et al.* claim that an empty light wave arrives earlier than the photons. Failure to detect earlier stimulated emission, which would have been induced by the wave, is the basis for their claim.

While it is true that the phase velocity exceeds the group velocity, the phase velocity does not apply to the arrival of the wave. This is true even for a monochromatic wave as long as the wave is not present at a given point until a particular time. In fact, only an infinitely long wave can be thought of as truly monochromatic. Of course, such as infinite wave does not "arrive" and so cannot be used in the present experimental context. Putting it another way, any gradient in the square of the waves moves with the group velocity. Thus, the region which changes from no wave to wave moves with the group velocity. So the wave arrives at the same time as the photons.

It seems that this experiment cannot be used to prove or disprove the existence of empty waves; that is, the experiment cannot be used to detect the wave.

5.5. *Martinolli–Gozzini*

Martinolli and Gozzini(27) tested the nature of the two-photon state by allowing it to fall on a semitransmitting mirror. They measured the statistics of the reflection process.

There are three possible results: both photons are transmitted, both photons are reflected, or one photon is transmitted and the other is reflected. An extensive analysis of the experiment by Heidmann and Reynaud(28) leads to probabilities of $\frac{1}{4}$, $\frac{1}{4}$, and $\frac{1}{2}$ for the cases above. These probabilities generally hold for the case of distinguishable particles.

Martinolli and Gozzini on the other hand obtained results compatible with probabilities of $\frac{1}{3}$, $\frac{1}{3}$, and $\frac{1}{3}$. These probabilities hold for identical photons.

Heidmann and Reynaud reach their results even though they consider the photons as indistinguishable bosons. This is accomplished through symmetrization between the reflected and transmitted modes. Their precollision wave packet

$$|1,2\rangle \tag{55}$$

and their postcollision state

$$\frac{1}{\sqrt{2}}(|1\lambda_T,2\lambda_R\rangle + |1\lambda_R,2\lambda_T\rangle) \tag{56}$$

(where T and R stand for transmitted and reflected modes) are the symmetrized representation of identical bosons.

The problem is that the states shown in Eqs. (55) and (56) are very different from $|II\rangle$. $|II\rangle$ is nonseparable (EPR-like),

$$|II\rangle \neq |1,2\rangle \tag{57}$$

The origin of this difference is that the photons in $|\mathrm{II}\,\rangle$ are, in fact, not bosons.[29,30] This also explains the peculiar properties of the squeezed states discussed in Section 2.2.

A fuller exposition of this experiment will be presented elsewhere.

5.6. *Conclusion*

By analyzing these and other experiments we conclude that the wave is unobservable. The reasons for this unobservability vary with each experiment. These difficulties in observation might be part of a pattern, like the difficulties observing violations of the uncertainty principle or of conservation of energy. That is, the wave may be unobservable in principle.

The principle, which as we will see is duality itself, manifests itself on the physical level in different ways, depending on the experimental situation. In this regard it is similar to other principles.

In looking into the nature of this principle we will be attempting to give a reason for the unobservability of the wave. To do this we will have to question the idea of objectively *coexisting* particles and waves, an idea to be found in writings from de Broglie to Selleri. If the wave and the particle do not coexist we will have to find different times or physical situations during which they exist separately. The only natural division in time in quantum mechanics is between before and after measurement.

The relation between measurement and duality occupies us in one way or another in the last section of this chapter.

6. DISCUSSION: MEASUREMENT AND DUALITY

6.1. *Locality*

We have attempted to show that the wave is unobservable. Our inability to directly observe the wave (as we can the particle) may in fact be a necessary limitation. We now wish to show that this unobservability of the wave produces observed locality, i.e., locality on the observed level.

This is particularly brought out by EPR. In EPR the inability to observe ψ, which is the probability amplitude, prevents nonlocality from being observable.

Consider an EPR experiment with two observers. When the second observer performs a measurement on the photon that arrives at his location, he does in fact obtain *some* information about the photon before measurement. He can, in the usual versions of the experiment, eliminate the possibility that before his measurement the photon had a particular polarization.

He could not do this before his measurement. But although some information about the photon survives the measurement, precisely that information which would violate locality is "filtered out" by the measurement.

Suppose the first observer is choosing between two measurement apparatuses U and V and basic sets $|u\rangle$ and $|v\rangle$ corresponding to a choice between plane or circularly polarizing filters. This *choice* is the message to be sent to the second observer.

Once this choice is made, and the first measurement is carried out, the photon arriving at the second observer is in an eigenstate of $\hat{u}$ or $\hat{v}$. This is a consequence of conservation principles, to which we will return. This "element of reality" contains nonlocal information which will not be present after measurement.

Assume that the basic sets $|u\rangle$ and $|v\rangle$ are related by a unitary transformation:

$$|u\rangle \mathrm{v} = \sum_{v} a_{\mathrm{uv}} |v\rangle \tag{58}$$

The measurement our second observer makes eliminates the possibility that the photon was in the eigenstate or orthogonal to the one obtained as a result of the measurement. However, the various probabilities that the photon was in one of the eigenstates of the *other* basis always add up to an equal probability since

$$\sum_{v \text{ or } u} a^{+}_{\mathrm{uv}} a_{\mathrm{uv}} = 1 \tag{59}$$

by unitarity. This "protection" of locality by the unitary transformation may be seen as a case of quantum mechanics saving relativity. We may consider this as the inverse of the famous Bohr–Einstein clock-in-the-box weighing experiment, where relativity was brought to the aid of quantum mechanics.

We should not be surprised if detecting the wave in space and time (where else?) should prove difficult.

Such a wave could not be superimposed over events, as the empty wave is often called on to do. We will have to find another place for the wave, other than where events take place.

Locality we conclude is a property confined to the observed level. It is present after, and a result of, measurement.

6.2. *Unitarity*

A central problem of quantum theory is to state the conditions under which a measurement is carried out. A few physicists have argued the necessity of involving the observer, as a conscious being. The majority of physicists have, however, rejected this approach as leading to a subjective interpretation of quantum mechanics. The alternative is to place measurement in an "irreversible act of amplification."

Aside from the fact that no program based on irreversibility has been convincing, the same difficulties stand in the way of such an approach. The Bohr–Wheeler approach to measurement (no elementary phenomenon is a phenomenon until it is an observed phenomenon) has been criticized as being ultimately referenced to subjective criteria such as observation and meaning.[(31)]

Such a fundamental irreversibility would involve a nonunitary transformation. The projection postulate itself selects one of the terms on the right of

$$\sum C_K|\psi_K\rangle|A_o\rangle \rightarrow \sum C_K|\psi_K\rangle|A_K\rangle \tag{60}$$

and should not be considered reversible.

($|A_o\rangle$ is the original position of the apparatus. The $|A_K\rangle$ are the final states of the apparatus which are correlated to $|\psi_K\rangle$ the eigenstates of an observable Q.)

We have just discussed the role of unitarity in preserving locality of the observed level. What is not always noted is that the unitary nature of measurement is also necessary to the Hermitian nature of observables. This was first pointed out by Wigner.[(32)]

Now we assume that the initial state of our "object" is $|\psi_K\rangle$ an eigenstate of Q. The measurement of this state (the interaction between the apparatus and object) carries out the transformation:

$$|\psi_K\rangle|A_o\rangle \rightarrow |\psi_K\rangle|A_K \tag{61}$$

Note that by carrying out our measurement on $|\psi_K\rangle$, an "element of reality," we avoid all problems associated with the projection postulate.

If we measure a different eigenfunction, we obtain

$$|\psi_l\rangle|A_o\rangle \rightarrow |\psi_l\rangle|A_l\rangle \tag{62}$$

The right sides of Eqs. (61) and (62) must be orthogonal because of the distinguishability of the pointer readings. If the transition represented by the arrows in Eqs. (61) and (62) is unitary, the left sides are also orthogonal, and therefore also the $|\psi_K\rangle$, the eigenfunctions.

But if the measurement process is nonunitary, if in Wheeler's words it is "brought to a close by an irreversible act of amplification," then we can no longer assume orthogonal eigenfunctions.

The coherent states themselves are nonorthogonal. They are often called classical states. Interestingly, they have been suggested as a substitute for the "observer" in Bohr's formulation of quantum theory.[33]

The nonunitary transformation leads to nonphysical observables. Suppose the distinguishable eigenvalues of our observable Q are given by

$$Q|\psi_i\rangle = q_i|\psi_i\rangle$$
$$Q|\psi_j\rangle = q_j|\psi_j\rangle \tag{63}$$

It makes no difference whether the q_i are real, as long as they are unequal. We then have

$$\langle\psi_i|Q\psi_j\rangle = q_j\langle\psi_i|\psi_j\rangle$$
$$\langle Q\psi_i|\psi_j\rangle = q_i\langle\psi_i|\psi_j\rangle \tag{64}$$

and

$$\langle Q\psi_i|\psi_j\rangle \neq \langle\psi_i|Q\psi_j\rangle$$

or

$$\langle\psi_i|Q^+\psi_j\rangle \neq \langle\psi_i|Q\psi_j\rangle \tag{65}$$

this following from the nonorthogonality of the eigenfunctions in Eqs. (64) and (65).

Note that we are only trying to show the consequences of avoiding the subjective aspects of measurement by making irreversibility a physical property. If we were forced to offer an example of a nonphysical observable we might choose a mental image or a perception. It is certainly observable and not obviously physical. It is the result of an irreversible "process" (consciousness) and not least in importance (according to the relation of unitary measurement to locality we discussed in Section 6.1), not located in space.

6.3. *Duality and Reality*

6.3.1. Failure of Local Realistic Theories. We see that unitarity leads on the one hand to an observable locality and on the other to physical observables. This assures that whatever we can detect will be real, i.e., local and objective. But we are left with a question: If ordinary, unitary measurement gives us local, real observables, why all this trouble with local realistic *theories*? The problem is that when we talk about a realistic *theory* we ask about the reality and locality of things *before* measurement. This is the EPR or element-of-reality approach, the criterion of which is *pre*diction *before* measurement. This is the realistic interpretation associated with Einstein. On the other hand, we have seen here that the familiar realistic properties apply only after measurement.

It is well known that we do not observe ψ. We have spent some time pointing out that we do not observe the wave in general. We infer the existence of the wave from the evidence of observations on particles. In the case of light we readily accept the existence of the wave. In the case of matter we are more hesitant. The difference lies in the fact that classical light is a wave.

This inability to directly observe the wave is due to the measurement process of quantum mechanics. Measurement brings into quantum theory the same dualism we are exploring. Here it is between the Schrödinger equation and the measurement (projection) process, between ψ and the observed values.

We can display this generalized duality as follows:

Postmeasurement	Premeasurement
Observable	Unobservable
Particle	Wave
Observables (real)	(Nonreal)
Locality	Nonlocality
Separability	Nonseparability

Other familiar properties could be added to this list. We could show that invariance holds only after measurement. By this we mean that the measured values are the same for all observers, while the objects before measurement, which may include elements of reality, are not invariant.

Temporal dualities such as "becoming versus being" as well as others can be brought into this relation.

The line here represents measurement. Duality offers a concept of measurement—or at least a way of singling it out: crossing the line.

But why *duality* rather than just a measurement upon the wave? One reason is that the wave is not in space. Another is, as we suggest above, that the duality is more general than wave versus particle.

There is no absolute place for the line. In Bohr's terms we would say that the measuring instruments define the line.

Our sense of the *complementarity* of physical properties is that they apply before and after measurement, respectively. In EPR (as we have seen) the (nonlocal) properties of the photon's polarization before measurement are determined by conservation principles. After measurement, only local information survives.

The EPR problem itself can be approached from the point of view of duality. One need suppose that the basis of measure on the right side, which is not spatial, is instead dependent on phase. The EPR photons would then be at the same "place" on the premeasurement side and so be capable of instantaneous communication.

Unobservability of the wave follows from the fact that it does not lie on the left, postmeasurement, observable side of the line. It does not coexist with the particle. The wave is always inferred from measurements on the particles by retrodicting from the time of measurement back across the line to the time before measurement.

"Realism" calls for the existence of this premeasurement, right side of the duality and so brings in nonlocality. Hence the failure of local realistic theories.

We can produce such a theory only by ignoring the right side. In Bohr's words, "concepts of space and time by their very nature acquire a meaning only because of the possibility of neglecting their interaction with the means of measurement." A Copenhagen interpretation means dealing only with the observations, with the left side.

6.3.2. A Historical Aside.[(34)] Our apparent local realism is facilitated by the unobservability of the wave, just as the existence of a previous reality, the celestial sphere, was facilitated by the inability to observe parallax.

The Ptolemaic reality was also dualistic, consisting of Earth on the one hand, and the celestial mechanisms, in their evolving historical forms, on the other. It can be argued that *all* widely accepted worldviews have been dualistic.

Similarly, unobservable entities are not new to physics. Virtual particles are unobservable but are inferred because they give rise to forces between real particles. In the same way unobservable waves give rise to interferences.

In the older, geocentric physics the region beyond the celestial sphere was unobservable. In more recent physics we have the example of cosmic censorship hypothesis wherein singularities are shielded from observation.

The prejudice among scientists against unobservables is due to the long history of religion, which contained the great Unobservable. Observability has consequently come to grant a special status, equivalent to being part of physical reality, both within and without physics itself. We have instead argued that observability is a particular state of affairs, one due in fact to the *unobservability* of the wave.

6.3.3. Conclusion. If we could observe the wave, with all its nonlocal effects, our conception of reality would have to change. We would have to accept either

1. That space-time is an illusion (not real), or
2. That space-time is not everything (i.e., that there is also a nonspatiotemporal reality)

Thus, duality is more than a question of whether the wave exists; it is a question about the nature of reality. Duality is a worldview with its place alongside the conventional worldview of reality. Locality only contradicts reality. It does not contradict duality. Duality by its very nature *includes* reality.

REFERENCES

1. M. SURDIN, *Lett. Nuovo Cimento* **31**, 86 (1984).
2. T. W. MARSHALL and E. SANTOS, Stochastic Optics, University of Santander preprint (1986).
3. A. BACH, *Lett. Nuovo Cimento* **42**, 443 (1985).
4. A. GARUCCIO and F. SELLERI, *Phys. Lett. A* **103**, 99 (1984).
5. F. SELLERI, *Found. Phys.* **17**, 739 (1987).
6. A. EINSTEIN, *Phys. Z.* **10**, 185, 817 (1909).
7. R. J. GLAUBER, *Phys. Rev.* **130**, 2529, **131**, 2766 (1963).
8. P. GORDON, *Phys. Lett. A* **117**, 447 (1986).
9. U. M. TITULAER and R. J. GLAUBER, *Phys. Rev.* **140**, B676 (1968).
10. P. GORDON and N. CHASE *Lett. Nuovo Cimento* **42**, 1975 (1985).
11. W. MÜCKENHEIM, *PHYS. REP.* **133**, 337 (1986).
12. D. YAO, *Phys. Lett. A*, **122**, 77 (1987).
13. M. HILLERY, *Phys. Lett. A*, **111**, 409 (1985).
14. Y. AHARONOV, D. FALKOFF, E. LERNER, and H. PENDLETON, *Ann. Phys.* **39**, 498 (1966).
15. P. GORDON, *Nuovo Cimento* **32**, 1548 (1964).
16. P. GORDON, *Nuovo Cimento* **29**, 935 (1963).
17. L. MANDEL, *Phys. Lett. A* **89**, 32 (1982); R. L. PFLEEGOR and L. MANDEL, *Phys. Rev.* **159**, 1084 (1967).
18. G. D. BLAKE and D. SCARL, *Phys. Rev. A* **19**, 1948 (1976).
19. P. GORDON, *Phys. Lett. A* **138**, 359 (1989).
20. V. L. GINZBURG, *Soc. Phys. Usp.* **26**, 713 (1983).
21. J. R. KLAUDER, *J. Math. Phys.* **8**, 2392 (1967).
22. M. BORN, W. HEISENBERG, and P. JORDAN, *Z. Phys.* **35**, 556 (1926).
23. P. GORDON, *Phys. Lett. A* **138**, 359 (1989).
24. F. SELLERI, *Phys. Lett A* **120**, 371 (1987).
25. F. SELLERI, *Lett. Nuovo Cimento* **1**, 908 (1969).
26. W. MÜCKENHEIM, P. LOKAI, and B. BURGHARDT, *Phys. Lett. A* **127**, 387 (1988).
27. A. GARUCCIO, in: *Open Questions in Quantum Physics* (G. TAROZZI and A. VAN DER MERWE, eds.), Reidel, Dordrecht (1985).
28. A. HEIDMANN and S. REYNAUD, *J. Phys.* **45**, 873 (1984).
29. A. BACH, *Phys. Lett. A* **121**, 1 (1987).
30. P. GORDON, *Nuovo Cimento* **29**, 935 (1963).
31. B. D'ESPAGNAT, in: *Quantum Implications* (B. J. HILEY *et al.*, eds.), Routledge, London, p. 151 (1987).
32. E. P. WIGNER, *Z. Phys.* **133**, 101 (1952).
33. P. STAPP, Ref. 31, p. 255.
34. For an excellent historical perspective, see P. W. MILONNI, *The Wave Particle Dualism* (S. DINER *et al.*, eds.), Reidel, Dordrecht (1985).

CHAPTER 5

Dualism within Dualism

Open Questions

Philippe Guéret

The general point of view of wave–particle duality proposed by Albert Einstein and Louis de Broglie represents all atomic objects such as photons, electrons, protons, etc. as consisting of the physical association of two entities: (1) a *wave-packet*, devoid of energy and momentum but nevertheless objectively real and propagating in space and time; (2) *energetic corpuscles* always localized inside the wave-packets.

Moreover, within this wave–particle duality exists another which appeared early in classical physics for the explanation of diffraction phenomena by means of the *Huygens–Fresnel principle*. This "dualism within dualism" retains all of its interest in the modern theory of light and serves as a basis to interpret the wave-mechanics in terms of the "double solution" hypothesis.

Our purpose is to expose this problem and draw from it some experimental and mathematical inferences coming into sight of open questions for all that.

Let us consider a pointlike source S of a wave phenomenon in a homogeneous isotropic medium and let Σ be a spherical wave-surface at a given time t (Figure 1). According to the Huygens hypothesis, each point M of Σ is regarded as the source of a secondary spherical wavelet the radius of which is $V\Delta t$ at a forthcoming instant $t + \Delta t$, V being the wave propagation velocity (phase velocity). At the same time $t + \Delta t$, the wavefront becomes Σ' with radius $V(t + \Delta t)$ and behaves as the envelope of all the wavelets emitted by the elements of Σ. Therefore, Huygens evidenced a progressive propagation mechanism of the wave phenomenon from a point M of space to another.

The Huygens construction has been completed by the Fresnel hypothesis of

Philippe Guéret • Institut de Mathématiques Pures et Appliquées, Université Pierre et Marie Curie, F-75230 Paris Cedex 05, France.

Wave–Particle Duality, edited by Franco Selleri. Plenum Press, New York, 1992.

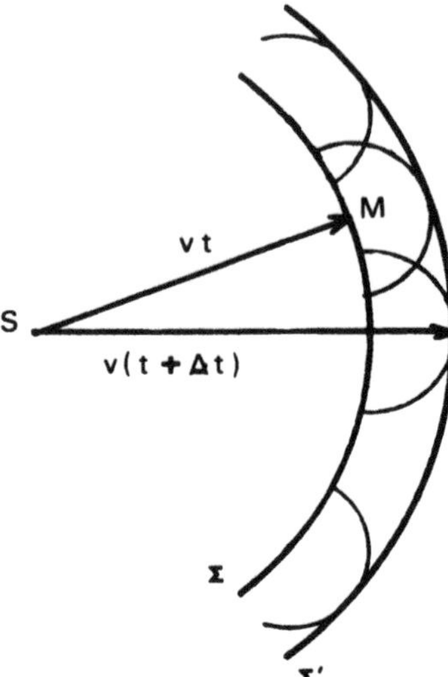

FIGURE 1. Σ is a spherical wave emitted by the source S. According to Huygen's hypothesis, each point M of Σ emits at its turn a spherical wavelet.

possible interferences between the wavelets themselves. Then, it is natural to admit that the secondary sources distributed on the Σ surface could have the same phase as the oscillating state of the wave front Σ. A more detailed investigation of this situation shows that actually the wavelets have a phase advance of $\pi/2$ before the Σ oscillations.

The Huygens–Fresnel principle does not only give a wave front propagating forward, but also a wave front Σ″ propagating in the opposite direction and which is another envelope of the wavelets (Figure 2). A mathematical analysis of this fact enables us to justify the Huygens–Fresnel principle and to eliminate the undesirable Σ″ wave inconsistent with empirical evidence. This was performed by Kirchhoff for scalar waves as acoustic waves in fluids. Kirchhoff built up an accurate formula including two terms adding or canceling each other according to whether one considers a point of Σ′ or a point of Σ″. The electromagnetic wave problem is more intricate because of the vectorial nature of the field. The expression of the Huygens–Fresnel principle requires three assumptions:

1. In a point M of space, the field must be the sum of all the fields sent by the elements of the wave front Σ. This holds also for the first derivatives of the fields.
2. The integral expression of the field, taken on Σ, must be equal to zero in each point located inside the spherical surface Σ (no back-waves).
3. The secondary wavelets are real electromagnetic waves.

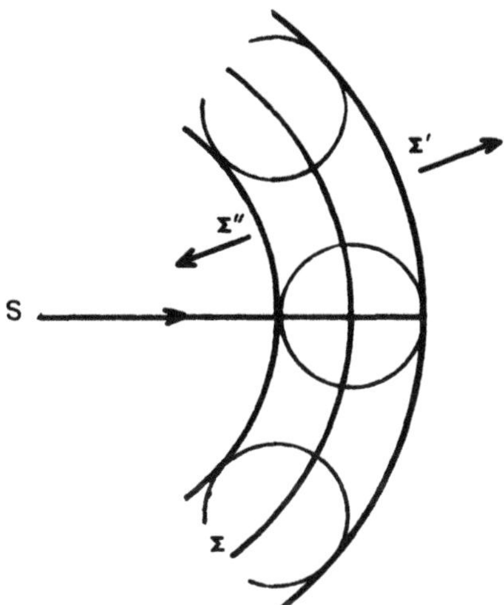

FIGURE 2. The wavelets have two envelopes: Σ', which propagates forward, and Σ'', propagating backward.

On this basis, one can deduce different formulas, equivalent to one another, but adjusted to the applications: in optics or for the transmission techniques of short waves, for instance.

The emission mechanism of light vibrations is not yet fully understood. However, it is known from the pioneer works of Einstein and Bohr that the energy exchanges between matter and electromagnetic radiation are carried out by quanta with energy $h\nu$. Particularly, when one atom emits one photon with energy $h\nu$, an electron of this atom passes from an energetic level E_2 to another of lesser energy E_1, the height difference of these levels being precisely equal to $h\nu = E_2 - E_1$. Then, in optics, it is reasonable to think that

1. The electromagnetic radiation is not emitted continuously but by limited wave-trains (or wave-packets) containing photons and coming from the different atoms from which the course is composed.
2. The energy density of these waves inside a small volume measures the probability of finding photons in this space region.
3. Electromagnetic waves surrounding photons emitted by different atoms can very likely interfere as shown by the Brown–Twiss or the Pfleegor–Mandel experiments.

Thus, the following representation occurs: the light source S being made of a very great number of atoms, the electromagnetic wave emitted at a given instant t, is the sum of individual wave-packets due to the different atoms. These emit only for a time τ corresponding to the mean life of their excited levels. After a fairly long time, the wave-packets emitted at the instant t will be kept away and new atoms take the relay by emitting new waves with new amplitudes and phases: there is a loss of coherence between successive emissions of radiation. A source is coherent with itself only providing that we compare waves emitted at very near instants, the time interval being clearly below τ.

The successive atomic emissions of the source S are made of wave trains limited in the propagation direction. The phase velocity V of the waves inside the wave train is related to the wavelength λ and the frequency ν by $V = \lambda\nu$ and the wave-train velocity v itself (the group velocity) is given by the Rayleigh formula

$$1/v = \partial(1/\lambda)/\partial\nu = \partial(\nu/V)\partial\nu \tag{1}$$

The wave-train length $L = v\tau$ (coherence length) is a *measurable size* by well-known methods of the interference techniques: for instance, by disappearing of the interferences by interposing transparent plates or by separating sufficiently the mirrors of a Michelson interferometer.

We can define τ as the "coherence time" of a wave train by remarking that at a given point M, it behaves as a monochromatic plane wave of the same length L as the wave train. Such an approximation meets its justification by applications which can be made to the calculations of interferences and diffraction phenomena.

We have experimental numerical data on the coherence length (and consequently on the coherence time τ) of the light wave-packets: for ordinary visible light of mean wavelength $\lambda = 0.5\ \mu\text{m}$, L is equal to about 1 m which corresponds to a coherence duration τ of about 10^{-8} s. Moreover, the time constant τ is connected to the spectral ray width of the light: one can easily devise that a long wave train is sufficiently near that of a sinusoidal wave to have a well-defined frequency whereas a short wave train admits a more spread spectrum.

These considerations about the structure of light emissions bring up more physical consistence to the Huygens–Fresnel principle. Thus, the wave-surface Σ corresponds to the geometrical distribution of the wave trains in space surrounding the source S. The wave-packets take the place of the Huygens wavelets. For a source of ordinary intensity, the great number of atomic emissions gives to the wave front an apparent "materiality" in the sense that the Σ wave can be regarded itself as responsible for the interference phenomena. On the contrary, for a source of very low intensity, emitting photons one by one, for instance, Σ is reduced to its statistical signification of geometrical function of repartition of wave-packets around the source and can be useful for calculations of probabilities.

Let us recall that Einstein propounded such a model for atomic emissions of radiation under the name of *Nadelstrahlung*. In the case of light, this model exhibits a *wave–wave dualism with the wave–particle dualism*.

The idea of a wave associated with a particle of matter occurred first to de Broglie[(1)] and was embodied in his well-known doctoral thesis of 1924.[(2)] His assumptions were that the particle has an internal vibration (the "de Broglie clock") and moves in the space-time of special relativity, where at every moment it is localized.

A free particle at rest has mass m_0 and energy $E = m0c^2$. If this energy is taken to equal one quantum $h\nu_0$, the internal particle vibration frequency is defined by $\nu_0 = m_0c^2/h$ and a wave function Ψ_0 can be associated with the vibration so that

$$\Psi_0 = a \exp(2\pi i\nu_0)\tau \tag{2}$$

τ being the proper time of the particle. Ψ_0 is uniform throughout space and does not serve in any way to locate the particle.

Now, if it is supposed that the same particle is moving freely with uniform velocity v in the $+x$ direction, a Lorentz transform gives

$$\tau = (t - \beta x/c)\sqrt{(1 - \beta^2)} \tag{3}$$

($\beta = v/c$) and, for an observer at rest, the wave function Ψ_0 becomes

$$\Psi(x,t) = a \exp\{[2\pi i\nu_0(t - \beta x/c)]\sqrt{(1-\beta^2)}\} \tag{4}$$

or, on taking $\nu_0/\sqrt{(1 - \beta^2)} = \nu$ and $c^2/v = V$

$$\Psi(x,t) = a \exp[2\pi i\nu(t - x/V)] \tag{5}$$

Note that, under the Lorentz transformation, the wave frequency becomes $\nu = \nu_0/\sqrt{(1 - \beta^2)}$, while the particle frequency becomes $\nu = \nu_0\sqrt{(1 - \beta^2)}$ according to the clock slowdown formula. In order to remove this contradiction, de Broglie lays down his "phase concordance principle": *the internal particle vibration and its associated wave remain in phase where the particle is located*. de Broglie waves consequently act as a guide for the particle motion.

According to Eq. (5), for an observer at rest, the particle motion is associated with the propagation of a plane wave of frequency ν and phase velocity $V > c$. On defining, as usual, the wavelength by $\lambda = V/\nu$, one gets for this phase wave

$$\lambda = c^2h/Ev = h/p \tag{6}$$

E and $p = |\vec{p}|$ respectively representing the energy and relativistic momentum of the moving particle.

On the basis of classical reasoning in wave theory, one can consider a wave-packet made up of a superposition of plane wave with closely related frequencies. Using $k = \nu/V$, such a packet can be written as

$$\Phi(x,t) = \int_{\Delta k} \{a \exp[2\pi i(\nu t - kx)]\}\, dk \tag{7}$$

To apply this formula to matter waves, one puts $k = p/h$ and $E = h\nu$ obtaining

$$\Phi(x,t) = \int_{\Delta p} \left\{ a \exp\left[\frac{i}{h}(Et - px)\right]\right\} dp \tag{8}$$

Let p_0 be a central value in the wave-packet and assume that E varies slowly enough with p to justify the Taylor expansion

$$E = E_0 + \frac{dE}{dp}(p - p_0) + 0(p - p_0)^2 \tag{9}$$

Substitution of Eq. (9) into Eq. (8) yields

$$\Phi(x,t) \sim a \exp\left[\frac{i}{h}(E_0 t - p_0 x)\right] \int_{\Delta p} \left\{ \exp\left[\frac{i}{h}\left(\frac{dE}{dp}t - x\right)(p - p_0)\right]\right\} dp \tag{10}$$

In Eq. (10) the exponential term outside the integral represents a plane wave moving with constant velocity, and the integral behaves like a wave-packet when

$$\frac{dE}{dp}t - x = 0 \tag{11}$$

One thus obtains for the wave-packet velocity U (group velocity)

$$U = \frac{dE}{dp} = \frac{x}{t} \tag{12}$$

and one notes that

1. *In the nonrelativistic case* (Schrödinger waves): $E = p^2/2m$, $p = mv$ and thus

$$U = \frac{dE}{dp} = \frac{d}{dp}(p^2/2m) = v \tag{13}$$

2. *In the relativistic case* (de Broglie waves): $E^2 = p^2c^2 + m_0^2c^4$, $p = m_0v/\sqrt{(1 - \beta^2)}$ and thus

$$U = \frac{dE}{dp} = \frac{d}{dp}(p^2c^2 + m_0^2c^4)^{1/2} = pc^2/E = \frac{\sqrt{1 - \beta^2)}m_0vc^2}{\sqrt{(1 - \beta^2)}m_0c^2} = v \tag{14}$$

In both cases the identification of the group velocity U of a wave-packet with particle velocity v holds. This coincidence is a common source of confusion between the relativistic de Broglie waves (phase velocity $V = c^2/v$, frequency

$\nu = E/h$) and the nonrelativistic Schrödinger waves (phase velocity $V = v/2$, frequency $\nu = mv^2/2h$). Moreover, the Schrödinger waves which are statistical featured waves without local concentration of energy (as the Born waves) cannot be likened to real physical waves. For this reason, de Broglie[3] expressed his "double solution hypothesis": *to each solution* $\Psi = a \exp(iS)$ *of the propagation equation, there must correspond a solution* $U_0 = f \exp(iS)$ *with the same phase S, but one whose amplitude f exhibits a singularity moving with the particle velocity.* The U_0 corresponds to an extended wave phenomenon centered on a very small region standing, strictly speaking, for the particle. The U_0 wave would be the solution of a yet unknown nonlinear equation and the Ψ wave the solution of its linear approximation, at least outside the singularity.

If we take $\Psi = a \exp(iS)$, the linear Klein–Gordon equation

$$\Box\Psi - (m_0^2 c^2/\hbar^2)\Psi = 0 \tag{15}$$

splits into

$$\text{(C):} \quad \nabla(a^2 \nabla S) = 0 \tag{15}$$

$$\text{(J):} \quad (\nabla S)^2 + m_0^2 c^2 = \hbar \Box a/a \tag{16}$$

(C) is a continuity equation and (J) a Jacobi equation.

In the right-hand side of Eq. (16,J) one recognizes the relativistic generation of the quantum potential introduced by de Broglie, which expresses the reaction of the wave deformed in the presence of obstacles to its propagation. In particular, the explicit calculation of this quantum potential in the two slit situation[4] shows how to obtain interference without the need to abandon the notion of well-defined trajectories.

Now, let us consider a wave-packet of *arbitrary shape* surrounding the particle and defined by

$$u(x,t) = R(\xi \Delta k) \exp(iS) \tag{17}$$

with $\Delta k = mc/\hbar$, $\xi = x - vt$, $S = (mc^2/\hbar)t - (mv/\hbar)x$, $m = m_0/\sqrt{(1 - \beta^2)}$. This wave-packet is the solution of a nonlinear two-dimensional Klein–Gordon equation

$$\Box u - \varkappa^2 u = \frac{\Box\sqrt{\rho}}{\sqrt{\rho}} u \tag{18}$$

with $\Box = \partial^\mu \partial_\mu = \partial^2/\partial x^2 - \partial^2/c^2\partial t^2$, $\varkappa^2 = m_0 c^2/\hbar^2$, $\sqrt{\rho} = (u\bar{u})^{1/2} = |u|$. This equation is the relativistic extension of the nonlinear Hasse equation[5] (a Schrödinger equation with a nonlinear term of "quantum potential" type) and describes propagation of kinks and solitons of arbitrary shape in the two-dimensional space-time.

Equation (18) stems from the Lagrangian

$$[(u/\bar{u})(\partial_\mu \bar{u})(\partial^\mu \bar{u}) + (\bar{u}/u)(\partial_\mu u)(\partial^\mu u)]/4 + [(\partial_\mu \bar{u})(\partial^\mu u)]/2 - \varkappa^2 u\bar{u} = \mathcal{L} \tag{19}$$

From this Lagrangian can be derived the same current as in the linear case, that is a condition required by the double solution theory.[3] We have

$$J_\mu = -(ie/\hbar c)\,[\bar{u}(\partial_\mu u) - u(\partial^\mu \bar{u})] \tag{20}$$

and for the energy–momentum tensor

$$\begin{aligned} T_{\mu\nu} = {} & [(u/\bar{u})(\partial_\mu \bar{u})(\partial_\nu u) + (\bar{u}/u)(\partial_\mu u)(\partial_\nu u) \\ & - (\partial_\mu u)(\partial_\nu \bar{u}) - (\partial_\mu \bar{u})(\partial_\nu u)] - \delta_{\mu\nu}\mathcal{L} \end{aligned} \tag{21}$$

In four-dimensional space-time, instead of (17), the nondispersive wave-packet associated with a particle of rest mass m_0 and traveling in the $+x$ direction are defined by

$$u(r,t) = R(r)\exp(iS) \tag{22}$$

with $r = |\vec{r}| = [(x - vt)^2/\sqrt{(1 - \beta^2)} + y^2 + z^2]^{1/2}$. They are solutions of

$$\Box u - \varkappa^2 u = (R_{,rr} + 2R_{,r}/r)\exp(iS) \tag{23}$$

But a simple calculation shows immediately that, for u given by Eq. (22),

$$\frac{\Box\sqrt{\rho}}{\sqrt{\rho}}u = (R)\exp(iS) = (R_{,rr} + 2R_{,r}/r)\exp(iS) \tag{24}$$

so that Eqs. (23) and (18) have the same form.

Among the solutions (22) there are solutions for which arise simultaneously

$$\Box u = 0 \quad \text{and} \quad m_0^2 + (\hbar^2/c^2)\frac{\Box\sqrt{\rho}}{\sqrt{\rho}} = 0 \tag{25}$$

i.e., the Eulerian differential equation

$$R_{,rr} + 2R_{,r}/r + \varkappa^2 R = 0 \tag{26}$$

which admits the general solution

$$u(r,t) = \left[A\frac{\sin\varkappa r}{\varkappa r} + B\frac{\cos\varkappa r}{\varkappa r}\right]\exp(iS) \tag{27}$$

where A and B are constants.

As Mackinnon has shown, the sine solutions express the relativistic covariance of the phase concordance principle.[6] These solutions are also well known to represent the superposition of two spherically symmetrical waves, one converging and the other diverging.[3] In electromagnetism, they behave as waves in a phase-locked cavity similar to those analyzed by Jennison[7] and have the inertial properties of classical particles.

The cosine solutions, unbounded at the center, can also be retained since, in a realistic scheme, particles are not pointlike and have a radius r_0 very small but different from zero. These solutions can express the very high value of the amplitude near the particle.

The previous elementary calculations give a schematic model of $U._0$ waves consistent with experimental data on the interferences of particles. Furthermore, as in the optical case, the coherence length L of matter waves is a *measurable quantity*: the electron longitudinal coherence length L used in interference experiments (Moellenstedt, Faget) is about 10^{-6} cm with a velocity on the order of 10^9 cm/s; for neutrons, L varies between 10^{-5} and 10^{-3} with a velocity on the order of 10^5 cm/s. It is important to underline the fact that an interference pattern is not dependent on the observer may be explained by de Broglie waves, but not by Schrödinger ones,[8] even in the nonrelativistic case.

Thus, a dualism between the Ψ wave and the U_0 waves appears within the wave–particle dualism for matter, exactly in the same way for light and the Nadelstrahlung model occurs again.

A satisfactory manner to avoid this problem would be to ascribe all the physical properties to the wave-packets and the statistical ones only to the Δ wave, in the Born way. de Broglie discussed this important problem in a very interesting but practically unknown article published in the *Cahiers de Physique*. Analyzing the Einstein model, de Broglie did not retain the above proposition. Indeed, he reminded the reader that, about 1920, in order to put the Nadelstrahlung to the test, Schrödinger suggested the following experiment: "with the help of mirrors, to have a try at doing interfere the light beams emitted in nearly opposite directions" and de Broglie wrote: "The experiment was performed and has given a positive result. As the photons were emitted one after the other by the atoms of the source, this result seems to give an evidence, contrary to the Nadelstrahlung hypothesis, that just when the emission occurs, the wave going out the atom and carrying a photon is a classical spherical wave,"[9] i.e., a Σ wave.

This experiment seems to have made a great impression on de Broglie's mind and led him to abandon Einstein's idea. But nowadays, one knows that it is not simple to build a source emitting photons *one after the other*, so that the Schrödinger experiment needs confirmation. If new experiments would negate the results of the former and uphold the Nadelstrahlung, a space to new inquiries about *individual* atomic processes would be created. The wave functions U_0 should describe a real phenomenon propagating in space-time as in the original de Broglie wave mechanics, the wave functions Ψ, elements of an abstract functional space, retaining all their quantum mechanical assignments. This view implies that, mathematically and experimentally, a few questions remaining open, as, for instance:

1. The transversal size of a wave-packet, i.e., the transversal coherence length.
2. The physical nature of the U_0 waves (in other words, what is waving?). For light, it appears clearly that Maxwell electromagnetic waves are associated with photons, in the case of massive photons introduced by de Broglie,[10] or for conventional zero-mass photons as shown, particularly, by the Majorana equation.[11] But the problem remains unsolved for the other particles: for instance, are protons and neutrons associated with the same "nucleonic" wave? A solution of this question is very likely tied to a realistic reinterpretation of the quantum field theory.
3. The "empty wave-packet" problem, i.e., the examination of the properties of wave-packets deprived of their particles. Empty waves are necessary to explain one particle interference with itself. The theoretical study of their possible intervention in the domain of neutronic interferometry[12] and their eventual ability to induce stimulated emissions of radiation[13] have already been examined. But other problems remain unsolved: for instance, are the empty waves absolutely devoid of energy? A negative answer should lead to a concrete description of how a photon grows old by repeated diffractions and, particularly, by taking in account this consideration, it is possible to obtain an available alternative explanation of the redshift of astronomical objects.[9] Another problem is: why an empty wave separated from an electrically charged particle, would remain sensitive to an electrostatic field?
4. The wave–wave dualism within the wave–particle dualism involves the existence of two kinds of quantum potential: one $\Box a/a$ built from the wave function $\Psi = a \exp(iS.)$ gives the trajectories of the wave-packet center,[4] the other, built from the wave function U_0, is connected with the position fluctuations of the particle inside the wave-packet. This scheme is consistent with stochastic theories, but without the need for a recourse to a hypothetical chaotic medium as the Dirac ether.
5. An important problem is the spin dependence of the U_0 waves. From its solution depends, namely, space-time specification of the polarization measurements as it occurs peculiarly in the EPR experiments.

REFERENCES

1. L. DE BROGLIE, in: *Wave Mechanics: The First Fifty Years*, Butterworths, London (1973).
2. L. DE BROGLIE, *Ann. Phys. (Paris)* **3**, 22 (1925).
3. L. DE BROGLIE, *Nonlinear Wave Mechanics*, Elsevier, Amsterdam (1960).

4. C. PHILIPPIDIS, C. DEWDNEY, and B. J. HILEY, *Nuovo Cimento B* **52**, 1 (1979).
5. R. W. HASSE, *Z. Phys. B* **37**, 83 (1980).
6. L. MACKINNON, *Found. Phys.* **8**, 157 (1978).
7. R. C. JENNISON, *J. Phys. A Gen. Phys.* **11**, 1525 (1978).
8. L. DE BROGLIE, *Cah. Phys.* **147**, 1 (1962).
9. L. DE BROGLIE, *Ondes électromagnétiques et photons*, Gauthier–Villars, Paris (1968).
10. R. MIGNANI, E. RECAMI, and M. BALDO, *Lett. Nuovo Cimento* **11**, 568 (1974).
11. C. DEWDNEY, P. GUÉRET, A. KYPRIANIDIS, and J. P. VIGIER, *Phys. Lett. A* **102** (7), 291 (1984).
12. F. SELLERI, *Found. Phys.* **17**(8), 739 (1987); A. GARUCCIO, P. GUÉRET, and F. SELLERI, *Found. Phys. Lett.* **1**(2), 139 (1988).

CHAPTER 6

RECENT CONTRIBUTIONS OF ELECTRON INTERFEROMETRY TO WAVE–PARTICLE DUALITY

FRANZ HASSELBACH

1. INTRODUCTION

Louis de Broglie's wave–particle duality hypothesis, published in his famous paper[1] in 1924, was verified for electrons only 3 years later by Davisson and Germer[2] and Thomson.[3] No interferometer was available at that time to measure the very short electron wavelengths predicted by the de Broglie relation. The idea underlying their experiment to prove the hypothesis was that if electrons are in some sense represented by waves, they should undergo diffraction from crystalline lattices in a way almost identical to X rays. Crystalline lattices of appropriate periodicity were at hand and therefore a diffraction experiment provided the first evidence of wave–particle duality.

By 1940 the electron microscope had developed to a point where it became possible to observe phenomena below 10 nm in scale and at the end of this decade an electron interferometer seemed to be within reach. Two schemes to produce two coherent electron waves and to recombine these waves were investigated: Amplitude division occurring upon crystalline diffraction by Marton[4] and wavefront splitting by an electron optical biprism by Möllenstedt and Düker.[5,6] Both schemes work with electrons. However, biprisms proved to have the greater utility for electrons. That is why they are used exclusively in today's electron interferometers. The single-crystal interferometer got to the top two decades later in neutron interferometry.[7] The most direct and spectacular experiments concerning wave–

FRANZ HASSELBACH • Institut für Angewandte Physik, Universität Tübingen, D-7400 Tübingen, Germany.

Wave–Particle Duality, edited by Franco Selleri. Plenum Press, New York, 1992.

particle duality were done with electron and neutron interferometers. However, there is of course—apart from these experiments—an overwhelming mass of evidence for de Broglie's hypothesis.

Conventional electron interferometers were constructed by electron microscopists according to customary principles approved in electron microscopy. In fact, they were most of the time just suitably modified electron microscopes. Their sensitivity to alternating magnetic stray fields and mechanical vibrations is therefore similar or, in interferometers with wide separation of the coherent beams, even higher than that of electron microscopes with atomic resolution. Electron interferometers therefore had to be located in special laboratories far from electric cables, far from any traffic and often additionally mounted on vibration isolation systems. To put such an interferometer on a rotating table, e.g., in order to perform a Sagnac experiment with electron waves seemed to be unimaginable. A closer look at the problem fortunately revealed that the special constructional requirements necessary for a rugged interferometer were disregarded in conventional instruments. We therefore dropped traditional constructive principles and developed a totally new design.[8] The resulting instrument is many orders of magnitude more insensitive to the disturbances just mentioned. The Sagnac experiment, for which the insensitivity of the instrument to vibrations is crucial, could be performed successfully as we will see later. While we focus here our interest on experiments done with this new interferometer, supplementary information may be found in review papers on conventional electron interferometry.[9–12]

2. THE NOVEL ELECTRON-OPTICAL BIPRISM INTERFEROMETER

An electron optical biprism interferometer, in principle, consists of an electron source which is illuminating the biprism (Figure 1a). The biprism is composed of a very fine metallized quartz filament (less than 1 μm in diameter) held at positive potential between two grounded electrodes. The incoming wave front is split into two partial waves when passing the biprism filament and—by the positive charge—the two partial waves are deflected toward each other. In analogy to the light optical biprism the electrons (partial waves) seem to emerge from the two virtual sources (marked by crosses). Interference fringes are formed in the region of superposition only if the spatial and the temporal coherence conditions are met. Both conditions can be satisfied very easily when the electrons emerging from the very fine virtual source of a field-emitter tip are used to illuminate the biprism.[13] When a single positively charged biprism is used, the widest separation of the coherent partial waves is on the order of the diameter of the biprism filament, i.e., a micrometer or less. A wider separation of the coherent electron waves is mandatory in many experiments, e.g., when a very small coil or superconducting tube carrying magnetic flux has to be inserted between the coherent waves in order to measure the Aharonov–Bohm phase shift [14–17] or the flux quantization in superconductors.[17,18] Likewise, an enclosed area between the

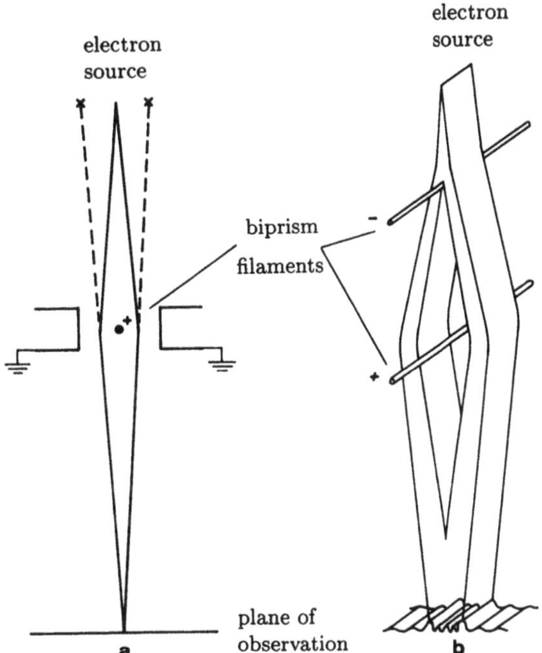

FIGURE 1. (a) Schematic diagram showing the path of the rays in an electron-optical biprism interferometer. (b) Schematical setup for achieving wide separation of the coherent electron waves by means of a dual biprism arrangement.

coherent beams which is as large as possible is needed for the Sagnac experiment. Wide separation of the coherent beams and in turn an enclosed area can easily be accomplished by multiple biprism arrangements as shown, e.g., for the case of two biprisms in Figure 1b. Here the first biprism is charged negatively and bends the partial waves apart. A second, positively charged biprism recombines them. The separation of the wave fronts depends on the negative voltage of the first biprism filament and the distance of the two filaments. It reaches its maximum in the vicinity of the second biprism filament.

A conventional electron interferometer operated under such "wide separation conditions" has an enormously increased sensitivity to vibrations and—due to the phase shifting action of magnetic fluxes enclosed by the coherent beams—to ac magnetic stray fields. The mechanical resonance frequency of a conventional instrument is low and, since it departs only slightly from the frequencies of the vibrations coming along the floor of the building it is excited easily. As a remedy it had to be the primary goal to make the new interferometer as rigid as possible, i.e., to raise the mechanical eigenfrequency of the whole assembly to values as high as possible; then external vibrations cause the interferometer to vibrate as a whole of course, but the relative positions of its components are not influenced.

In turn, the visibility of the interference fringes is not impaired. The consequence of these considerations is that the dimensions as well as the weight of the interferometer have to be reduced drastically. Mechanical alignment of the interferometer, while operating, has to be abandoned in favor of prealigned high-precision electron optical components. Fine alignment has to be done exclusively by electromagnetic deflection systems.

The practical realization of the interferometer, the electron-optical setup, and a beam path are given in Figure 2a,b,c. The total length of the interferometer is only 30 cm, the diameter of the electron-optical components 28 mm, and the total mass less than 1 kg. In the setup shown, up to three biprisms can be used according to the individual requirements of the experiment. Fine alignment is achieved by the deflection elements and by the coils. The homogeneous magnetic field created by the coils allows rotation of the directions of the wave fronts. Inevitable slight rotational misalignments of the biprism filaments relative to each other can be compensated in this way. The Wien filter incorporated in our instrument is a novel component in electron interferometry. It is obligatory—as we will see later—in multiple biprism interferometers working with low-energy

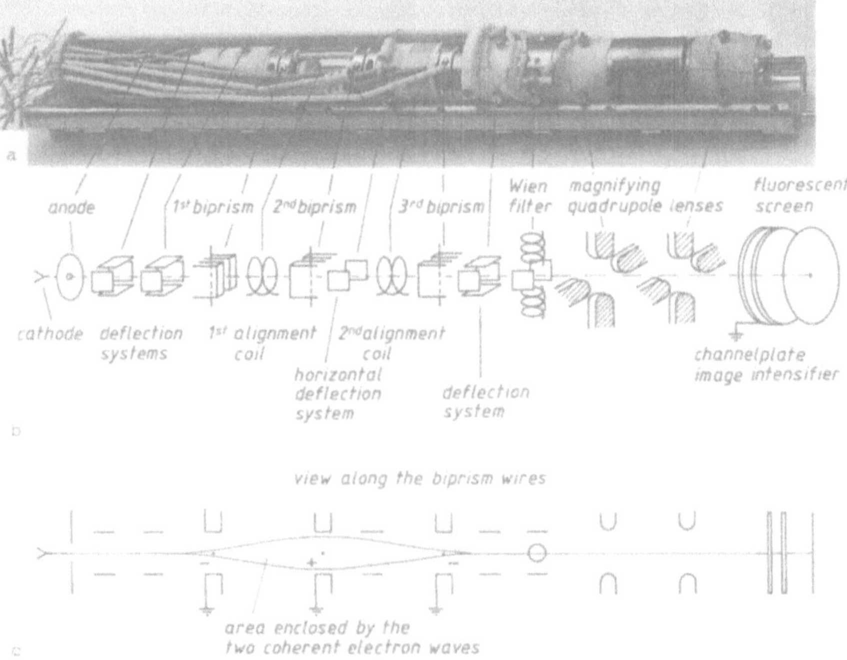

FIGURE 2. (a) Technical realization of the triple biprism interferometer. (b) Electron-optical setup. (c) Beam path.

electrons in the range of 150 eV up to a few keV. Figure 2c gives the beam path in a three-biprism $(-,+,-)$ arrangement with an enclosed area. The third, negatively charged biprism deflects the beams so that they intersect at a smaller angle in order to increase the width of the fringes in the interference pattern.

3. ELECTRON INTERFEROMETRIC VERIFICATIONS OF WAVE–PARTICLE DUALITY

3.1. *Early Experiments: Diffraction at an Edge, Electron Biprism Interferences, and Diffraction by Slits*

Electron diffraction on macroscopic objects was observed for the first time by Boersch[19–21] in 1940 in an electron microscope. He observed contour fringes on an edge in out-of-focus electron micrographs and identified these as Fresnel diffraction fringes.

As an example of electron biprism interferences, a series of interferograms taken at an energy of 2.5 keV with our new interferometer are given in Figure 3a. The potential of the biprism filament was chosen in the range 0.0–1.6 V. In the uppermost panel of Figure 3a the shadow of the biprism filament is visible in the middle. The Fresnel diffraction fringes of both edges of the filament are clearly visible. With increasing positive voltage applied to the biprism filament the partial waves begin to overlap. With further increasing angle of superposition—corresponding to an increasing lateral distance of the two virtual sources—more and more fringes become visible.

Microminiaturization was launched in Tübingen at the end of the 1950s.[22,23] Slits about 0.3 μm wide in a thin copper foil were produced with this new technique. Single-, double-, up to ten-slit diffraction patterns were observed by Möllenstedt and Jönsson in 1959[24,25] and diffraction by a transmission grating by Holl in 1969.[26] In Figure 3b, single-, double-, and five-slit diffraction patterns taken from Jönsson's Ph.D. thesis[27] are given. The single slit interference pattern is complementary to the shadow image of the biprism filament given in Figure 3a and demonstrates Babinet's theorem.

3.2. *Novel Experiments*

3.2.1. Buildup of an Interference Pattern out of Single Events. One of the most impressive experiments which directly shows quantum mechanics at work is to observe the buildup process of an electron biprism interference pattern by accumulating the arrival sites of single electrons on a photographic plate or, even more impressive, in the memory of an image processing system.[28] While with the photographic method the buildup process can be seen just after the developing process is finished, with the image processor the buildup process can be visualized

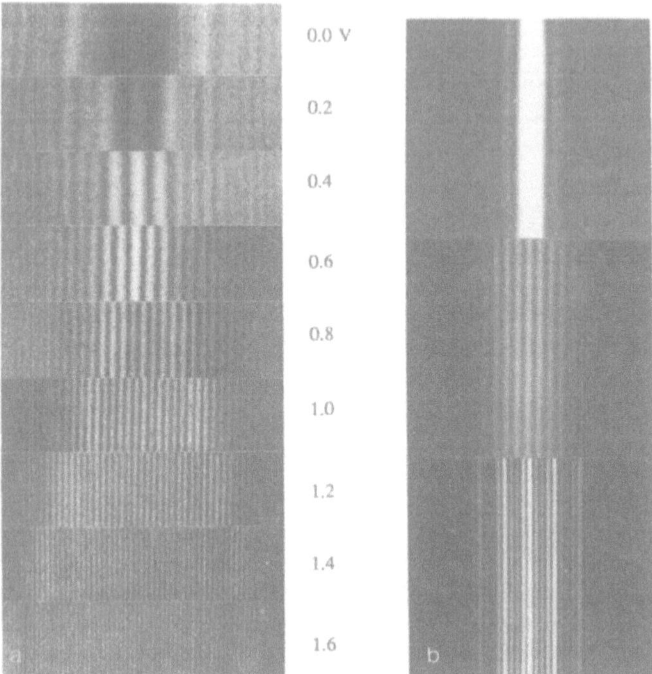

FIGURE 3. (a) Electron biprism interference patterns taken at an electron energy of 2.5 keV. The voltage applied to the (single) biprism filament is given at the right. Fresnel diffraction fringes on both sides of the filament are clearly visible especially when no voltage is applied to the biprism filament. (b) Single-, double-, and five-slit electron diffraction patterns taken by C. Jönsson in 1959. The freestanding slits had a width of about 0.3 μm.

in real time. The fringe pattern, which has been accumulated in the memory of the image processor, is simultaneously displayed with the electrons incoming in every moment. In order to be able to discriminate between the incoming electrons and the accumulated fringe pattern, the brightness of the dots on the cathode ray tube, corresponding to the momentarily incoming electrons, is enhanced by a suitable program routine. Unfortunately, this dynamic buildup process cannot be demonstrated in a book. We must be content here with a static demonstration as given in Figure 4.

In order to obtain these micrographs the emission current of the cathode of our interferometer was adjusted to such a low value and the gain of the image intensifier to such a high level that the sites of incidence of single electrons become visible as tiny bright spots on the fluorescent screen of the image intensifier. In the series of micrographs the exposure time has been doubled from micrograph to micrograph resulting in an increasing density of the bright spots. While fringes

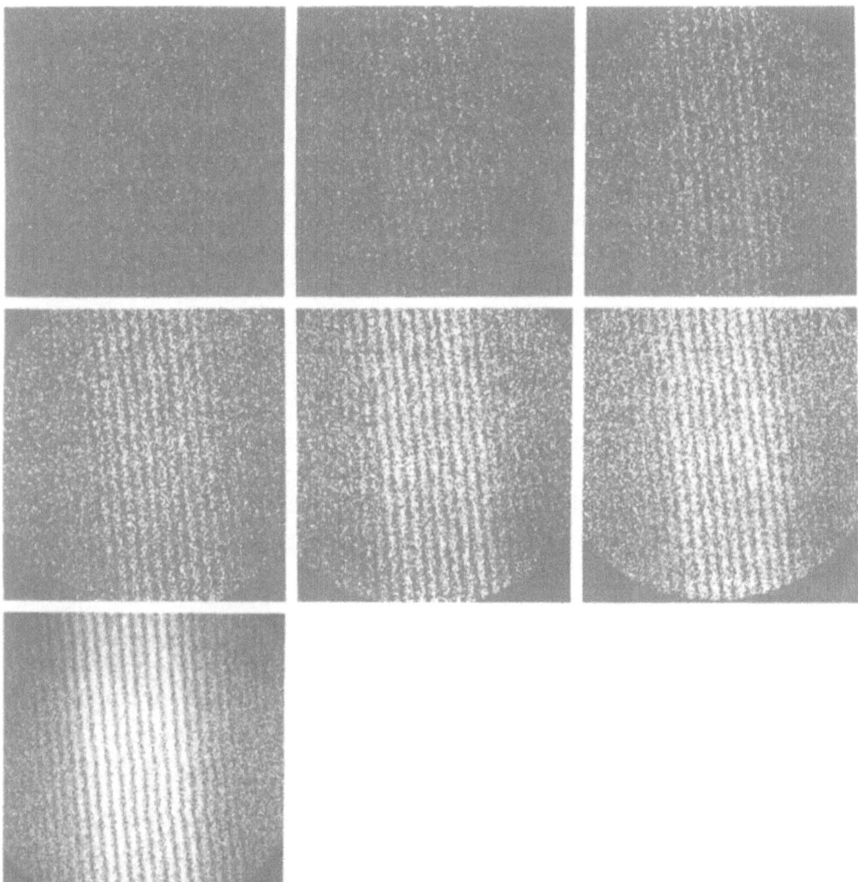

FIGURE 4. Interference patterns obtained at extremely low emission current. The exposure time (starting from 1/8 s) has been doubled from micrograph to micrograph. With increasing integration time the fringe visibility becomes better and better. The bright spots in the micrographs show the arrival sites of single electrons and demonstrate the corpuscular character of the electrons, the arrangement to fringes their simultaneously present wave character.

cannot be seen at all when one is observing the fluorescent screen and in the first micrograph which has been taken with an exposure time of 1/8 s, they are well marked in the last ones. The appearance of well-localized bright spots demonstrates the corpuscular nature of the electrons and the arrangement of the spots to fringes the simultaneous presence of their wave nature.

A prerequisite for this demonstration of particle–wave duality was the availability of image intensifiers with a gain sufficient to visualize single electrons. The first micrographs showing the statistical nature of the formation of interference fringes were taken by Merli *et al.* in 1976[29] followed by Wohland[30] and Matteucci and Pozzi.[31] The micrographs presented in Figure 4 were taken in the first test phase of the new interferometer and presented at the 1979 meeting of the German Electron Microscopical Society.[32]

3.2.2. The Wien Filter as a Device to Shift Wave Packets Longitudinally. What is a Wien filter, and what is the salient point of a Wien filter in an electron interferometric instrument? A Wien filter consists of crossed electric and magnetic fields (Figure 5). It is in its compensated state when the electric force on the electrons is just compensated by the magnetic force, that is, the electrons travel through the Wien filter without any deflection rectilinearly. Let us assume that the two coherent wave packets enter into the Wien filter Δx apart from each other and that the condenser plates of the Wien filter are on a potential of $-U$ and $+U$, respectively. The wave packet on the right-hand side travels through the Wien condenser in a region of positive potential with respect to that on the left-hand side. That is, the wave packet on the right has a higher group velocity in the Wien filter than that on the left. Consequently, the wave packets leave the Wien filter shifted longitudinally Δy relative to each other. The acceleration and deceleration of the wave packets happens in the fringing electric fields of the Wien filter. With increasing excitation of the compensated Wien filter, the longitudinal shift increases, and for sufficiently high excitation, the two wave packets leave the Wien filter one behind the other. They do not overlap any more, and the contrast of the interference fringes vanishes.

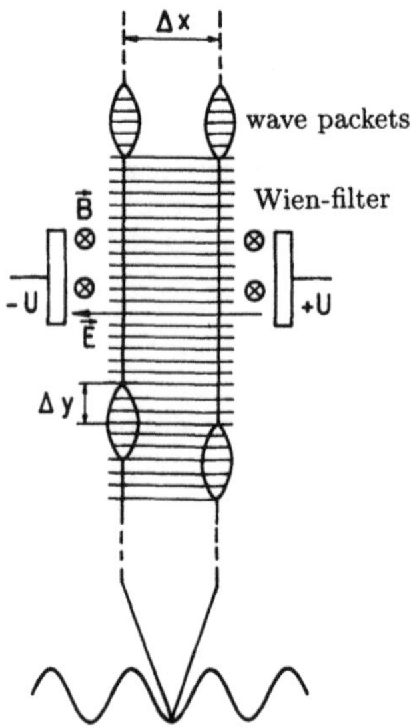

FIGURE 5. Influence of a Wien filter in its compensated state on two spatially separated electron wave packets and on the phase of the waves. The wave packets are shifted longitudinally, the phase velocity is not affected (see text). Therefore, the positions of the horizontal lines, which symbolize the crests of the waves, are not shifted at all by the electromagnetic fields inside the Wien filter.

It is noteworthy that in all compensated states of the Wien filter, the electron-optical index of refraction equals one in the nonrelativistic limit. Consequently: (1) the phase velocity of the electron waves is not affected at all by the presence of a Wien filter in its compensated state and (2) the order of the interference is not increased without regard to the fact that the wave packets are shifted longitudinally.

3.2.3. Coherence Lengths. The ability of a Wien filter to shift wave packets of charged particles longitudinally was discovered by Möllenstedt and Wohland; this feature was used to perform the first coherence length measurements of electron waves in 1980.[(33,34)] They increased the excitation of the Wien filter until the fringe contrast vanished and calculated the corresponding coherence length from the geometrical dimensions of their Wien filter and the electric field strength for vanishing fringe contrast.

In order to overcome the large errors of more than 10% inherent in the measurement method just mentioned, we refined it substantially in the following way[(35)]: The electric and magnetic field are no longer increased simultaneously. We increase in a first step the electric field only. The Wien condenser then works as a deflection element. Let us assume that the interference fringes are deflected, e.g., by 3 fringe widths to the left on the fluorescent screen. This is due to the fact that the wave packet traveling in the more negative region of the Wien condenser is slower and loses three wavelengths. We now increase the magnetic field until the deflection due to the electric field is just compensated. This state of the now again compensated Wien filter corresponds to the following physical situation: Both beams travel rectilinearly through the Wien filter but the left-side wave packet is shifted longitudinally by three wavelengths with respect to the right one in the Wien filter. For measuring the coherence length this procedure is repeated while counting the total number of fringes until the contrast in the fringe field has decreased to $1/e$. We define the coherence length by twice this number of fringes times the wavelength of the electrons. The factor limiting the precision of this method is given by the precision with which the contrast of the fringes can be determined densitometrically. This is a question of counting statistics only.

Let me note here that for this measurement method it is not necessary to know anything about the geometry, the field strengths and homogeneities of the electromagnetic fields in the Wien filter, not to mention the fringing fields. The only thing we need to know is the de Broglie wavelength of the electrons.

In order to simulate different energy widths of our electrons and in turn different coherence lengths, we superimposed to the extraction voltage of 4 kV of our field emitter a triangular-shaped voltage of variable amplitude. The result of our coherence length measurement is given in Figure 6. The natural width of the field emission spectrum plus 3-V triangular-shaped voltage peak to peak results in a coherence length of 25 nm. By reducing the amplitude to 0.5 V it increases to 120 nm. The natural energy width of the field emitter of 0.36 eV corresponds to a coherence length of 280 nm.

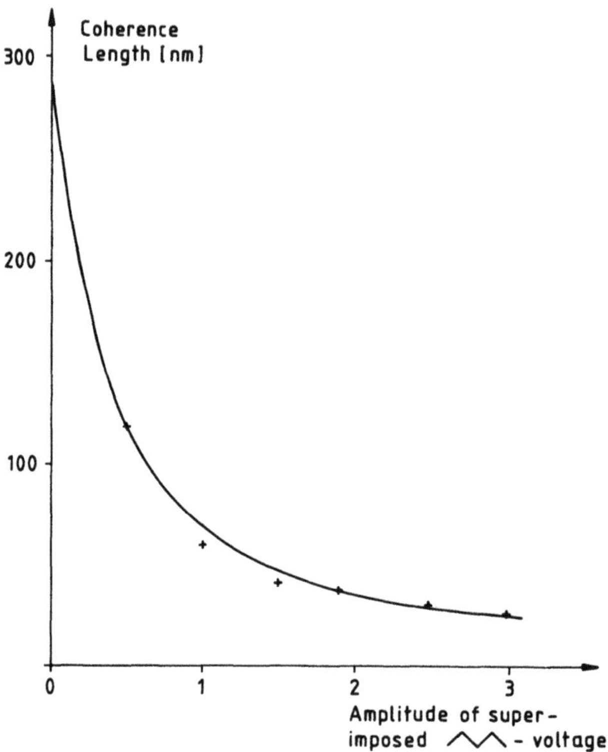

FIGURE 6. Coherence length as a function of energy spread of the electron beam. Different energy spreads were artificially simulated by superimposing a triangular-shaped voltage of 0–3 V to the extraction voltage of 4 kV of the field emitter. The natural energy width of the field emission of 0.36 eV corresponds to a coherence length of 280 nm.

In this chapter, no distinction has been made between coherence length and the length of the wave packet even though the length of de Broglie wave packets increases due to their intrinsically dispersive propagation. The question arises, did we measure the coherence length or the longitudinal shape of the electron wave packet which is varying with the distance from its origin. It is beyond the scope of this contribution to present the theory here or even to discuss the problem in detail. The result is: In spite of the fact that the wave packet spreads, the coherence length remains equal to its initial value, which is exclusively determined by the wave-number spread of the beam. Irrespective of the spread of a wave packet, the fringe visibility depends on the coherence length only.[36–38]

3.2.4. Fourier Spectroscopy of Electron Waves. For a quantitative Fourier spectroscopic measurement of an electron energy distribution,[39] it is not sufficient just to count the number of fringes until a certain decrease of the contrast is reached. Here we must record quantitatively the contrast in the whole interference field. Such an interference field consists, e.g., for a beam energy of 2.5 keV with an energy spread of ≈ 0.36 eV of about 20,000 fringes. We recorded the contrast in this whole interference field in sets of, e.g., ten fringes successively with a television camera.

In order to use all information contained in a two-dimensional fringe pattern and in turn to reduce the noise, the fringe intensities were integrated along the fringe direction. The result is a one-dimensional low-noise densitometer trace across a single set of fringes. The digitized data of all sets of fringes are put together with matched phases in a personal computer. The data are then Fourier-analyzed in a VAX computer. For the spectrum of the field emitter we obtained a full width at half maximum of 0.6 eV instead of the theoretical width of about 0.36 eV (Figure 7a). The spectrum, which was obtained under unfavorable experimental conditions, demonstrates in this very first test of the new method a resolution of better than 0.4 eV. In a second experiment we superimposed a square wave of 30 V amplitude to the extraction voltage of our field emitter. This simulates an energy spectrum containing two discrete lines 30 eV apart from each other (Figure 7b). The two lines are clearly visible, the peak in the middle is an artifact.[39]

3.2.5. The Sagnac Effect of Electron Waves. The Sagnac effect[40–43] is one of the most faceted gemstones of physics: With respect to theoretical physics it links classical and relativistic physics in a unique way. It helped to clarify many interrelations between classical mechanics, the theory of relativity, nonrelativistic and relativistic quantum mechanics. On the other hand, it is not only of interest from a theoretical point of view: today's state-of-the-art navigation systems, e.g., in the latest generation of civil airplanes, are based on the light optical Sagnac effect. Additionally, the most promising developments in the area of rotation sensing are based on the Sagnac effect of matter waves,[44] i.e., of ions, molecules,[45,46] and last but not least of superfluid helium.[47] With the latter, the most sensitive detector for absolute rotation sensing seems feasible. The first experi-

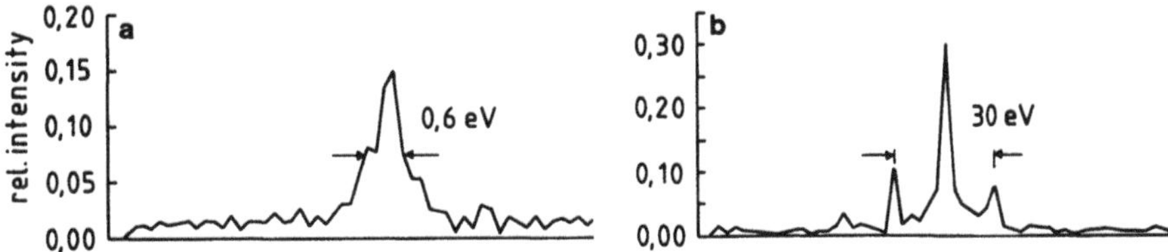

FIGURE 7. Spectra obtained by Fourier analysis. (a) Spectrum of a field emission cathode; accelerating voltage 2.5 kV (see text). (b) In order to simulate a spectrum containing two discrete lines, a square wave voltage of 30 V in amplitude was superimposed to the extraction voltage of the field emitter. The two peaks are clearly visible. The central peak is an artifact.

mental confirmations of the Sagnac effect for matter waves were undertaken by Zimmermann and Mercereau in 1964[48] for Cooper pairs in a rotating superconducting quantum interferometer (SQUID) and—following a proposal of Anandan[42]—by Werner *et al.* in 1979 for neutrons.[49]

In a Sagnac experiment (Figure 8) two signals or wave packets run in opposite directions around an enclosed area A. The whole experimental setup rests on a disk and is set into rotation with respect to the laboratory frame. For the co-moving observer the signals run with equal velocity v around their paths and arrive at the detector after the time $\pi R/v$ where R is the radius of the circle. In the inertial system with respect to which the disk is rotating, the clockwise signal has a speed of $v + \Omega R$ and the counterclockwise, $v - \Omega R$. When we calculate the arrival time of the signal at the detector, this difference in speed is just compensated by the difference in the distances that the signals have to travel. The signals reach the detector at the same instant of time in the rotating and the laboratory system. In the classical Galilean invariant theory, there is no observable effect of the rotation. However, in the relativistic theory as well as in Schrödinger's nonrelativistic quantum mechanics the Sagnac phase shift for matter waves is predicted.

Dieks and Nienhuis[50] discuss the question, how is it possible that a nonrelativistic theory yields a phase shift at all and, moreover, the correct one? Their conclusion is that nonrelativistic quantum mechanics contains some relativistic elements in the sense that the quantum mechanical Galilei group "is not an invariance group in exactly the same way as is the Galilei group for classical theories." This group theoretical aspect of the Sagnac effect has been profoundly discussed by Anandan.[43] In essence, the Sagnac effect is purely a relativistic phenomenon: The difference in the arrival time of the two signals at the detector is a consequence of the fact that there exists no absolute time according to the theory of relativity. Thus, an adequate derivation of this time difference or the corresponding Sagnac phase shift is possible only within the framework of the

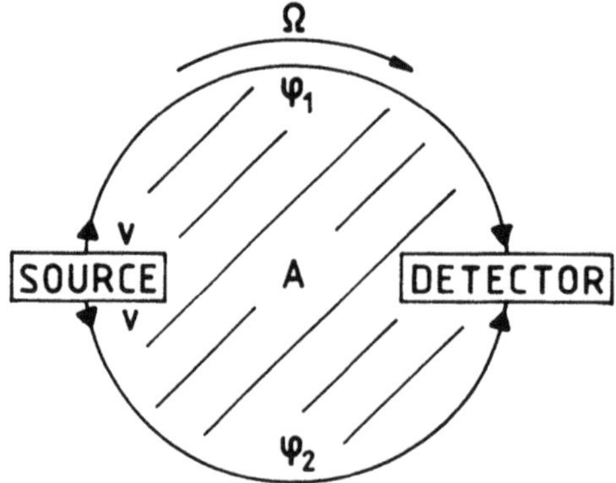

FIGURE 8. Principle of the Sagnac interferometer.

special theory of relativity.[42,43,51] The special theory of relativity predicts for the Sagnac phase shift for electromagnetic and matter waves:

$$\varphi_1 - \varphi_2 = \frac{2m}{\hbar} A\Omega = \frac{2}{\hbar c^2} EA\Omega$$

where E denotes the total energy of the particles or photons, h is Planck's constant, $\hbar = h/2\pi$, c the velocity of light, A the enclosed area of any shape and Ω the angular velocity of the rotating system, and m the relativistic mass and not the rest mass m_0. In the nonrelativistic limit $m \rightarrow m_0$, the Sagnac phase shift of matter waves, in contrast to electromagnetic waves, becomes independent of the wavelength.

The following should be noted: (1) The total energy of an electron is at least 511 keV compared to 2 eV of an optical photon. Consequently, an electron Sagnac interferometer should be more sensitive by a corresponding factor of 250,000 compared to an optical one. However, while it is relatively easy to realize an enclosed area comparable to a soccer field with light waves—let me mention that Michelson and Gale realized such a large area in their famous experiment to detect the earth's rotation[52]—it is very hard to attain enclosed areas on the order of 1 cm^2 in electron interferometers. (2) The Sagnac phase difference does not depend on the speed of the signals or wave packets, but on the angular velocity of the rotating system only. Thus, the application of the Sagnac effect to detect rotation is obvious.

In the present experimental observation of the Sagnac effect of electron waves,[51,53,54] two of the three electron biprisms (Figure 2) were used to realize an enclosed area of about 4 mm^2. The small enclosed area necessitates relatively high rotation rates on the order of 1 rev/s to obtain detectable Sagnac phase shifts of about 3% of a fringe. These high rotation rates lead to centrifugal effects, such as minute bending of the vacuum chamber causing fringe shifts comparable, or even larger than the expected fringe shifts due to the Sagnac effect. In order to avoid any influence of centrifugal forces, we measured the phase differences between successive alternating clockwise and counterclockwise rotations with exactly the same rotation rates.

The Wien filter incorporated into our Sagnac interferometer (Figure 2) has proved to be indispensable in restoring the temporal coherence of the electron waves arriving in the plane of interference. In low-energy electron biprism interferometers the electrostatic deflection elements for fine alignment almost always reduce or totally destroy the longitudinal coherence in the plane of interference. Typically, the coherent electron wave packets travel laterally separated, in regions differing in their electric potential through the deflection fields resulting in an unwanted longitudinal shift. Because of the low electron energies used in our interferometer, resulting in a correspondingly short coherence length, the longitudinal shifts in the numerous deflection systems usually add up to a value greater than the coherence length. The two wave packets arrive in the interference plane one after another. No fringe contrast is observed. With the Wien filter we routinely create a longitudinal shift that is exactly compensating that caused by the deflection elements.[55] In the interference plane, longitudinal coherence and in turn maximum fringe contrast is reestablished. This is demonstrated in Figure 9. Without excitation of the Wien filter the incoherent overlap of the arriving wave packets leads to a stripe of enhanced intensity in the middle of the uppermost micrograph (a). With increasing compensation of the relative delay of the wave packets by the Wien filter, temporal coherence is reestablished (b,c). The interference fringes reappear.

The phase information was extracted by image processing from the interference pattern. The pixel columns of the television camera are aligned mechanically parallel to the interference fringes. The information in all pixels of one column is integrated. The result is a low-noise densitometer trace across our

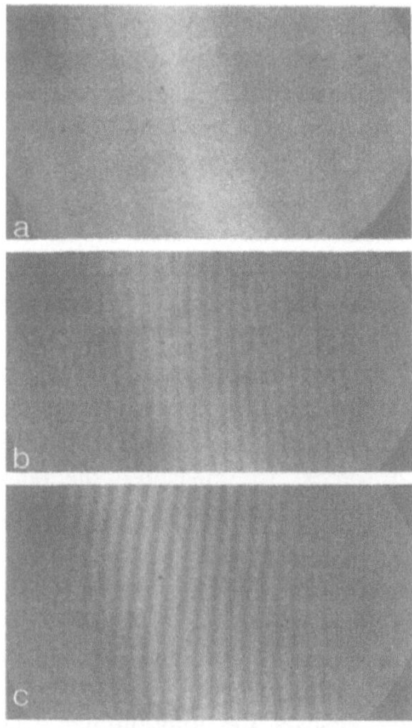

FIGURE 9. Reestablishing temporal coherence by means of a Wien filter. Without excitation of the Wien filter the wave packets arrive in the plane of interference one after another (a). With increasing excitation of the Wien filter the interference fringes appear (b). The excitation is increased until maximum contrast of the interference fringes is observed corresponding to a full overlap of the wave packets (c). Only after this step is our low-voltage biprism interferometer fully adjusted and ready for an experiment.

interference field. This trace is Fourier-analyzed and the phase information is calculated via the arctan of the Fourier components. The phase error achieved with this method is less than 1%.

A high mechanical and electronic long-term stability is required. The wavelength of the 1.5-keV electrons is about 0.03 nm. One percent of a fringe corresponds to a length of 0.0003 nm. The entirety of all disturbances, i.e., due to mechanical vibrations, electronic instabilities, and instabilities of the field emission electron source must not exceed the equivalence of this length. The stability needed in this experiment is comparable to that of an electron microscope with atomic resolution. But, while exposure times of a few seconds are usual in such microscopes, we need this stability for at least 10 min because of the relatively long time needed for accelerating and decelerating of the whole apparatus.

The results of our measurements of the Sagnac phase shift are given in Figure 10 and agree well with theory. The error limits of about 30% are due to long-term drifts of the electronic supplies and instabilities of the field emission.

Let us now consider the striking formal analogy between the formula for the Sagnac phase shift and the Aharonov–Bohm phase shift in the nonrelativistic case:

$$\varphi_1 - \varphi_2 = \frac{2m}{\hbar}\int \vec{\Omega}\, d\vec{\sigma} \qquad \varphi_1 - \varphi_2 = \frac{e}{\hbar c}\oint \vec{A} d\vec{s} = \frac{e}{\hbar c}\int \vec{B} d\vec{\sigma}$$

The Sagnac phase shift is given on the left and the Aharonov–Bohm phase shift on the right hand side. The surface integrals are over the oriented enclosed area, $\vec{\Omega}$ is the angular velocity vector, $\vec{B}$ the magnetic field vector, and $\vec{A}$ the vector potential. $2m\vec{\Omega}$ in the Sagnac formula corresponds to $(e/c)\vec{B}$ in the equation for the Aharonov–Bohm effect. The angular velocity corresponds to the magnetic field. Hendriks and Nienhuis[56] derive the Dirac, Klein Gordon, and Schrödinger equation in the rotating frame of reference and show that the rotation has the same effect on the Schrödinger equation, as an electromagnetic field, described by a

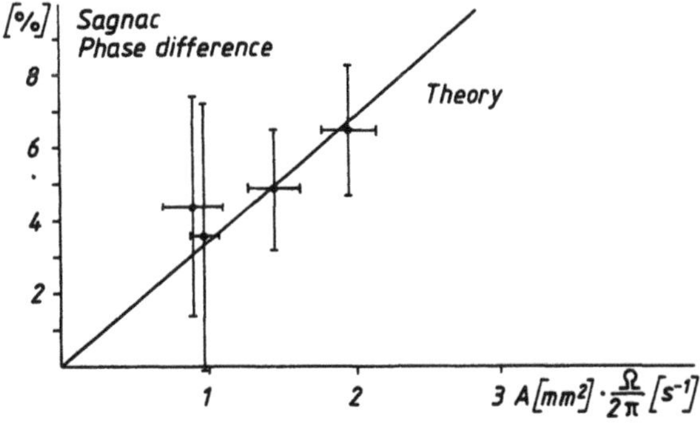

FIGURE 10. Sagnac phase shift as a function of the product of rotation frequency times enclosed area.

vector potential. That is, the Sagnac effect is the mechanical counterpart of the Aharonov–Bohm effect. Consequently, since the Aharonov–Bohm phase shift is an example for a geometric phase or Berry phase,[57] the same holds for the Sagnac effect. Bernstein and Phillips[58] discuss the Aharonov–Bohm effect in the context of the mathematical theory of fiber bundles and vividly demonstrate that the Aharonov–Bohm experiment can be modeled by a parallel transport of a vector on a truncated cone capped with a spherical dome. The region of the spherical dome corresponds to the enclosed magnetic field and this in turn to the total curvature enclosed between the paths. In this model the Aharonov–Bohm phase shift equals the angular excess when a parallel transport of a vector over a closed curve around the cone is performed. In the case of the Sagnac effect the corresponding phase shifting field is likewise caused by a velocity-dependent potential. The study of the Sagnac phase shift in the framework of the fiber bundles/geometric phase model will shed new light on quantum physics, the theory of relativity, and their interrelations.

ACKNOWLEDGMENTS. I thank Prof. Dr. G. Möllenstedt for the constant support during my development of the new type of interferometer. I thank my colleagues H. Gauch, A. Schäfer, and especially M. Nicklaus for the fruitful cooperation and many discussions. The Sagnac experiment was supported by the Deutsche Forschungsgemeinschaft (Ha 1063/2-1,2,3).

REFERENCES

1. L. DE BROGLIE, *C. R. Acad. Sci.* **177**, 507–510 (1923); *Philos. Mag.* **47**, 446 (1924); *Ann. Phys. (Paris)* **3**, 22–128 (1925).
2. C. J. DAVISSON and L. H. GERMER, *Phys. Rev.* **30**, 705 (1927).
3. G. P. THOMSON, *Proc. R. Soc. London Ser. A* **117**, 600 (1928); *Nature* **120**, 802 (1927).
4. L. MARTON, *Phys. Rev.* **85**, 1057–1058 (1952).
5. G. MÖLLENSTEDT and H. DÜKER, *Naturwissenschaften* **42**, 41 (1954).
6. G. MÖLLENSTEDT and H. DÜKER, Z. Phys. **145**, 377–397 (1956).
7. H. RAUCH, W. TREIMER, and U. BONSE, *Phys. Lett. A* **47**, 369–371 (1974).
8. F. HASSELBACH, *Z. Phys. B: Condensed Matter* **71**, 443–449 (1988).
9. A. SIMPSON, *Rev. Mod. Phys.* **28**, 254–260 (1956).
10. G. MÖLLENSTEDT and H. LICHTE, in: *Neutron Interferometry, Proceedings of an International Workshop* (U. BONSE and H. RAUCH, eds.), Clarendon Press, Oxford, pp. 364–388 (1979).
11. G. F. MISSIROLI, G. POZZI, and U. VALDRÈ, *J. Phys. E* **14**, 649–671 (1981).
12. A. TONOMURA, *Rev. Mod. Phys. 59*, 639–669 (1987).
13. W. BRÜNGER, *Naturwissenshaften* **55**, 295–296 (1968).
14. R. G. CHAMBERS, *Phys. Rev. Lett* **5**, 3–5 (1960).
15. W. BAYH, *Z. Phys.* **169**, 492–510 (1962).
16. H. SCHMID, Ph.D. thesis, Universität Tübingen (1985).
17. A. TONOMURA, *Physica B* **151**, 206–213 (1988).
18. H. WAHL, *Optik* **30**, 508–520, 577–589 (1970).
19. H. BOERSCH, *Naturwissenschaften* **28**, 709 (1940).

20. H. BOERSCH, *Phys. Z.* **44**, 202–211 (1943).
21. H. BOERSCH, *Phys. Z.* **44**, 32–38 (1943).
22. G. MÖLLENSTEDT and R. SPEIDEL, *Phys. Bl.* **16**, 192–198 (1960).
23. G. MÖLLENSTEDT and C. JÖNSSON, *Z. Phys.* **155**, 472–474 (1959).
24. C. JÖNSSON, *Z. Phys.* **161**, 454–474 (1961).
25. C. JÖNSSON, D. BRAND, and S. HIRSCHI, *Am. J. Phys.* **42**, 4–11 (1974).
26. P. HOLL, *Optik* **30**, 116–137 (1969).
27. C. JÖNSSON, Ph. D. thesis, Universität Tübingen (1960).
28. M. NICKLAUS, To be published in *Am. J. Phys.*
29. P. G. MERLI, G. F. MISSIROLI, and G. POZZI, *Am. J. Phys.* **44**, 306–307 (1976).
30. G. WOHLAND, Diploma thesis, Universität Tübingen (1977).
31. G. MATTEUCCI and G. POZZI, *Am J. Phys.* **46**, 619–623 (1978).
32. F. HASSELBACH, *19. Tagung der Deutschen Gesellschaft für Elektronenmikroskopie in Tübingen*, Abstract 7L1, p. 90 (1979).
33. G. MÖLLENSTEDT and G. WOHLAND, in: *Electron Microscopy 1980* (P. BREDORO and G. BOOM, eds.), Vol. 1, pp. 28–29.
34. G. WOHLAND, Ph.D. thesis, Universität Tübingen (1981).
35. I. DABERKOW, H. GAUCH, and F. HASSELBACH, *Joint Meeting on Electron Microscopy*, Antwerp 1983, Program and Abstract book p. 100.
36. H. KAISER, S. A. WERNER, and E. A. GEORGE, *Phys. Rev. Lett.* **50**, 560–563 (1983).
37. A. G. KLEIN, G. I. OPAT, and W. A. HAMILTON, *Phys. Rev. Lett.* **50**, 563–565 (1983).
38. G. COMSA, *Phys. Rev. Lett.* **51**, 1105–1106 (1983).
39. F. HASSELBACH and A. SCHÄFER, in: *Proc. 12th Int. Congress for Electron Microscopy*, Seattle 1990 (L. D. PEACHEY and D. B. WILLIAMS, eds.), San Francisco Press. San Francisco, Vol. 2, pp. 110–111.
40. G. SAGNAC, *C. R. Acad. Sci.* **157**, 708–710, 1410–1413 (1913).
41. E. J. POST, *Rev. Mod. Phys.* **39**, 475–493 (1967).
42. J. ANANDAN, *Phys. Rev. D* **15**, 1448–1457 (1977).
43. J. ANANDAN, *Phys. Rev. D* **24**, 338–346 (1981).
44. C. V. HEER, *Bull. Am. Phys. Soc.* **6**, 58 (1961).
45. F. HASSELBACH, German patent No. DBP 3504278C2.
46. J. F. CLAUSER, *Physica B* **151**, 262–272 (1988).
47. R. Y. CHIAO, *Phys. Rev. B* **25**, 1655–1662 (1982).
48. J. E. ZIMMERMANN and J. E. MERCEREAU, *Phys. Rev. Lett.* **14**, 887–888 (1965).
49. S. A. WERNER, J. L. STAUDEMANN, and R. COLELLA, *Phys. Rev. Lett.* **42**, 1103–1106 (1979).
50. D. DIEKS and G. NIENHUIS, *Am. J. Phys.* **58**, 650–655 (1990).
51. M. NICKLAUS, Ph.D. thesis, Universität Tübingen (1989).
52. A. A. MICHELSON and H. G. GALE, *Astrophys. J.* **61**(3), 137–145 (1925).
53. F. HASSELBACH and M. NICKLAUS, *Physica B* **151**, 230–234 (1988).
54. F. HASSELBACH and M. NICKLAUS, submitted for publication to *Phys. Rev. A*.
55. F. HASSELBACH and M. NICKLAUS, submitted for publication to *Phys. Rev. A*.
56. B. H. W. HENDRIKS and G. NIENHUIS, *Quantum Opt.* **2**, 13–21 (1990).
57. M. V. BERRY, *Proc. R. Soc. London Ser. A* **392**, 45–57 (1984).
58. H. J. BERNSTEIN and A. V. PHILLIPS, *Sci. Am.* **245**(1), 95–109 (1981).

CHAPTER 7

The Aharonov–Bohm Effect from the Point of View of Local Realism

Dipankar Home* and Franco Selleri

1. INTRODUCTION

In 1959 Aharonov and Bohm pointed out that in quantum mechanics, unlike in classical physics, there can be observable effects on charged particles (such as electrons) confined to an electric and magnetic field-free space when there is an enclosed magnetic field in a region inaccessible to the electron wave function. Prediction of this striking effect, known as the Aharonov–Bohm (AB) effect, stimulated intensive investigations (more than 300 journal articles over the past 30 years) which culminated in the beautiful experiments using electron holography and microlithography that unambiguously verified the existence of this effect (these experiments have been reviewed by Tonomura[1] in a separate contribution to this volume; see also the pertinent references cited therein).

It is somewhat surprising that though there have been considerable discussions about the implications of the AB effect, its significance from the point of view of local realism has not yet received proper attention. The present chapter seeks to fill this gap from the perspective of Einstein's conception of what he called *Gespensterfelder* ("ghost" waves, devoid of momentum and energy, guiding the atomic and subatomic entities) which will be referred to as "empty" waves in the present chapter. It is important to recall that in his "Reply to Criticisms," Einstein[2] stated that the problem of wave–particle duality was "probably the most interesting subject" to discuss. In the same book, Bohr[3] referred to Einstein's "use of such picturesque phrases as ghost waves (*Gespensterfelder*)"

Dipankar Home and Franco Selleri • Dipartimento di Fisica, Università di Bari, I-70126 Bari, Italy. • *On leave from the Department of Physics, Bose Institute, Calcutta 700 009, India.

Wave–Particle Duality, edited by Franco Selleri. Plenum Press, New York, 1992.

which, Bohr emphasized, "implied no tendency to mysticism, but illuminated rather a profound humor behind his piercing remarks."

Almost all the discussions concerning the significance of the AB effect give the impression that its existence (now experimentally proved) should necessarily imply a nonlocal effect. However, in their original paper, Aharonov and Bohm(4) had written: "Two possible directions are clear. First, we may try to formulate a non-local theory in which, for example, the electron could interact with a field that was a finite distance away. . . . Secondly, we may retain the present local theory and, instead, we may try to give a further new interpretation to the potentials." Later, Aharonov(5) advocated the outlook viewing the AB effect as an example of a nonlocal phenomenon, since the value of the vector potential at a point is not gauge invariant and therefore not measurable. The interpretation of the AB effect as a manifestation of a kind of nonlocality in quantum mechanics was also discussed by Spasskii and Moskovskii,(6) Van Kampen,(7) Olarin and Popescu.(8) Philippidis *et al.*(9) analyzed the meaning of the AB effect using the quantum potential approach and concluded that nonlocal aspects of quantum mechanics come into play as a result of "global" determination of the quantum potential which, in their words, provides "an intuitive understanding in the context of the AB effect of what may physically underlie Bohr's notion of the wholeness of the form of the experimental conditions and the content of the experimental results." On the other hand, Yang(10) asserted: "The AB effect is the result of a local equation of motion in the Heisenberg representation, but with non-commuting dynamical variables. Electrodynamics is not non-local in quantum mechanics." It is also relevant to note that the recent analysis of the time-dependent AB effect (the effect of a time-varying enclosed magnetic flux on the AB phase-shift)(11,12) has revealed that in a dynamic situation the magnetic phase-shift is not determined solely by the enclosed flux but also depends on the time of switching of the flux. Brown and Home(12) show that the effect has to be attributed to the vector potential acting locally at the electron's position as the center of the wave packet moves along the interfering stationary-action trajectories from the source to the screen. In such a time-dependent case, Stokes's theorem ceases to be applicable; hence, Brown and Home(12) conclude: "There are no global aspects and this implies that their invocation in the static limit reflects a mathematical accident rather than a fundamental physical feature."

Experimental investigations of the AB effect carried out in the early 1960s(13) claimed to have confirmed the existence of this phenomenon. However, it was later pointed out(14) that since these experiments used current-carrying solenoids or magnetic whiskers of finite length, the observable effects on the electrons could be accounted for by the inevitable leakage of magnetic field through the ends. Furthermore, some authors(15) even questioned the theoretical justification for the AB effect on the issues related to the use of Stokes's theorem in the multiply-connected space to fix the vector potential and also on the ground of an incompatibility with energy and momentum conservation. The experiments by Tonomura[1] settled the controversy decisively. These experiments used tiny toroidal magnets (a few microns in diameter) avoiding completely the leakage effects. In the light of these experiments one is compelled to accept the physical reality of the AB effect whose deeper understanding promises to provide new insights into conceptual foundations of quantum mechanics. In the following section we recapitulate briefly the key elements in the theoretical treatment of the AB effect.

2. THEORY OF THE AB EFFECT

It is well known that Maxwell's equation can be expressed in the following elegant form in terms of the 4-vector potential $A_\mu(x) = (\vec{A}(x), i\Phi(x))$ at every spacetime point $x = (\vec{x}, ict)$:

$$\Box A_\mu(x) = -(4\pi/c)j_\mu(x) \tag{1}$$

where $j_\mu(x) \equiv (\vec{j}(x), ic\rho(x))$ is the 4-current and A_μ satisfies the Lorentz condition

$$\frac{\partial A_\mu}{\partial x_\mu} = 0 \tag{2}$$

We now note that the 4-vector potential A_μ can undergo what is known as gauge transformations of the second kind:

$$A_\mu \rightarrow A'_\mu = A\mu + \frac{\partial \Lambda}{\partial x_\mu} \tag{3}$$

where Λ is a scalar function of x that satisfies the homogeneous D'Alembertian equation

$$\Box \Lambda = 0 \tag{4}$$

It is because of Eq. (4) that both Eqs. (1) and (2) remain unchanged when subjected to Eq. (3). Let us now consider the electromagnetic tensor $F_{\mu\nu}$, which is antisymmetric in μ, ν and comprises six independent quantities, viz. the components of the electric field $\vec{E}$ and of the magnetic field $\vec{B}$. It is easily seen that its well-known expression in terms of the 4-potential, given by

$$F_{\mu\nu} = \frac{\partial A_\nu}{\partial x_\mu} - \frac{\partial A_\mu}{\partial x_\nu} \tag{5}$$

also remains unaltered by the transformation (3) subjected to (4). In classical physics the 4-potential A_μ is regarded as a mathematical construct, devoid of any physical significance in itself, but a useful aid in computing the fields which generate physically observable effects by acting on the charges, accelerating them and thereby affecting their physical attributes such as energy and momentum.

The ingenuity of the AB example lies in that it provides an instance where a physically observable quantum mechanical effect on charged particles is ascribable to $A\mu$ even when $F_{\mu\nu} = 0$ (zero-field condition). Note that the fields deduced from (5) turn out to be

$$\vec{E} = -\vec{\nabla}\Phi - \frac{1}{c}\frac{\partial\vec{A}}{\partial t}$$
$$\vec{B} = \vec{\nabla} \times \vec{A} \tag{6}$$

It is then evident that sufficient conditions for $\vec{E} = \vec{B} = 0$ are: (i) $\vec{\nabla}\Phi = 0$, (ii) $\vec{\nabla} \times \vec{A} = 0$, and (iii) $\partial\vec{A}/\partial t = 0$; i.e., $\vec{A}$ be time-independent. However, it is not necessary for $\vec{A}$ or Φ to vanish in order to satisfy these conditions. The AB example in which $\vec{A} \neq 0$ but still $\vec{E} = \vec{B} = 0$ is the following:

We consider a very long and closely wound current-carrying cylindrical solenoid. Then one has essentially a confined magnetic field $\vec{B}$ trapped inside the solenoid. However, $\vec{A} \neq 0$ outside the solenoid because it has to satisfy the following condition because of Stokes's theorem:

$$\Phi = \int \vec{B}\cdot d\vec{S} = \oint \vec{A}\cdot d\vec{x} \tag{7}$$

where Φ is the total magnetic flux through every closed circuit containing the symmetry center of the solenoid.

Now let us consider a beam of electrons split into two parts (Figure 1), each going on opposite sides of the solenoid, but avoiding it. The beams are recombined within the region F where at any point, say, P the wave function can easily be shown to be given by

$$\Psi = \Psi_1^0 \exp(-iS_1/\hbar) + \Psi_2^0 \exp(-iS_2/\hbar) \tag{8}$$

where

$$S_1 = (-e/c) \int_{\text{ABP}} \vec{A}(\vec{x}')\cdot d\vec{x}' \tag{9}$$

and

$$S_2 = (-e/c) \int_{\text{ACP}} \vec{A}(\vec{x}')\cdot d\vec{x}' \tag{10}$$

with the integrals extending over the trajectories associated with the first and the second wave packet, and Ψ_1^0, Ψ_2^0 are the wave functions when $\vec{A} = 0$.

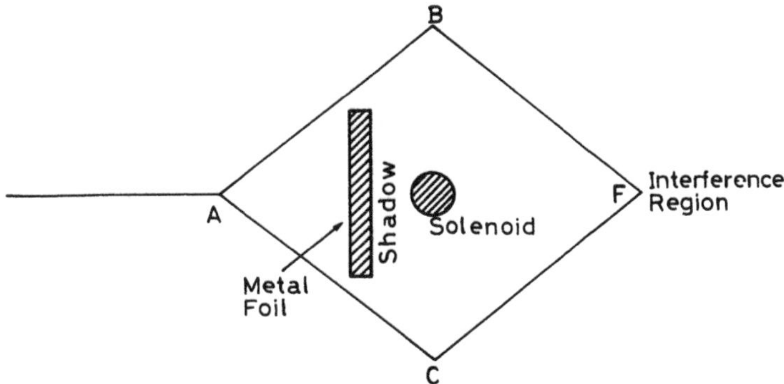

FIGURE 1. A beam of electrons going around a solenoid on opposite sides (ABF and ACF) are recombined within the region F.

The interference pattern observed in the region F will depend on the phase difference

$$(S_1 - S_2)/\hbar = (e/\hbar c)\oint \vec{A}\cdot d\vec{x} = (e/\hbar c)\Phi \tag{11}$$

which is determined by the vector potential $\vec{A}$ through the gauge invariant closed loop integral $\oint \vec{A}\cdot d\vec{x}$ where the closed path of integration is ABFCA of Figure 1. We have, therefore, an observable phase difference which results in the shift of the interference pattern depending on Φ. There are two most important features of this effect that need to be stressed:

A. The magnetic field is confined to a region that should be totally *inaccessible* to electrons that propagate in a different region of space where electric and magnetic fields are rigorously *zero*.
B. The vector potential $\vec{A}$ must instead be *nonvanishing* in the region where the electrons propagate.

3. LOOKING BEYOND THE MATHEMATICAL FORMALISM

If one follows the standard interpretation espoused by the Bohr–Heisenberg school, the question of any further "understanding" of the AB effect does not arise. According to this viewpoint, quantum mechanics is to be regarded as a package of mathematical recipes by means of which one can predict the statistical results obtainable under specified experimental conditions; the possibility of providing a consistent space-time causal description of the individual quantum events (accounting for the observed statistical results) is ruled out even in principle. On the other hand, there exists a different school of thought that believes that the need for a space-time causal description of "underlying physical reality" in terms of an unambiguous well-defined model is of paramount importance, a point of view eloquently articulated in the following words of Einstein[(16)]: "Physics is the attempt at the conceptual construction of a model of the real world, as well as its lawful structure." We now know of rigorous theoretical models that provide in a quantitative way a causal description in space and time of individual microevents underlying many quantum phenomena (for an up-to-date overview see Selleri[(17)]). There is no difficulty, for example, in reproducing in such a way the quantum mechanical predictions for the famous double-slit experiments. There is even a movie, produced with a computer by C. Dewdney, in which individual particle trajectories can be seen to reproduce the statistical interference pattern.

The main idea here is to take seriously the Einstein–de Broglie[(18)] picture of the microphysical objects, according to which an individual atomic entity is actually a localized particle accompanied by an extended, objectively real wave $\Phi(x,y,z,t)$ (proportional to the Schrödinger wave function $\Psi(x,y,z,t)$) which "guides" the particle through its trajectory. The action of the wave on the particle is represented by a suitably defined "quantum potential," say, Q (which has qualitatively new features vis-à-vis the so-called classical potentials) whose exact expression can be deduced from the solution of the Schrödinger equation written in the form $\Psi = R\exp(iS/\hbar)$:

$$Q = -\frac{\hbar^2}{2m}\frac{\nabla^2 R}{R} \tag{12}$$

The upshot of the various calculations using the "quantum potential" approach is the realization that the quantum phenomena can be consistently interpreted in terms of the dual aspects of the wave and the particle which together compose a quantum object, both behaving causally in space and time. One can therefore conclude that the way for a deeper probing of the nature of physical reality underlying the quantum phenomena has opened up and the barrier presented by the so-called "impossibility" or "no-go theorems" has been overcome. It is against the background of this scenario that we now proceed to analyze the ramifications of the AB effect from the point of view of "realist" descriptions adhering to the "locality" (no action at a distance) condition. As everybody knows, the quantum potential Q is strictly local only for the case of a single particle. The Schrödinger wave function describing two particles is instead

$$\Psi = \Psi(x_1,y_1,z_1;\, x_2,y_2,z_2;\, t) \tag{13}$$

and depends on the coordinates of both particles. As a consequence of this, Bohm and Hiley showed that the quantum potential Q acting e.g. on particle "1" is

$$Q_1 = -\frac{\hbar^2}{2m}\frac{\nabla_1^2|\Psi(x_1,y_1,z_1,;\, x_2,y_2,z_2;\, t)|}{|\Psi(x_1,y_1,z_1;\, x_2,y_2,z_2;\, t)|} \tag{14}$$

and depends in general on the instantaneous position of particle "2." This is clearly a form of nonlocality. Several comments should, however, be added. First, it has never been shown that violations of Bell's inequality occur as a consequence of (14), and our feeling is that one deals here more with a formal than with an essential type of nonlocality. Second, the validity of the two-body Schrödinger equation has been checked experimentally only for particles at a very short distance (e.g., in the case of the He atom). Third, it can be hoped to deduce the many-body Schrödinger equation in configuration space from the single-particle equation in ordinary space, and attempts in this direction have been made, e.g., by Andrade e Silva.[19] No impossibility theorem has ever been formulated against the success of such an idea and so it appears that nonlocality implied by (14) will be eliminated if one can find a formulation of the Schrödinger equation for n particles in ordinary three-dimensional space.

4. LOCAL REALIST INTERPRETATION OF THE AB EFFECT

We begin by recalling that the basic motivation for introducing the concept of fields is to enable physical effects on charged particles to be understood as local phenomena. The central puzzle posed by the AB effect is that it defies such an understanding in terms of local fields. One has to inevitably take resort to the concept of vector potential, but then the value of vector potential at a given space-time point is apparently devoid of any physical reality because it is not a gauge invariant quantity. This difficulty readily dissolves if one adopts the "quantum potential" approach. Note that the quantum potential Q (as defined in Eq. (12)) at a given space-time point is gauge invariant. In the regions surrounding points B and C of Figure 1 we have the wave functions for the electron given by

$$\Psi_1 = \Psi_1^0 \exp(-iS_1/\hbar) \tag{15}$$

and

$$\Psi_2 = \Psi_2^0 \exp(-iS_2/\hbar) \tag{16}$$

respectively, where Ψ_1^0 and Ψ_2^0 are the wave functions corresponding to $\vec{A} = 0$. The quantum potential Q depends only on $|\Psi_1| = |\Psi_1^0|$ and on $|\Psi_2| = |\Psi_2^0|$ in the two said regions and is therefore gauge invariant.

In the region surrounding point P of Figure 1 we have according to (8), the wave function given by

$$|\Psi|^2 = |\Psi_1^0|^2 + |\Psi_2^0|^2 + 2|\Psi_1^0||\Psi_2^0| \cos[(S_1 - S_2)/\hbar] \tag{17}$$

which is gauge invariant since $|\Psi_1^0|$, $|\Psi_2^0|$, and $(S_1 - S_2)$ are so. Q being a function only of $|\Psi|$ is also therefore gauge invariant.

As shown by Philippidis *et al.*[9] one can calculate explicitly the trajectories (Figure 2a and b) in the AB effect situation and explain the fringe shift as arising from the interaction of the electrons with the quantum potential. The quantum potential, therefore, mediates the local effect of the gauge-independent part of the vector potential on the electrons. There is no need to invoke the idea of an "immediate nonlocal action" by the enclosed magnetic flux on electrons. Philippidis *et al.*, however, restrict their attention in interpreting their calculations as an illustration of nonlocal aspects of quantum mechanics manifested through the so-called "global" determination of the quantum potential. In what follows we discuss the relevant significance from a local realist point of view.

The local action of the physically real empty wave $\Phi(x,y,z,t)$ on an electron at a space-time point is mediated through the value of the quantum potential at that point, which in turn is determined by the solution $\Psi(x,y,z,t)$ of the Schrödinger equation, along with the relevant boundary conditions. In the AB example, since Ψ contains a phase factor involving the preexisting vector potential $\vec{A}(x,y,z)$, the action of an empty wave on each individual electron changes its trajectory (compared to the situation $\vec{A} = 0$) and results in the overall shift of the interference pattern. A significant element inherent in this interpretation of the AB effect is that momentum (and eventually energy) of each individual electron gets altered by this action of the empty wave, though the empty wave in itself is devoid of momentum (and energy) and since $F_{\mu\nu} = 0$, the electromagnetic momentum (and energy) density also vanishes in the region accessible to the electrons (classically as well as quantum mechanically), because ignoring the zero point fluctuations of the field strengths, quantum electrodynamics satisfies the correspondence principle, a point discussed in the standard treatise by Heitler.[20] It is, therefore, evident that the momentum and energy conservation laws do not hold for the individual electron trajectories interpreted in terms of local realism. This is, of course, an inevitable feature associated with the local realist interpretation of quantum interference phenomena in terms of empty waves and not limited to the AB effect. For instance, in the double slit experiment, one gets diffraction pattern-like distribution of particle trajectories emerging from the slits corresponding to an incident uniform distribution of particles. Here again, the empty waves act on the individual particles modifying their momentum even though there is no other source with which they can locally exchange momentum. One could think that "in practice" momentum is exchanged with the screen on which the slits are pierced and hence no violation of the conservation laws need be invoked. But this can at most explain the particle deviations occurring in the single slit (diffraction) experiment. In a double-slit experiment the relevant wave interference is essentially active in a region several centimeters away from the screen and in such

⟶

FIGURE 2. (a) Particle trajectories for the two-slit arrangement calculated using the quantum potential approach; (b) particle trajectories for the AB effect situation with the shifts arising due to the presence of the enclosed magnetic flux.

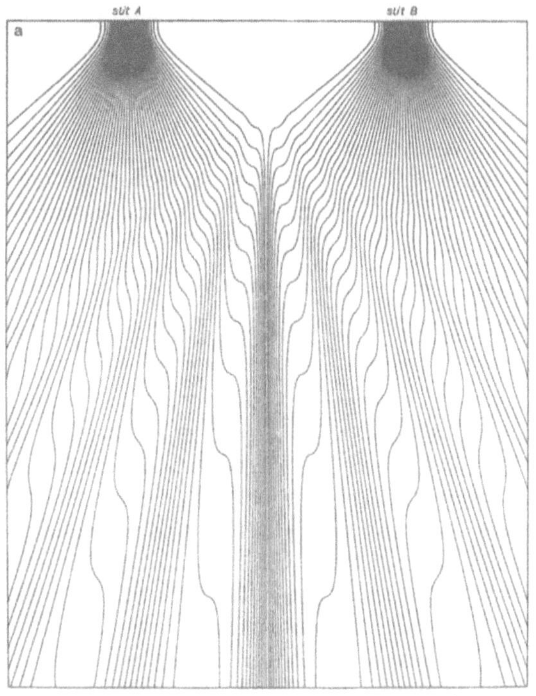
slit A
slit B
a

b

conditions it is obvious that no local interaction with the screen can explain the modified momentum distribution of the particles.

The speciality of the AB effect lies in that here, even at the statistical level, the momentum (and energy) conservation does not hold from the point of view of local realism. This is easily seen by considering that $\langle\Psi|-i(\partial/\partial\vec{x})\Psi\rangle$ has different values for $\vec{A} = 0$ and $\vec{A} \neq 0$ (AB example) because of the presence of the phase factor $\exp(-ie/c)\int\vec{A}\cdot d\vec{x}$ in Ψ when $\vec{A} \neq 0$, which leads to an additional contribution to the momentum expectation value. Note that

$$\langle\Psi|-i\frac{\partial}{\partial\vec{x}}|\Psi\rangle_{\vec{A}=0} - \langle\Psi|-i\frac{\partial}{\partial\vec{x}}|\Psi\rangle_{\vec{A}\neq 0} = \langle\Psi_0|\frac{\partial S}{\partial\vec{x}}|\Psi_0\rangle \tag{18}$$

where Ψ_0 is the wave function when $\vec{A} = 0$ and $S = (-e/c)\int\vec{A}(\vec{x})\cdot d\vec{x}$.

It is therefore evident in the AB effect-type situation the expectation value of momentum (and energy) changes even though locally the electromagnetic field has zero momentum (and zero energy). One possible argument[(21)] at this stage could be that the reflection of the electrons at the impenetrable boundary of the enclosed magnetic flux provides the source for this momentum (and energy) change. However, this argument can be readily circumvented by considering a localized wave packet whose spreading during the time of transit is made negligibly small (by taking an initially highly localized packet with sufficiently high energy so that its time of transit is quite small) so that it remains confined to a region away from the boundary of the trapped flux. Another point that might be raised is that a moving electron produces a magnetic field which overlaps with the enclosed magnetic field giving rise to an additional source of interaction energy. However, in the experiments by Tonomura *et al.*, the enclosed magnetic field is shielded by a superconducting barrier which (due to the Meissner effect) excludes the possibility of such an overlap. The only way one can, therefore, salvage the momentum (and energy) conservation laws in the AB example is by invoking "nonlocal" influence of the enclosed magnetic flux on electrons. For those believing in local realism the physical reality of the AB effect has, therefore, the important message that the locality condition and momentum (energy) conservation laws appear to be incompatible for such quantum mechanical phenomena. This is extremely worrying for anyone who believes in a realistic and rationalistic approach to physics, given the enormous amount of empirical evidences that overwhelmingly support the conservation laws. This deep problem, perhaps, hides an equally profound truth that should be brought to light. One possible conjecture is that the zero point fluctuations of the field strengths could provide locally the missing balance of momentum and energy for the particle trajectories in the AB effect-type situations. The validity of this conjecture needs to be carefully assessed by more detailed considerations, but we would like to stress that the reality of the quantum zero point field of the vacuum, strikingly manifested in the attraction between neighboring uncharged metallic plates in vacuum (known as the Casimir effect), should be taken seriously. The fluctuations of the field strengths about the mean value zero in the vacuum need not necessarily be random and it could well be that the momentum and energy changes involved in the local action of the quantum potential on particles are mediated through these zero point fluctuations.

ACKNOWLEDGMENTS. One of the authors (D.H.) gratefully acknowledges the fellowship provided by the Commission of the European Community which enabled the collaboration work leading to this paper. We thank E. Santos for helpful suggestions.

REFERENCES

1. A. TONOMURA, in the present volume.
2. A. EINSTEIN, in: *Albert Einstein: Philosopher-Scientist* (P. A. SCHILPP, ed.), Open Court, La Salle, Ill. (1970).
3. N. BOHR, in Ref. 2.
4. Y. AHARONOV and D. BOHM, *Phys. Rev.* **115**, 485 (1959).
5. Y. AHARONOV, in: *Proc. Int. Symp. on Foundations of Quantum Mechanics*, Tokyo, 1983 (S. KAMEFUCHI *et al.*, eds.), Physical Society of Japan, Tokyo (1984).
6. B. I. SPASSKII and A. V. MOSKOVSKII, *Sov. Phys. Usp.* **27**, 273 (1984).
7. N. G. VAN KAMPEN, *Phys. Lett. A* **106**, 5 (1984).
8. S. OLARIN and I. I. POPESCU, *Rev. Mod. Phys.* **57**, 339 (1985).
9. C. PHILIPPIDIS, D. BOHM, and R. D. KAYE, *Nuovo Cimento B* **71**, 75 (1982).
10. C. N. YANG, quoted by M. PESHKIN and A. TONOMURA, *The Aharonov–Bohm Effect*, Springer-Verlag, Berlin, (1989) p. 97.
11. T. TROUDET, *Phys. Lett. A* **111**, 274 (1985).
12. R. A. BROWN and D. HOME, *Locality and Causality in Time-Dependent Aharonov–Bohm Interference, Nuovo Cimento B* **107** (1992).
13. H. A. FOWLER *et al.*, *J. Appl. Phys.* **32**, 1153 (1961); G. MÖLLENSTEDT and W. BAYH, *Phys. B I* **18**, 299 (1962); R. G. CHAMBERS, *Phys. Rev. Lett.* **5**, 3 (1966).
14. P. BOCCHIERI, A. LOINGER, and G. SIRAGUSA, *Nuovo Cimento A* **51**, 1 (1979); S. M. ROY, *Phys. Rev. Lett.* **44**, 111 (1980); D. HOME and S. SENGUPTA, *Am. J. Phys.* **51**, 942 (1983).
15. P. BOCCHIERI and A. LOINGER, *Lett. Nuovo Cimento* **35**, 469 (1982).
16. A. EINSTEIN, *Letter to M. Schlick*, Nov. 28, 1930; quoted by A. FINE, *The Shaky Game—Einstein, Realism and the Quantum Theory*, University of Chicago Press, Chicago, (1986) p. 97.
17. F. SELLERI *Quantum Paradoxes and Physical Reality*, Kluwer, Dordrecht (1990).
18. A. EINSTEIN, *Ann. Phys. (Leipzig)* **18**, 639 (1905); L. DE BROGLIE and J. ANDRADE E SILVA, *Phys. Rev.* **172**, 1284 (1968).
19. J. L. ANDRADE E SILVA, *Thése de Doctorat*, Gauthier–Villars, Paris, 1960.
20. W. HEITLER, *The Quantum Theory of Radiation*, Oxford University Press, London, (1944) p. 58.
21. M. PESHKIN and A. TONOMURA, *The Aharonov–Bohm Effect*, Springer-Verlag, Berlin, (1989) p. 19.

CHAPTER 8

ARE TWO-BEAM SELF-INTERFERENCES MASS-INDEPENDENT?

NOT THOROUGHLY KNOWN (?) ROLE OF THE MASS

YUJIRO KOH

1. INTRODUCTION AND HISTORICAL BACKGROUNDS

The symmetric two-beam self-interference of single particles, which is one of the most elementary quantum processes and named after Dirac, is usually understood to exhibit the same normalized interference pattern under the same geometric condition for both the wavelength and slit-detector system, no matter how different the properties* of the particles are (abbreviated as *the same-geometry-interference patterns* of various particles).

Although the above assumption (abbreviated as *the Dirac assumption*) contributed to simplifying quantum mechanics at its beginnings, I think that the experimental basis of the assumption, or the evidence as to whether there is any difference among self-interferences of various particles (abbreviated as *the particle-independence concept* in the Dirac assumption) or not, was at that time (and is even now) *disproportionately weak*, in contrast to its important roles in theories, e.g., that for the wave function reduction in measurement theories. At the stage of 1925–1928, when the early construction of quantum mechanics had finished, the experimental data of the self-interferences of non-massless particles which might be available to test the Dirac assumption were the diffraction of electrons by crystal only.

The discovery of the electron crystal diffraction, however, could not support

*For example, the mass, the charge, the inner structure, and the history experienced before superposition of the wave functions, etc.

YUJIRO KOH • Department of Physics, Ibaraki University, Bunkyo 2-1-1, Mito 310, Japan.

Wave–Particle Duality, edited by Franco Selleri. Plenum Press, New York, 1992.

the Dirac assumption experimentally, in spite of its contribution to the wave–particle dualism. The reason is that, at the technical level in those days, it was impossible to ascertain evidently the particle-independence concept between electron and X-ray photon, by observation only of too complicated crystal diffraction patterns due to multiple diffraction under exposure to the background reflection. Since, at that time, the simplest two-beam self-interference (e.g., double-slit diffraction) patterns of various particles could not be realized, it was actually impossible to test the Dirac assumption experimentally by such simple same-geometry-interference patterns.

In describing the actual history, it can be pointed out that the particle-independence concept in the Dirac assumption has played the same role as *a principle of simplicity*. The subsequent accumulation of a vast amount of quantum-physical information including the neutron crystal diffraction has not changed the situation basically, but has extended the particle-independence assumption from the simple two-beam self-interference to the whole of interference phenomena. It must, however, be remembered that in the history of science there are examples where even the important basic laws based on the simplicity principle, such as the mass-conservation law, could be shown to hold only approximately, not exactly.

The two-beam self-interferences of non-massless particles have been realized with great technical difficulties since 1952 (~25 years after the appearance of quantum mechanics), as shown in Table I. In 1969[(6)] I noticed that the electron biprism techniques could be modified to test the Dirac assumption, and in 1985[(7)] the neutron double-slit diffraction as well. Since then, in order to compare the two-beam self-interferences of the X-ray photon, electron, proton, and neutron, my colleague and I have been investigating the possible modifications of the experimental techniques used in the area of the above-named particles. Our aim is to make the self-interference patterns so distinct, like the ideal same-geometry-

TABLE I
Two-Beam Interferences of Non-massless Particles

Electron[a]
1952, L. Marton,[(1)] three thin copper films
1955, G. Möllenstedt and H. Düker,[(2)] biprism by coulomb field

Neutron[a]
1974, H. Rauch, W. Treimer, and U. Bonse,[(3)] LLL interferometer
1976, A G. Klein and G. I. Opat,[(4)] ferromagnetic domain boundary
1981, A. Zeilinger, R. Gähler, C. G. Shull, and W. Treimer,[(5)] double slits

Charged[a]	(μ, π),	$\underline{p,}$	$\underline{d,}$	$\underline{H_2^+,}$	t,	$^3He^+$,	HD^+,	etc.
Neutral	(e–e^+,	H,	D,	H_2,	T,	^3He,	HD,	etc.)

Increasing mass →

[a]The underlined particles are the known or immediate objects for reexamination of the Dirac assumption; d and H_2^+ will be useful for comparing the interferences of the mass centers of systems. Technical problems for the particles in parentheses are hardly solvable at the present time.

interference patterns, as to be able to distinguish whether the particle-independence concept is correct or not.

Of course, the wide range of the rest masses of these interfering particles creates difficult technical problems. For example, the condition of *the same wavelength* in the same-geometry-interference pattern is crucial. However, we do not think that it is hopeless to utilize the simple structure in the two-beam setups and reexamine the so far ignored role of the particle's history before superposition. Then, it seems possible for us to make proposals of the modified two-beam self-interference experiments, which satisfy the above-mentioned aim, realizable by using continuous developments of existing techniques (Section 2).

The fundamental problem left to be overcome in the proposed experiments is estimation of machine times in the particle source necessary for accumulating data to obtain reliable information as to whether the Dirac assumption will get more exact or only transitionally approximate (Section 2.1).

In the former case which is advantageous to the Copenhagen interpretation, the first experimental evidence for the particle-independence assumption will be brought forward. Other new weak points will also be pointed out (Section 3).

In the latter case the estimation seems to be impossible unless the behaviors of the particles (possibly slow) between splitter and detector are described in a slightly different way from the usual one, considering the particle-independence assumption as the first approximation which has to be improved in the next stage. Then, the long-pending debates about some propositions in the Copenhagen interpretation may be reexamined (Section 4).

It should be emphasized, however, that the meaning of the proposed experiments is ambivalent for the respective interpretations of both cases.

2. PROPOSITIONS OF NEW TEST EXPERIMENTS

In the modified experiments which are proposed to test the particle-independence concept in the Dirac assumption, the points of modification for the existing techniques are as follows: (A) Extension of the biprism techniques from electron to proton or heavier ions. (B) Macroscopic separation of the two split beams of the two-beam interference setup, by an artificially controllable separator of macroscopic Total Separated Path Lengths* (abbreviated as TSPL, according to

*TSPL can be said to be a special kind of the particle's parametrized history prior to superposition. TSPL should not be confused with conventional path difference which is comprehended in the vagueness of the TSPL concept itself.

The fact (which is liable to be misunderstood) that TSPL play no role in the two-beam interference has been experimentally ascertained only for two (not necessarily split) beams of macroscopic or many-particle waves (e.g., two laser beams of slightly different frequencies), but never for quantum-mechanical two split beams of single particles. It is, I surmise, unsure even for visible single photons, since beam intensities high enough for too long TSPL of two split beams of single photons will be difficult technically; though possible for two laser beams.

Ref. 6) which are elongated between the splitter and the exit slits in such a way that the separated beams cannot be superposed in front of the exit slits. (Examples of TSPL in existing techniques are illustrated in Figure 1a and 1b.) In both A and B the restriction of the same-geometry-interference pattern must be maintained between the exit slits and the detectors.

In order to parametrize the normalized two-beam interference pattern, the visibility

$$V = (I_{max} - I_{min})/(I_{max} + I_{min}) \tag{1}$$

introduced by Michelson to measure the degree of unobscurity of the interference fringes, is convenient. In short, the proposed experiments are measurements of the TSPL versus V curves of the double-slit diffractions of X-ray photon, neutron,

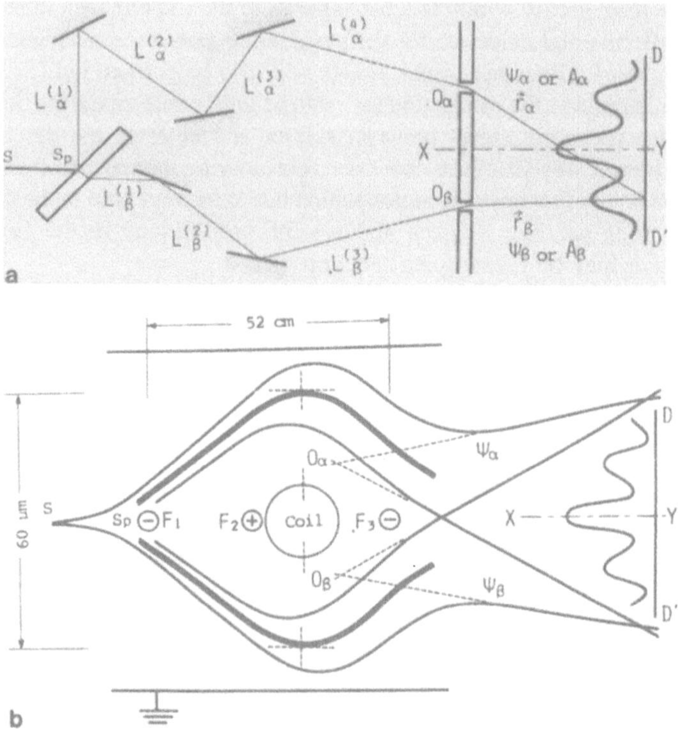

FIGURE 1. (a) Two-beam interferometer for visible photon, shown schematically. Total Separated Path Lengths (TSPL) = $\sum_i L^{(i)}_{\alpha \text{ or } \beta}$.

(b) Möllenstedt–Bayh's electron biprism,[9,10] shown schematically. F_1, F_2, and F_3 are the fiber electrodes. The complex electrode structure is designed so that a tiny coil can be inserted into the loop composed of the two split beams without disturbance from the beam current. TSPL = lengths of the solid lines. (Sections 2.4 and 5.2.)

and the ion-biprism interferences under the restriction of as close to the same-geometry-interference as possible and comparison among them, although the Dirac assumption as *a simplicity principle* asserts all the same V = constant line without any experimental basis.

2.1. Modified Neutron Double Slits

Figure 2a and 2b show respectively the neutron double slits and their clear interference fringes of Zeilinger *et al.*[5] The 100-μm boron wire can be regarded as the splitter and the separator of TSPL ≃ 100 μm at the same time.

Figure 3 shows a proposed setup modified from that shown in Figure 2a, according to modification point B of Section 2. The separator of the artificially

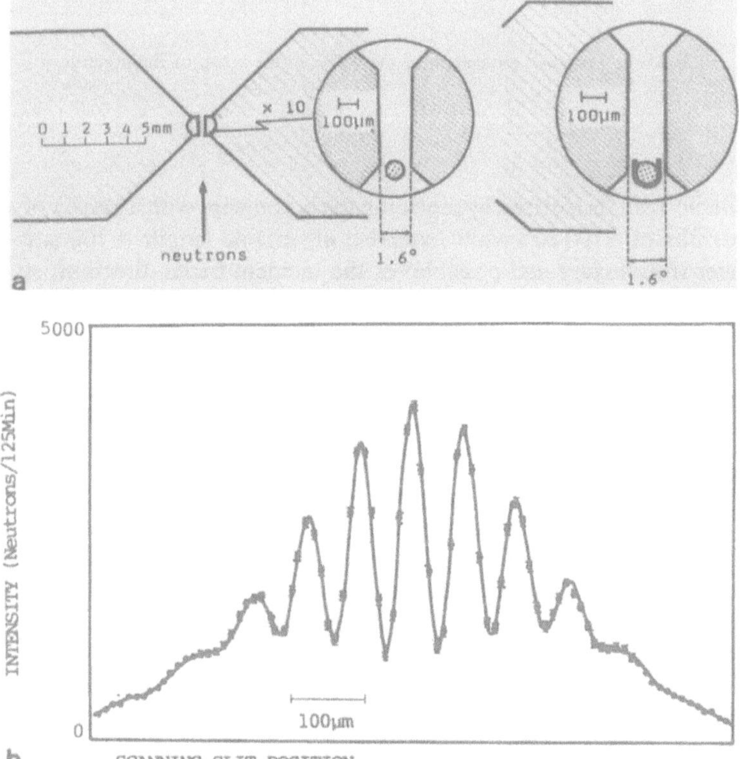

FIGURE 2. (a) Double-slit arrangement of Zeilinger *et al.*[5] The boron wire is mounted in the gap between the two neutron-absorber edges. The solid line shows the TSPL.

(b) Neutron diffraction pattern of double slits with the curve calculated from the Schrödinger equation. λ_n = 18.45 Å, v_n = 211.4 ms^{-1}, slit–counter distance = 5 m. From Zeilinger *et al.*[5]; illegible parts of the original figure have been retouched by the author.

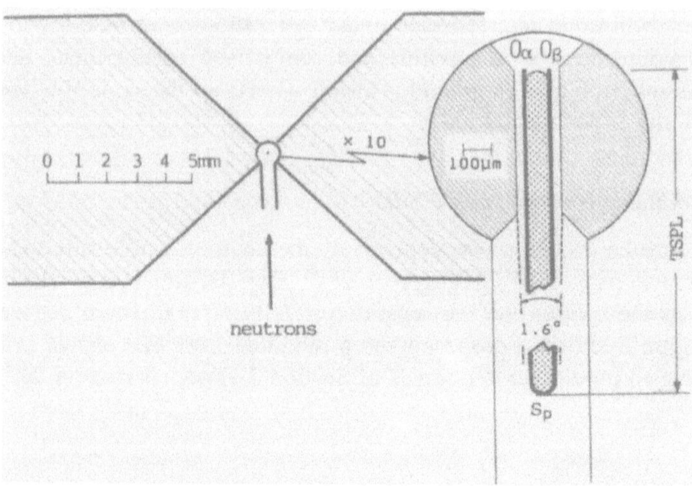

FIGURE 3. Proposed elongation of the TSPL in the setup of Zeilinger *et al.*[5]

controllable TSPL is realized by replacing the boron wire with a barrier of neutron-absorber film of ≃100 μm width extended in variable length ≃ 100 μm–100 cm (or longer if necessary and possible) of the incident beam direction, so that the split beams cannot be superposed with each other in front of the exit slits. The superposing space between the exit slits and the detector plane is kept unchanged for the variable TSPL. The normalized constant-λ wave function in the above space is exactly the same, as long as the incident beam can be regarded as parallel, no matter how long the TSPL are.

The data of Figure 2b correspond to a point (TSPL ≃ 100 μm, $V = 0.60$) on the TSPL–V plane. For the setup of Figure 3, the Dirac assumption predicts a $V = 0.60$ straight line in a possible range of TSPL as an example of the TSPL–V curve. The technical upper bounds (≃ order of 1 m) of TSPL will be limited by the reliability of the location techniques on a huge optical bench on which the slits, splitter, and separator must be mounted exactly.

There is no essential problem which cannot be overcome technically. But Zeilinger *et al.*[5] needed the machine time of 7.5 days to obtain just Figure 2b. In practice, we are unable to design any plan for accumulating enough data to measure a reliable TSPL–V curve, unless we can estimate beforehand total machine time necessary and sufficient for an image of the whole experiment plan (Section 4, point 4).

In any case, either independence or dependence of the visibility upon TSPL will have important physical meanings.

2.2. *TSPL versus V Curve*

In observation of the TSPL–V curves, there is no conflict with the Heisenberg uncertainty principle. In fact, the Dirac assumption assumes the V = constant lines, as examples of the TSPL–V curves, for all particles, as long as their two-beam self-interferences are possible. The possibility to observe the TSPL–V curve is compatible with the impossibility of simultaneous observation of orbit and interference fringes. This compatibility should not be regarded as puzzling according to the uncertainty principle.

Nevertheless, we cannot but feel it puzzling that the same-geometry-interference patterns of all two-beam self-interfering particles with variable TSPL have the same normalized fringe shape and their TSPL–V curves are the same V = constant line, no matter how different the particles' properties are. This feeling of puzzlement is illustrated in Figure 4 with "Another Question of the Symbol Cat of the International Workshop on Matter Wave Interferometry, Wien, 1987" which is different from the conventional question about her career or orbit just after splitting.

Since the physics during the time interval = (TSPL)/(speed), both experimental and theoretical, is scarcely known, it does not seem unnatural for us to expect that in the TSPL–V curves of the same-geometry-interference patterns

FIGURE 4. The cat cannot be sure about whether or not all will have the same future (interference pattern), though they have obeyed the order (uncertainty principle) of Heisenberg well and been under the same environment (geometric condition).

there may be distinguishable differences dependent on the properties of the particles in the following ways:

A. *The interference fringes should become more obscure for heavy masses as the TSPL become longer.*
B. The particle-independence concept in the Dirac assumption should be treated as the first approximation.

2.3. *Effect of Source Size*

Next, before the TSPL–V curves of various particles are discussed, an effect of source size on observation of the visibility must be investigated. Excluding the two exceptional cases of neutron double-slit diffraction and electron biprism interference, it is technically not easy for us to decrease the source size effect which is injurious to the apparent fringe clearness or visibility.

The actual two-beam interference fringes, of both double slits and biprism, can be calculated as the weighted average of fringes from each point-source (weighted weight ∝ point-source intensity) over the tiny source domain spreading near the plane of symmetry of the two-beam interference setup. The domain can be either an assemblage of real points emitting actually interfering particles or only a focal region of incident beam through a lens device.

Drahoš and Delong[8] estimated such averaged visibility* of biprism fringes

$$V_{\mathrm{D-D}} = (\Delta/\pi b)(a/W)\cdot\sin\{\pi b/\Delta)(W/a)\} \tag{2}$$

where a/b = (source-fiber distance)/(fiber-observation distance), W is the source width perpendicular to incidence, $\Delta = \lambda(a + b)$/(separation of virtual sources) is the spacing of fringes in the observation plane perpendicular to incidence.

$$V_{\mathrm{D-D}}\{(W/a)\rightarrow 0\}/V_{\mathrm{D-D}} = 1/V_{\mathrm{D-D}} = C_{\mathrm{D-D}} \tag{3}$$

*Assuming that the fringe intensity at x, $i(x,s)$, of each particle radiated from a point source at s, is

$$i(x,s) \propto \sin^2\{x/\Delta - s/(a\Delta/b)\}$$

and the weighted average of the fringe intensity at x is

$$I(x) = \int_{-W/2}^{W/2} i(x,s)\cdot\sigma(s)\,ds = \sigma(0)\int_{-W/2}^{W/2} i(x,s)\,ds$$

where x and s are deviations from the plane of symmetry, respectively, on the observation and source planes perpendicular to the incidence. The source intensity $\sigma(s)$ is uniform.

Drahoš and Delong ascertained that in their electron-biprism experiment, as in the old visible-photon biprism, according to variation of the factor $\sin\{(\pi b/\Delta)\cdot(W/a)\}$ for increase of (W/Δ), positions of the fringe maxima and minima are alternated or the visibility becomes zero at $(W/\Delta) = a/b$, $2a/b$, ... This fact seems to show enough reliability of their assumptions.

can be regarded as a correction factor in order to calculate the corrected visibility for a point source on the symmetry plane from the observed visibility of symmetrical fringes for sources distributed in a domain of small size W.

Their assumption from which $V_{\mathrm{D-D}}$ and $C_{\mathrm{D-D}}$ have been derived cannot be said to be well-grounded. But the oscillatory dependence of $V_{\mathrm{D-D}}$ on (W/Δ) has been ascertained experimentally. $C_{\mathrm{D-D}}$ and $V_{\mathrm{D-D}}$ seem to be usable as the correction factor and basic data for designing a future experimental setup.

The neutron double-slit diffraction of parallel incidence is under specially advantageous conditions because of the correction factor $C_{\mathrm{D-D}}(W/a \rightarrow W/\infty) \simeq 1$ almost independent of W, high source intensity, and high detection efficiency, compared with the other neutral particles in Table I. For the other neutral particles, however, such advantages are ineffective.

2.4. *Modified Ion Biprism*

Structures of electrodes of the electron biprism are simple enough that extension of the biprism techniques from electron to proton or heavier ions, adding the beam separator of adjustable TSPL, is realizable under the restriction of nearly equal wavelengths.

The term *realizability* is used here in the meaning of *possibility* to be made real, in a scope of continuous development from existing techniques, without both technical and financial difficulties which cannot be overcome essentially. The realizable experiment should be distinguished from the so-called Gedanken experiment: only the experimenter's intention determines whether the former will be carried out or not, while the latter may be often impossible in practice, though possible theoretically.

Energies and $\beta = v/c$ values for important ions in Table I and X-ray photon are estimated in order to get the feeling of their orders of magnitude in a case where their wavelengths are unified to 0.02 Å of the same order as that in the electron-biprism experiment of Möllenstedt and Bayh[9,10]; 620 keV and 1 for X-ray photon, 290 keV and 0.77 for electron, 210 eV and 0.66×10^{-4} for proton, 105 eV and 0.33×10^{-4} for deuteron, etc.

As a necessary condition for the wire electrode of the biprism in order to observe apparently clear interference fringes, we can derive the condition (Bayh's Winkelkohärenzbedingung[10]) that

$$bW/a\Delta \simeq 2\theta(W/\lambda) = (d/a)(W/\lambda) \rightarrow 0 \tag{4}$$

where $d = 2\theta a$ is the wire diameter, since

$$V_{\mathrm{D-D}}(\pi bW/a\Delta) \simeq 1 - \pi^2(bW/a\Delta)^2/(3!) \approx 1 - \pi^2(2\theta W/\lambda)^2/(3!)$$

In the case of the electron biprism, very small values of $(d/a)(W/\lambda)$ have been achieved as a result of modern electron-microscope techniques, especially due to the exceptionally small $W \simeq 200 \sim 50$ Å of the pointed hot emission filament or the cold emission tip. The very clear fringes of the experiment of Möllenstedt and Bayh[9,10] correspond to $(d/a)(W/\lambda) = (1\ \mu\text{m}/245\ \text{mm})(200\ \text{Å}/0.06\ \text{Å}) \simeq 0.0136$ and $V_{\text{D-D}} \simeq 0.99974$. Even for $d = 20\ \mu$m, Bayh[10] reported still recognizable fringes, which correspond to $(d/a)(W/\lambda) \simeq 0.272$ and $V_{\text{D-D}} \approx 0.883$.

The TSPL = 52 cm shown in Figure 1b is a record long TSPL obtained as a by-product of the beam's lateral separation as wide as possible. It has, however, not yet been investigated as to how the visibility changes for TSPL > 52 cm. The maximum length 52 cm reported by Möllenstedt and Bayh[9,10] may be a significant feature or a symptom of the in-flight transition (Section 4, point 3).

For the biprism of protons and heavier ions, it is hopeless for us to invent the same type source as that of electrons. Instead, it seems hopeful that focal regions of their incident beams through some ion-optical lens devices will play roles of virtual image sources of small W. Expression (4) is useful as a necessary condition for the smallness of W. For example, $0.3 > (d/a)(W/\lambda) = (1\ \mu\text{m}/250\ \text{mm})(W/0.02$ Å) gives 1500 Å $> W$. For W not small enough, however, Drahoš–Delong's assumptions for (2) become inappropriate.*

In spite of these difficulties, I think that the extension of the biprism techniques from electrons to ions, including a preacceleration for detection if necessary, is realizable in the meaning of this Section within the scope of the continuous development of modern ion optics, until the very low energies of ions make application of the ion optics inappropriate.

3. IS THE DIRAC ASSUMPTION APPROXIMATE OR NOT?

The possibilities of experimental trials testing the reliability of the Dirac assumption by measuring the various TSPL–V curves have been treated in the preceding sections. In the foundations of such trials, I cannot deny there are some dubieties about the belief that the Dirac assumption cannot be approximate because of its great contribution. Reasons for the dubieties can be itemized from the viewpoint of the proposed experiments as follows.

*T. Sasaki is constructing a low-energy proton source, which has an extremely small width, by means of contraction of the two tandem magnets of iron-core (the image contractor). Its main design factor, Winkelkohärenzbedingung, is

$$(d/a)(W/\lambda) = (1\ \mu\text{m}/250\ \text{mm})(0.2\ \mu\text{m}/2.9\ \text{Å}) \simeq 0.27$$

This design factor is on the same order as that of Bayh,[10] resulting in still recognizable fringes. Compared with $d \simeq 100\ \mu$m in Figure 2a, $d \simeq 1\ \mu$m in Sasaki's factor is so small that the beam separator of the absorber-film type, as exemplified in Figure 3, will be impossible. Beam separation other than that of the electromagnetic method, as exemplified in Figure 1b, will be difficult.

1. As discussed in Section 2.2, the expected mass-dependence of the TSPL–V curves.

2. In the neutron double-slit diffraction of Zeilinger *et al.*,[5] the neutron wavelength $\lambda_n = 18.45$ Å corresponds to the neutron speed $v_n = 211.4$ ms^{-1}. The flight time interval from the splitter to the detector $T_{SD} = 5$ m/$v_n = 23.65$ ms. The extremely low speed is impressive and these values are on the order of macroscopic daily life ($v_n <$ speed of the supersonic airplane). When biprisms of protons and heavier ions are realized, the more impressive situations where the lower speeds of the heavier ions are on the daily-life order will be reached. The flight time interval T_{SD} will be elongated without limit by use of the splitter–detector distance elongated easily. We should note that such situations will be realizable in the meaning of Section 2.4 and, in part, have already been realized (neutron double-slit diffraction), compared with practically impossible Gedankenexperimente (such as often used in measurement theory).

Although the indeterministic distribution of final states of the two-beam interference cannot but be accepted as an experimental fact, it seems to be too unnatural and incomprehensible that the image of behavior of the *possibly heavy and slow* interfering particle in the *unlimitedly long* T_{SD} is abandoned on principle, except description by the superposed wave functions. Similarly so it does that its classical orbit in the *whole* interval of T_{SD} appears suddenly after being disturbed by some observation. The heavier or the longer they are, the more unnatural the situation is.

3. Figure 5 shows the realizable constant-λ ion biprisms which have six artificially controllable functions: mass number of the ions themselves, variable voltage V applied to the electrodes, beam splitter as the separator of adjustable TSPL, position of the observation plane, and adjustable two exit slits.

When one beam of the split beams is shut out by the corresponding exit slit, the ion biprism is transformed into the beam deflector. By gradually and continuously opening and closing one of the exit slits for the beam from image source O_α, the final particle distribution on the observation plane is converted from the pattern of quantum-mechanical two-beam interference fringes, F_{ap} in (B_{ap}), to that of the classical deflector slit image, $D_{v\beta}$ in ($S_{v\beta}$), and conversely, back and forth. $D_{v\beta}$ itself, however, has the edge diffraction of the wave function, as calculated by the Fraunhofer diffraction theory of classical optics. In more detail, the particle distribution as the form, $F_{ap} + R(s) \times D_{v\beta}$, where the mixing ratio, $R(s)$, is dependent upon the degree, s, of symmetry of the two split beams. There is no reason why it should be asserted that the so-called wave function reduction can take part in only the interference phenomenon resulting in the interference pattern F_{ap}. The wave function reductions take place in both the two-beam interference for the biprism (B_{ap}) and the edge diffraction for the deflector ($S_{v\beta}$).

For both setups of ($S_{v\beta}$) and (B_{ap}), the behaviors of ions in T_{SD} can be described by the respective wave functions which are determined by the respective conditions; the former in ($S_{v\beta}$) is made so as to be different from the latter in (B_{ap})

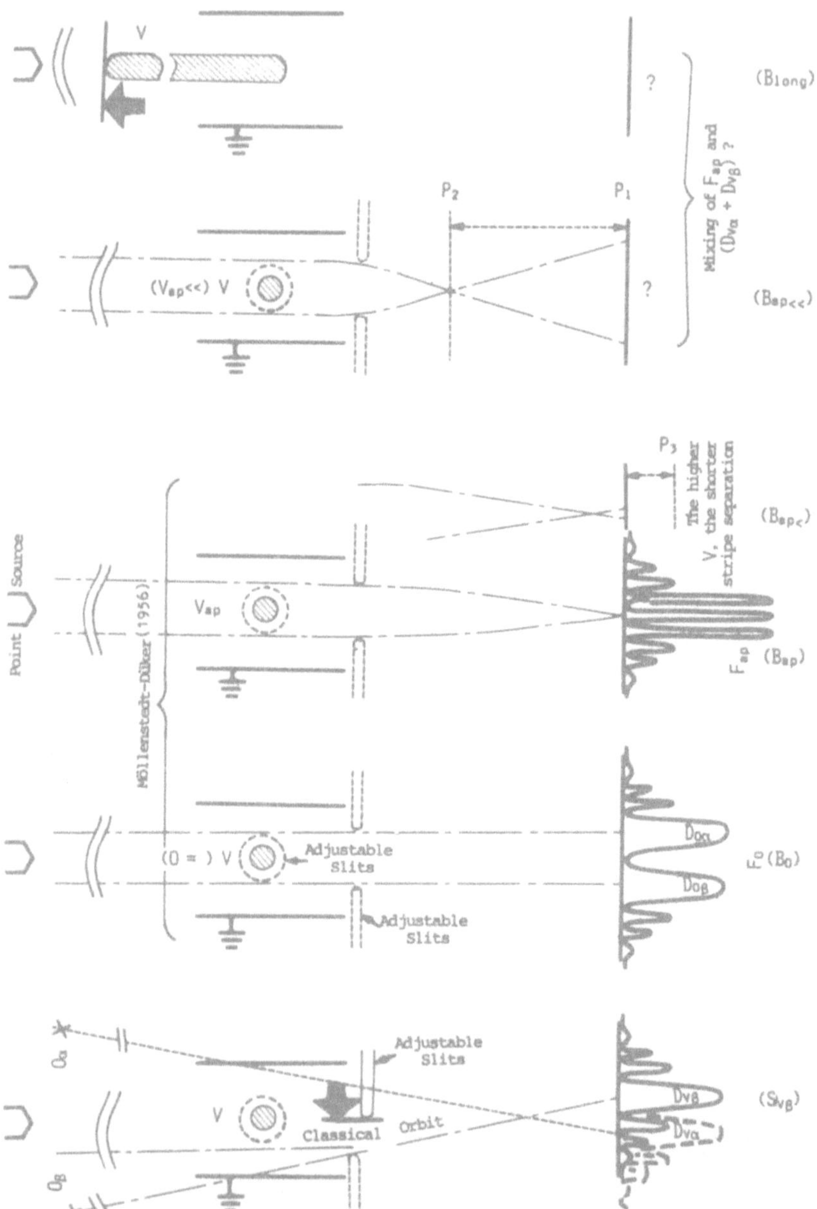

FIGURE 5. Combination of ion biprisms with the splitter capable of being elongated and the classical deflector. (B_{long}), biprism with elongated splitter; ($B_{ap<<}$), biprism with suitably high V; ($B_{ap<}$), biprism with slightly high V; (B_{ap}), biprism with V_{ap} appropriate for interference; (B_0), biprism with $V = 0$; ($S_{v\alpha \text{ or } \beta}$), single-slit or classical deflector. For the range restricted by the brace, Möllenstedt-Düker[11] published a beautiful photograph of interference patterns (F_0), (F_{ap}), and ($F_{ap<}$) which correspond to gradual variation of the voltage V applied to the electrodes.

with the artificial operation that the beam from the virtual image source O_α is restricted by means of the controllable slit. When the slit is closed perfectly, the description by the classical orbit, as in the geometrical optics for visible photon, also becomes possible for the behavior in T_{SD} for $(S_{v\beta})$. However, the more widely the slit is opened, the less meaning the classical orbit has. For the symmetric two split beams, the classical description plays no role.

Although it is correct to describe $(S_{v\beta})$ as classical, it is not correct to describe only (B_{ap}) as quantum-mechanical since both are describable with similar wave functions. We have, I think, unreasonable convention that we treat the quantum-mechanical description and the classical one as being too isolated.

4. If the speed of the interfering particle is slow enough compared with that of closing the exit slit, a movement of closing the slit for conversion $F_{ap} \rightarrow D_{v\beta}$ will be able to start and finish in the middle of T_{SD}, as a realizable process in the meaning of Section 2.4. Then, how will the *possibly heavy and slow* particle behave in the *unlimitedly long* T_{SD}?

According to the uncertainty principle, the classical orbit cannot exist before closing the slit. After closing, however, where will the classical orbit start from? Is that (a) the point on the classical orbit at the instant, t_{cl}, when the slit will have been closed, (b) at the instant, t_{ar}, when the particle will arrive at the closed slit $(t_{cl} \leqslant t_{ar})$, and (c) at another instant for which such classical descriptions are inappropriate? [The above (B_{ap}) with the partially opened slit is essentially equivalent to the delayed choice experiment. See Section 5.2.]

4. HYPOTHETICAL MODEL: THE IN-FLIGHT TRANSITION BEFORE ARRIVAL AT DETECTOR

In short, all the dubieties in Section 3 have come from the main dubiety about the abandoned *behavior* in the *unlimitedly long* flight time interval T_{SD} of each interfering particle passing through the symmetric two-beam interference setup, like that of Figure 2a or (B_{ap}) in Figure 5. (Notice that the word *behavior* is often used instead of the conventional word *orbit*. The *behavior* includes the motions of all possible types: the motion of the coherent channel mentioned in point 1 below, that of the incoherent channel, the in-flight transition in point 3 below, the classical motion, etc.) Both the probabilistic distribution of final states and the *behavior* in T_{SD} of the interfering particle should not be abandoned.

It is too conventional that we insist on the continuation of the classical orbit in the interference phenomenon, but on the other hand it seems too unnatural that we consider the *behavior* of each interfering particle should be abandoned at all instants in T_{SD} (except the initial and final instants) according to the Copenhagen interpretation. In the middle of both extreme viewpoints, we should search for another viewpoint.

Now, as discussed in Section 2.2, it seems natural for us to expect that *the longer the TSPL become, the more obscure the two-beam interference fringes should become for heavy masses*. In other words, the heavier the masses are, the steeper declivity the TSPL–V curves should have. A reason for the expectation is an assumption that, for heavy masses, the interference will hardly take place in far distance from the splitter.

Experimentally, if such setups as those of Figure 3 or (B_{long}) in Figure 5 can be constructed, the above expectation will be able to be tested, and the same expectation seems to be explained (or, more correctly, speculated) by use of the following series of hypothetical models:

1. The so-called traditional wave-packet reduction for observation cannot start until the instant, T_{SD}, when the particle, starting from the splitter or scatterer, arrives at the detector. A new type of spontaneous transition would have to occur at an instant ($\ll T_{SD}$ for a heavy particle) from a two-beam coherent state (i.e., a coherent channel) to either one of two incoherent single-slit-beam states (i.e., two incoherent channels), in order that such dubieties as stated in Section 3 will be solved. It would occur at a flight distance, R, from the splitter or scatterer, i.e., during the flight of the interfering particle and before its arrival at the detector.

2. If superposition of the two split beams which are coherent together is possible before the spontaneous transition mentioned in point 1, interference between the two beams can occur. If the two beams have a beam separator of TSPL longer than R, the superposition and the interference will be impossible and only a weighted sum of two single-beam patterns (weight $\propto$ the spontaneous transitions probability from the coherent channel to each incoherent channel) will be observed on the observation plane.

3. For introducing the new type of spontaneous transition, a minimum of necessary reformation of the theoretical picture should be tried.

If we temporarily postpone the question of the dynamical details of the interaction or of the origin of the new transition, an R-dependent scattering phenomenon about each of the scatterer particles in the splitter may play a role essentially equivalent to the splitting. It may be a microscopic and spontaneous transition of a completely new type from the outgoing spherical wave to one of the final mutually-incoherent plane waves, satisfying the momentum conservation laws and taking place at variable R ($\leqslant$ particle speed $\cdot\, T_{SD}$). This transition would occur probabilistically during the flight of the scattered particle (abbreviated as "the *in-flight* transition," according to Ref. 7), before the so-called observation starts in the detector.

Now, in general, a parallel incident beam of a usual scattering experiment or an initial plane wave isolated from its source, maintaining the 4-dimensional linear and angular momenta, may be prepared through such an in-flight transition.

4. The in-flight transition would have a lifetime, τ_{in}, as a kind of spontaneous decay. Then, it seems natural that we assume the TSPL–V curve has an expression

$$
\begin{gathered}
V \propto \exp(-\kappa \times \text{TSPL}) \\
\kappa(\text{mass, interaction, etc.}) = (\text{particle speed} \cdot \tau_{\text{in}})^{-1}
\end{gathered} \tag{5}
$$

Expression (4) is a result obtained from the speculated in-flight transition, but not a logical conclusion. It should be regarded as *a kind of working hypothesis*.

Of course, it is interesting to know what factors κ and τ_{in} are dependent upon. Let us, however, postpone this problem together with the origin of the new transition, as in the beginning of point 3 above. The most important at this stage is whether expression (5) can be verified experimentally or not.

As treated in Section 2.1, for reliable measurements of the TSPL–V curves the very long machine times are presumed, these being dependent very strongly on the magnitudes of κ. The work, describing Expression (5) as a *kind of working hypothesis*, means the work of estimation of the machine times.

Reliable methods to estimate κ are not yet known. However, all possible methods should be attempted, even if only prospectively. A method treated in Section 5.1 represents a trial example.

5. THEORETICAL PERSPECTIVES

The hypothesis of the R-dependent scattering composed of the in-flight τ_{in} decay seems to have much circumstantial evidence, but no positive proof at present. If the hypothesis is shown to be correct with increasing proof in the future, its effects to circumstances will be influential. Such situations may resemble those of the Avogadro hypothesis at its beginnings. Although the origin of the in-flight transition is not necessarily treated systematically in this chapter, emphasis will be laid on future perspectives and any possible theoretical clue to estimation of κ will be shown as follows.

5.1. Irreversible Microscopic Process

Consider an isolated microscopic system composed of two particles. If each particle is scattered by the other and followed by the spontaneous in-flight transition with lifetime τ_{in}, it is difficult for us to consider in the expression describing the R-dependent scattering that the direction of their time flow from the past to the future can be inverted.

Hence, study of the microscopic process including at least one in-flight transition seems to have the same purpose as that of the *Microscopic Theory of Irreversible Process in Quantum Systems* by Prigogine's Brussels group,[(12)] which introduces the irreversibility into the quantum-mechanical process in a sense similar to thermodynamics.

For example, the in-flight transition may be an example of the microscopic irreversible scattering in Prigogine's sense. He has introduced the new concept, *the internal time*, of the scattered particle. However, Prigogine himself explained it as *an imperfect concept under calculation*. On the other hand, τ_{in} of the in-flight transition is only a constituent of *a working hypothesis* at the present time. It seems natural that both imperfect concepts have the same order,[13]

$$\text{Order}(\tau_{in}) \simeq \text{Order}(\textit{internal time}) \tag{6}$$

We will be able to use relation (6) for estimation of κ.

5.2. *Effect on EPR and Delayed Choice Experiments*

Physical phenomena where the in-flight transitions play dominant roles may be common between those just after the decays observed in EPR experiments and those just after the scatterings treated in the preceding sections. Then, EPR experiments may be realizable, only before the in-flight transition takes place. In other words, it seems natural[7] that measurement of the TSPL–V curve is the experiment in which we observe "From how long a separation do the scattered and scatterer particles become separable, in the sense of the EPR phenomenon, and following the real time process after scattering?" $\kappa^{-1} = (\text{TSPL})_{\text{bound}}$ may be interpreted as the upper boundary of TSPL, within which the two particles are inseparable or the EPR phenomenon is possible.

Wheeler[14] pointed out that the delayed choice experiment mentioned in 1931 by von Weizsäcker[15] was the first type of EPR experiment. The problem of the dubiety discussed in point 4 of Section 3 is essentially the same as the delayed choice problem.

A delayed choice experiment for electrons, using the setup shown in Figure 1b, will be possible by inserting, into the record TSPL = 52 cm of electrons, an adjustable slit as exemplified in ($S_{v\beta}$) of Figure 5. However, the delayed choice phenomena of electrons in such setup may become difficult for long TSPL $\gg$ 52 cm and wide lateral separation $\gg$ 60 μm. This represents a perspective from this chapter, in which a possible restrictive effect on the EPR phenomenon from the in-flight transition can be forecasted.

5.3. *Are Roles of the Mass "Not Thoroughly Known"?*

If it should be generally accepted that the macrophysical laws will be derived approximately from the microphysical laws, the existing perspective that the derivation will be able to be completed on the basis of the combination of the Schrödinger equation, the $\Psi\Psi^*$ law, and the wave packet concept and its reduction seems to be too optimistic. The reason is that the derivation of the macrophysical determinism, as a special case of extremely large probability, from the microphysical indeterminism on the basis of the above-mentioned combination only seems to be hopeless, unless the microphysical indeterminism becomes restricted gradually with an increase of the mass.

The in-flight transition may be an example of such a source of restriction. The relation of κ and τ_{in} in (5) seems to suggest an unknown role of the mass for the in-flight transition.

We should say that the roles of the mass are not yet thoroughly known, spreading from the macroscopic space-time curvature to the microscopic in-flight transition. It seems, at least, to be worthy of testing and not irrelevant that we search for new methods of estimation of κ in relation to unknown roles of the mass, as discussed in Section 4.

6. EPISTEMOLOGICAL REMARKS

The motivation for daring to introduce the new in-flight transition into quantum evolution (added to the detection process at the end of the scatterer $\rightarrow$ detector flight time interval T_{SD} which can be unlimitedly long, or approach the order of daily life) is to search for compatibility between the probabilistic distribution of the scattered particle and its real behavior in the real interval T_{SD}. Both will be regarded as experimental data independent of intervention of the observer. It is significant for us whether or not such data should be regarded as symptoms of the physical realities. This, I think, depends on our personal viewpoint of the respective epistemologies. The important point is which view is the more advantageous to understand any future physical truth. There is a viewpoint which denies the existence of any new truth in the interval T_{SD}. This chapter belongs to a viewpoint which permits us to search for the existence of it.

The hypothetical model of the in-flight transition is only phenomenological at present; in other words, is supported by circumstantial evidence only. But it would be worthwhile searching for some positive proof both experimental and theoretical. The purposes of this chapter are proposal of the hypothesis and discussion of estimation in preparation for its experimental verification.

ACKNOWLEDGMENT. The author acknowledges helpful discussion with Dr. Kayoko Awaya, Nagoya University.

Note added in proof. Although study of the origin of the in-flight transition is postponed in this chapter, some clues to a future theory which will be useful for further κ-estimation or replace the tentative hypothesis should be searched for. Measurement of the TSPL–V curves will be the experimental clue. Another theoretical clue may be, I presume, given by Lie-algebraic reconsideration.

REFERENCES

1. L. MARTON, *Phys. Rev.* **85**, 1057–1058 (1952).
2. G. MÖLLENSTEDT and H. DÜKER, *Naturwissenschaften* **42**, 41 (1955).
3. H. RAUCH, W. TREIMER, AND U. BONSE, *Phys. Lett. A* **47**, 369–371 (1974).
4. A. G. KLEIN and G. I. OPAT, *Phys. Rev. Lett.* **37**, 238–240 (1976).
5. A. ZEILINGER, R. GÄHLER, C. G. SHULL, and W. TREIMER, in: *International Symposium on Neutron Scattering, Argonne 1981*, AIP Conf. Proc. No. 89, pp. 93–99.
6. Y. KOH, *Mass Spectrom.* **16**, 303–313 (1968); **17**, 464–473 (1969).
7. Y. KOH and T. SASAKI, in: *Microphysical Reality and Quantum Formalism* (A. VAN DER MERWE, G. TAROZZI, and F. SELLERI, eds.), Vol. 2, pp. 197–206, Kluwer Academic, Dordrecht (1988).
8. V. DRAHOŠ and A. DELONG, *Opt. Acta* **11**, 173–181 (1964).
9. G. MÖLLENSTEDT and W. BAYH, *Naturwissenschaften* **48**, 400 (1961).
10. W. BAYH, *Z. Phys.* **169**, 492–510 (1962).
11. G. MÖLLENSTEDT and H. DÜKER, *Z. Phys.* **145**, 377–397 (1956).
12. I. PRIGOGINE, *From Being to Becoming*, Freeman, New York (1980); I. PRIGOGINE and C. GEORGE, *Proc. Natl. Acad. Sci. USA* **80**, 4590–4594 (1983); C. GEORGE, F. MAYNÉ, and I. PRIGOGINE, *Adv. Chem. Phys.* **61**, 223–299 (1985).
13. Y. KOH and T. SASAKI, *Physica B* **151**, 362–365 (1988).
14. W. A. MILLER and J. A. WHEELER, in: *Proc. 1st ISQM Symp. Tokyo 1983*, Phys. Soc. Japan, pp. 140–152.
15. K. F. VON WEIZSÄCKER, *Z. Phys.* **70**, 114–130 (1931).

CHAPTER 9

Wave Mechanics and Relativity

Georges Lochak and Regis Dutheil

Many papers have already been published concerning de Broglie's ideas on matter waves, including papers by one of us (G.L.) and D. Fargue in an earlier collective book devoted to the foundations of wave mechanics.[1] One can find, in the latter, some important points such as the *law of phase accordance*, considered by de Broglie as his basic idea, the hypothesis of a *permanent localization of the particle in the wave*, upon which de Broglie based his theory of *measurement*, and the attempts to find a *nonlinear wave equation* which would be able to describe the wave–particle dualism in a synthetic way.

In the same papers, the importance of relativity for the development of wave mechanics was already strongly emphasized. Nevertheless, we shall come back again, with new arguments, to this unique problem in the present chapter, because we think, as Louis de Broglie, that relativity is the cornerstone of wave mechanics and that every attempt to better understand this theory must be based on it.

It must not be forgotten that, when de Broglie found his famous connection between the principles of Maupertuis and Fermat, his approach was purely relativistic and historically, it was even the first application of the principle of relativistic covariance in quantum mechanics.[2] But two years later, in one of his famous papers,[3] Schrödinger obtained some of the basic formulas of wave mechanics (especially wave length and group velocity) making use only of classical mechanics. He was especially very satisfied that group velocity could be obtained without relativity, contrary to what de Broglie did in his thesis. Later on, following Schrödinger, de Broglie himself often introduced, in some of his books, these formulas from classical mechanics.[4–6] But it was for him only a heuristic process which made it possible to shorten an introduction: afterwards, all of his fundamental works were based on relativity.

Georges Lochak and Regis Dutheil • Foundation Louis de Broglie, F-75006 Paris, France.

Wave–Particle Duality, edited by Franco Selleri. Plenum Press, New York, 1992.

However, many theoreticians seem to consider that despite a historical role played in the birth of wave mechanics, relativity has not a crucial importance for its foundations. In particular, almost all works about causal interpretation do not make use of relativity: among many examples, see for instance Refs. 7–9. de Broglie was the only one who always started from Klein–Gordon and Dirac relativist equations in this kind of work. He considered the Schrödinger equation only as an approximation, excepting for the problem of systems of particles.

For these reasons, we would like to show, in this chapter, that relativity is really unavoidable when we want to obtain some fundamental formulas of wave mechanics.

1. CAN WE REACH WAVE MECHANICS STARTING FROM CLASSICAL MECHANICS?

In this section, we shall examine (in a slightly different form) some calculations of de Broglie and Schrödinger, based on classical mechanics. But first of all, we are going to recall briefly one of de Broglie's initial relativistic reasonings.

Introducing relativity directly he set up a phase four-vector of the wave, which he likened to the energy–momentum four-vector of the particle. He used Planck's law:

$$E = h\nu \tag{1}$$

in order to identify time components ν/c and E/c of the two four-vectors. From this followed, by relativistic covariance, the identity of the space components $1/\lambda$ and p; hence the famous formula of wavelength:

$$\lambda = \frac{h}{mv} \tag{2}$$

Owing to the identification of the two four-vectors, de Broglie deduced the identity of Maupertuis's and Fermat's principles and he got the famous group-velocity theorem. Note that in the preceding formula, m must be considered as the relativistic mass (depending on v).

Now suppose that we ignore relativity. Consequently, we cannot use the above reasoning because we cannot introduce the four-vectors and thus we have to proceed differently. In this paragraph, we are going to do almost the same thing as Schrödinger and de Broglie in Refs. 3–6: we shall start form the identification of Maupertuis's and Fermat's principles (as Hamilton did, though in a different form, which we shall examine at the end of this chapter).

Let us then postulate the identity of the two extremal formulas:

$$\delta \int_A^B mv\, ds = \delta \int_A^B \frac{ds}{\lambda} = 0 \tag{3}$$

However, contrary to what Hamilton did, we shall not consider the particle and wave descriptions as alternative. In accordance with de Broglie, we assume an association between wave and particle within the same physical object. It will be understood that we consider only the case of a particular and its wave in the vacuum, in the absence of an external field.

The left-hand integral in (3) is taken along the particle's trajectory and the right-hand one along a radius of the associated wave: we postulate, with de Broglie, that these two curves coincide. *In order that this identity should hold for any motion, the two integrants must be identical up to a universal constant, which implies de Broglie's formula* (2) (now m is a constant because we are in classical mechanics).

In saying that, we assume implicitly that the constant of proportionality between the two sides of (3) is the Planck constant, but doing so we are only anticipating experimental results which prove it, without violating the logic of the reasoning.

Let us introduce phase velocity V through the formula

$$\lambda = \frac{V}{\nu} \tag{4}$$

Identifying (2) and (4) we get

$$V = \frac{h\nu}{mv} \tag{5}$$

Since we are in the case of inertial motion, the particle energy is entirely kinetic and according to Planck's law and classical mechanics, we obtain

$$E = h\nu = \tfrac{1}{2}mv^2 \tag{6}$$

Introducing (6) in (5), we find a relation between the phase velocity V of the wave and particle velocity v:

$$V = \frac{v}{2} \tag{7}$$

Oddly enough, this trivial relation is never written. And yet it is interesting because it is very different from the de Broglie formula:

$$V = \frac{c^2}{v} \tag{8}$$

This is a crucial point to which we shall return. Nevertheless, (7) shows a dispersion of phase waves, because according to (6), v becomes a function of frequency and therefore V too. Now, let us consider the group velocity U of this wave through Rayleigh's classical formula:

$$\frac{1}{U} = \frac{d}{d\nu}\left(\frac{1}{\lambda}\right) = \frac{d}{d\nu}\left(\frac{\nu}{V}\right) \tag{9}$$

Introducing (5) and (6) in (9), we find

$$\frac{1}{U} = \frac{1}{h}\frac{d}{d\nu}(mv) = \frac{d}{v\,d\nu}(v) = \frac{1}{v} \tag{10}$$

Consequently,

$$U = v \tag{11}$$

We see that, starting from a purely classical calculation, equivalent to Schrödinger's, we reach a result that de Broglie obtained using relativity: the group velocity of waves is equal to the particle velocity. Schrödinger underlined the fact that this relation "*does not necessarily follow from relativity, but remains valid for any conservative system of classical mechanics.*"(3) We can see that, remaining in the framework of classical mechanics and wave theory, postulating Planck's law and identifying Maupertuis's and Fermat's principles, we obtain:

1. de Broglie's wavelength (2)
2. A relation between phase velocity of the wave and particle velocity
3. Group velocity (11)

Among these relations, which ones can be experimentally checked? As long as we observe only material particles, like electrons, only the wavelength will be accessible and for low energies, formula (2), with constant m, holds. But we shall not be able to check the formula (11) of group velocity, because experiment only gives particle velocity, which does not prove that particle velocity is equal to group velocity. Moreover, energy E, phase velocity V, and frequency ν still remain inaccessible.

We cannot even claim verify Planck's formula, because E and ν might be wrong; likewise we cannot check equality (6) since v is measured, but E and ν are not; and above all we cannot say anything about the relation (7) between phase velocity V and particle velocity v. Schrödinger was satisfied to obtain the same group velocity as de Broglie, but this result proves nothing about phase velocity. We observe also that by measuring λ, (4) gives only the ratio V/ν: we must therefore measure one of these two quantities experimentally in order to deduce the other.

Since neither V nor ν is measurable directly for a material particle, let us introduce one of de Broglie's essential ideas which was at the basis of his work: wave mechanics must be a unifying theory of matter and light. In fact, his starting point was a mechanical theory of the photon and not a wave mechanics of matter.* Moreover, since Fermat's principle derives from optics and Maupertuis's principle from mechanics, unifying them involves a synthesis between these two fields of science, which we will assume from now on.

Can we apply to light the formulas deduced above? This can be answered experimentally because with light, and contrary to material particles, phase and group velocity (and even frequency) can be measured. Now these velocities are equal in vacuum, which invalidates (7). Besides, if (7) had been correct, it would have been known as far back as the 18th century, because the light velocity measured by Römer would have been twice the one measured by Bradley: indeed, Römer based himself on the occultation of Jupiter's satellites and measured group velocity, whereas Bradley measured phase velocity through astronomic aberration.[11,12] Of course, the distinction between these two velocities was unknown at the time because it was hidden by the equality of results in vacuum: it was observed later in refracting media.† To conclude, since (7) is incorrect, we cannot start from classical mechanics to find wave mechanics.

2. MINIMAL CONDITIONS OF WAVE–PARTICLE DUALISM: THEY ARE INCOMPATIBLE WITH CLASSICAL MECHANICS AND REQUIRE RELATIVITY

Let us now leave classical mechanics, but without introducing relativity yet. Still remaining within inertial motion, let us define wave–particle dualism with the following postulates:

1. Identity (3) of Maupertuis's and Fermat's principles
2. Planck's law (1)
3. Equality (11) between group velocity of waves and corpuscle velocity
4. Leaving classical mechanics and consequently (6), we replace it by the following general formula (in which we recall Planck's law already postulated):

*This means that, for de Broglie, the photon is an ordinary particle. Its mass is nonzero but only very small. Remember that the masslessness of the photon is not a physical fact but only a theoretical consequence of the principle of gauge invariance. Experiment cannot prove that a mass is zero but can only give an upper bound for it, which is the case for the photon.

†Recall that all of the methods that rely on the measurement of a light signal give the group velocity (Fizeau's cogwheel, Kerr's cell, Foucault's rotating mirror, Michelson's interferometer). Besides astronomical aberration, phase velocity appears in the refraction index and in the ratio of the electric and magnetic units.[13] It can also be measured in electromagnetic resonant cavities.[11]

$$E = h\nu = m(v)F(v) \tag{12}$$

Here v is the particle velocity. Thus, we only suppose that energy is proportional to mass, but we suppose that the latter—and the unknown factor F—may be dependent on velocity. Observe that in a homogeneous space and in the absence of field, if mass is not a constant it can only depend on v.

Let us underline that postulates 1 and 3 which describe the essential part of wave–particle dualism were, in de Broglie's work, theorems that were deduced from relativity. But it cannot be the case here, because relativity is not yet included in our hypotheses. However, we will see that relativity is the only possible mechanics, provided an essential condition, without which we could not calculate the functions $m(v)$ and $F(v)$, is fulfilled: *the principle of invariance of light velocity in vacuum*.

One could assert that this principle is not a relativist postulate but an independent physical fact deduced, for instance, from the observation of double stars, as was shown by De Sitter. All the same, this is the principle which paved the way for relativity.

According to the previous paragraph we know that the first two postulates we introduced lead to de Broglie's wavelength (2) and to the relation (5) we shall need.

Taking into account postulates 1 and 3 and using (12), we introduce (1) and (11) in Rayleigh's formula (9). Hence:

$$\frac{1}{v} = \frac{1}{U} = \frac{d}{d\nu}\left(\frac{1}{\lambda}\right) = \frac{d(mv)}{d(h\nu)} = \frac{d(mv)}{d(mF)} \tag{13}$$

from which we find a first equation between the unknown function $m(v)$ and $F(v)$

$$\frac{d(mF)}{dv} = v\frac{d(mv)}{dv} \tag{14}$$

Notice that if we supposed mass was not dependent on v, we would find:

$$m = C^{nt} \Rightarrow \frac{dF}{dv} = v \Rightarrow F = \tfrac{1}{2}v^2 \tag{15}$$

and we would come back to the case of classical mechanics, which we excluded earlier. Therefore, we will admit that $m(v)$ is really a function of v.

Let R_0 be a reference frame where the particle is at rest and R a reference frame with velocity v with respect to R_0. Space and time are denoted (x_0, t_0) in R_0 and (x,t) in R. Consider a stationary wave in R_0:

$$\Psi = \exp 2i\pi\, \nu_0 t_0 \tag{16}$$

With respect to R, it will be a propagating wave with a phase velocity V and a frequency ν:

$$\psi = \exp 2i\pi\nu\left(t - \frac{x}{V}\right) \tag{17}$$

V is given by (5) which, taking account of (12), yields:

$$V = \frac{F}{v} \tag{18}$$

Insert this formula in (17) and identify (16) and (17). This gives us the following transformation law for the time variable:

$$t_0 = \frac{m(v)F(v)}{h\nu_0}\left(t - \frac{vx}{F(v)}\right) \tag{19}$$

In order to find the space transformation, imagine a light signal which travels along the interval (x_0, t_0) in R_0 and the corresponding interval (x, t) in R. The velocity of light being the same in both reference frames (this is where the postulate interferes), the time necessary to cover the distance will be equal to

$$t_0 = \frac{x_0}{c} \text{ in } R_0 \text{ and } t = \frac{x}{c} \text{ in } R \tag{20}$$

Inserting (20) in (19), we get the law of space transformation:

$$x_0 = \frac{m(v)F(v)}{h\nu_0}\left(x - \frac{vc^2t}{F(v)}\right) \tag{21}$$

It is easily seen that the transformations (19) and (21) form a group and that their composition law implies the following composition law of velocities which already looks like the relativistic one:

$$\frac{v''}{F(v'')} = \frac{v/F(v) + v'/F(v')}{1 + [vv'c^2/F(v)F(v')]} \tag{22}$$

As v is the translation velocity of reference frame R with respect to the reference frame R_0, if we compose two transformations of opposite velocities v and $v' = -v$, we should obtain the unit transformation (with zero velocity); this implies that the function $F(v)$ is even, because:

$$\frac{v}{F(v)} + \frac{v'}{F(v')} = \frac{v}{F(v)} - \frac{v}{F(-v)} = 0 \text{ from which } F(v) = F(-v) \tag{23}$$

Then, inverting transformations (19) and (21), we find

$$t = \frac{h\nu_0}{m(v)F(v)} \frac{1}{1 - (v^2c^2/F^2)} \left(t_0 + \frac{vx_0}{F(v)} \right);$$
$$x = \frac{h\nu_0}{m(v)F(v)} \frac{1}{1 - (v^2c^2/F^2)} \left(x_0 + \frac{vc^2t_0}{F(v)} \right) \tag{24}$$

These inverse formulas should coincide with those which obtained by changing v in $-v$ in (19) and (21); as $F(v)$ is even, this gives a second equation between $m(v)$ and $F(v)$:

$$\frac{m(v)F(v)}{h\nu_0} = \frac{h\nu_0}{\mathrm{m}(v)F(v)} \frac{1}{1 - (v^2c^2/F^2)} \tag{25}$$

Thus:

$$m^2F^2 = h^2\nu_0^2 + m^2v^2c^2 \tag{26}$$

From which follows, taking the derivative:

$$mF\frac{d(mF)}{dv} = c^2mv\frac{d(mv)}{dv} \tag{27}$$

And taking account of (14):

$$F(v) = c^2 \tag{28}$$

This entails, in virtue of (18), de Broglie's relation between phase velocity and particle velocity:

$$Vv = c^2 \tag{29}$$

According to postulate 3 this is also, by definition, the relation which links phase and group velocity.

Then according to (28), the composition law (22) for velocities gives the relativistic formula:

$$\frac{v''}{c^2} = \frac{v + v'}{1 + (vv'/c^2)} \tag{30}$$

On the other hand, the general law (12) becomes Einstein's law:

$$E = mc^2 \tag{31}$$

If we introduce (28) into (26) we get:

$$mc^2 = \frac{h\nu_0}{\sqrt{1 - (v^2/c^2)}} \tag{32}$$

From this, using (31) and defining the rest mass by applying Planck's law in the system at rest, there follows the usual relativist dependence of mass with respect to velocity:

$$m = \frac{m_0}{\sqrt{1 - (v^2/c^2)}} \tag{33}$$

Finally, introducing (28) and (32) in (19) and (21), we recover the Lorentz transformation:

$$t_0 = \frac{t - (vx/c^2)}{\sqrt{1 - (v^2/c^2)}}; \; x_0 = \frac{x - vt}{\sqrt{1 - (v^2/c^2)}} \tag{34}$$

Let us summarize the argument:

1. We first defined wave–particle dualism through three postulates which were nothing else than Planck's and de Broglie's laws.
2. We then postulated the proportionality between energy and mass in every inertial system.
3. This form admits classical mechanics and relativity as particular cases but it could *a priori* correspond with many conservative dynamics, which would be in accordance with de Broglie's law for the group velocity. But we have shown that there are actually only two dynamics compatible with all of these four postulates: classical mechanics and relativity. Now the theory of light compels us to reject classical mechanics: thus, there remains only relativity as de Broglie always asserted.

3. A NOTE ABOUT A REASONING OF HAMILTON

It is well known that Hamilton having defined the surfaces of constant action,

$$S = C^{nt} \tag{35}$$

noticed that the length of the vector:

$$p = mv = \nabla S \tag{36}$$

varies as the inverse of the velocity V (i.e., the phase velocity) of the wave front defined by (35). Therefore, he called it the vector of normal slowness of the front.[10,14]

This remark led Hamilton to set

$$V = \frac{1}{v} \tag{37}$$

Relation (37) lies at the basis of his analogy between optics and mechanics: specifically, the relationship between corpuscular and wave optics.

Of course, in Hamilton's work group velocity is not mentioned, as the concept was unknown at that time. But we shall see that relation (37) which at first looks like de Broglie's relation (8) is in fact inconsistent with Planck's law and in any case could not lead to wave mechanics. Indeed, Hamilton seemed to consider (37) as a direct consequence of the identification of principles of Maupertuis and Fermat. But this cannot be the case because this would lead to set in (3) and (4):

$$m = C^{nt} \text{ and } v = C^{nt} \tag{38}$$

Actually, introducing (4) into (3) and eliminating the now superfluous factors m and v, the identity of extremal principles could be written under the form:

$$\delta \int_A^B v\, ds = \delta \int_A^B \frac{1}{V} ds = 0 \tag{39}$$

This equality (which by the way is inhomogeneous) leads indeed to relation (37) (which is not homogeneous either). But conditions (38) cannot be accepted because:

1. Already in classical mechanics (the Hamilton one) the second equality (38) contradicts Planck's law since it contradicts the equality (6).
2. In relativist mechanics, equalities (38) go against both Planck's law and equivalence of mass and energy.

Therefore, it would be wrong to assert that to derive formula (8) it is enough to identify Maupertuis's and Fermat's principles. However, we have shown at the beginning of this chapter that identification, when properly formulated, leads to the wavelength formula. But as we have seen, the true problem is the one of phase velocity, which forces us to choose relativity.

REFERENCES

1. *The Wave–Particle Dualism* (S. DINER, D. FARGUE, G. LOCHAK, and F. SELLERI, eds.), Reidel, Dordrecht (1984).

2. L. DE BROGLIE, Thesis (1924).
3. E. SCHRÖDINGER, *Ann. Phys. (4)* **79**, 489 (1926).
4. L. DE BROGLIE, *Théorie de la quantification dans la nouvelle mécanique*, Hermann, Paris (1932).
5. L. DE BROGLIE, *Nonlinear Wave Mechanics*, Elsevier, Amsterdam (1960).
6. L. DE BROGLIE, *Les incertitudes d'Heisenberg et l'interprétation probabiliste de la mécanique ondulatoire*, Gauthier–Villars, Paris (1982).
7. D. BOHM, *Phys. Rev.* **85**, 166, 180 (1952).
8. A. BELINFANTE, *A Survey of Hidden Variables Theories*, Pergamon, Oxford (1973).
9. A. BARUT and M. BOZIC, *Ann. Fond. L. de Broglie,* **15**, 67 (1990).
10. W. R. HAMILTON, *Mathematical Papers*, Cambridge University Press, London (1940).
11. P. DRUDE, *Précis d'Optique*, Gauthier–Villars, Paris (1911).
12. G. BRUHAT, *Optique*, Masson, Paris (1965).
13. M. BORN and E. WOLF, *Principles of Optics*, Pergamon, Oxford (1964).
14. V. ARNOLD, *Les méthodes mathématiques de la mécanique classique*, "Mir", Moscow (1976).

CHAPTER 10

UNSHARP PARTICLE–WAVE DUALITY IN DOUBLE-SLIT EXPERIMENTS

PETER MITTELSTAEDT

1. INTRODUCTION

The quantum mechanical double-slit experiment was discussed from an epistemological point of view in the Bohr–Einstein debate.[1] According to Bohr the particle–wave duality comes from the fact that the measuring instruments for the measurement of a particle (the path) and for the measurement of the wave (the interference pattern) are mutually exclusive. This was demonstrated by Bohr for many Gedanken experiments by means of the uncertainty relation between momentum p and position x. The same uncertainty relation can, however, also be used to show that in spite of the strict particle–wave duality, the path (momentum p) and the interference pattern (position x) can at least be measured simultaneously in an approximate sense. This surprising result was first demonstrated by Wootters and Zurek[2] within a quantum mechanical reformulation of one of the Gedanken experiments discussed by Bohr and Einstein.[1]

A more rigorous treatment of unsharp quantum mechanical measurements must make use of the general theory of unsharp observables in quantum mechanics, which replaces the sharp observables, which can be described by projector-valued measures, by effect-valued measures which correspond to unsharp observables (see Busch[3]). If in the double-slit experiment the observables of the path and the interference pattern are replaced by unsharp observables of this kind, unsharp joint measurements of the path (momentum p) and the interference pattern (position x) turn out to be possible. Unsharp measurements of this kind must not be confused with inaccurate ones since the unsharpness expresses the

---

PETER MITTELSTAEDT • Institut für Theoretische Physik, Universität zu Köln, D-5000 Köln 41, Germany.

Wave–Particle Duality, edited by Franco Selleri. Plenum Press, New York, 1992.

objective indeterminacy of the path and of the interference pattern, but not the observer's ignorance about these properties.

Several attempts have been made to realize unsharp joint measurements experimentally. A photon split-beam experiment has been performed using a modified Mach–Zehnder interferometer.[(4)] This experiment provides simultaneous unsharp wave and particle knowledge in full agreement with the theoretical predictions. More sophisticated experiments of this kind were proposed, but have not yet been realized.[(5)] Other kinds of experiments which were interpreted in the sense of unsharp joint measurements are neutron interference experiments, which were performed by several authors.[(6–8)] However, these experiments use a slightly different experimental setup which does not allow interpreting them as unsharp measurements, but rather as experiments with "unsharp preparation." This conceptual distinction can also be realized experimentally by extended experiments which are briefly discussed here.

2. THE PHOTON SPLIT-BEAM EXPERIMENT

As an illustration of the Bohr–Einstein debate we consider the photon split-beam experiment shown in Figure 1. In the experimental setup, a Mach–Zehnder interferometer, the incoming photon-state $|\varphi\rangle$ is split by the first beam splitter $(BS)_1$

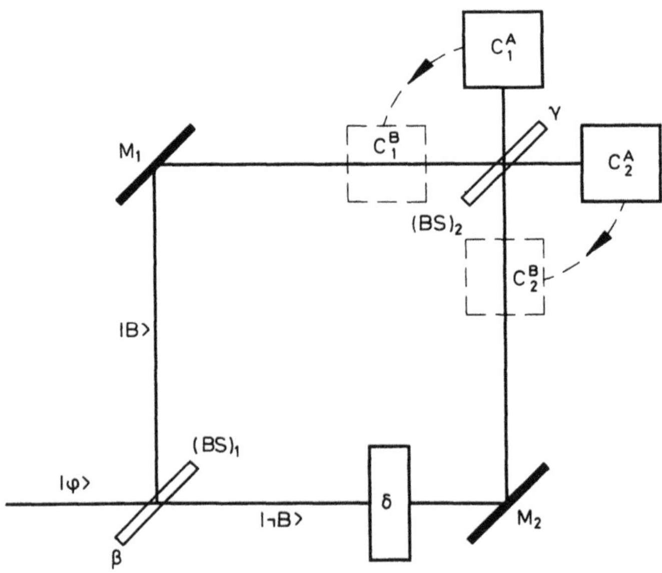

FIGURE 1. Photon split-beam experiment. Beam splitters $(BS)_1$ and $(BS)_2$ with transparencies $\beta = 1/2$, $\gamma = 1/2$, two fully reflecting mirrors M_1 and M_2, a phase shifter, and two photon counters C_1 and C_2.

into two components described by the orthonormal states $|B\rangle$ and $|\neg B\rangle$, respectively. The two parts of the split beam are reflected at the two mirrors M_1 and M_2 and recombined with a phase difference δ at the second beam splitter $(BS)_2$. In this experiment there are two mutually exclusive measuring arrangements: If the photon counters C_1, C_2 are in the position (C_1^B, C_2^B), one observes which way (B or $\neg B$) the photon came, and if the counters are in the position (C_1^A, C_2^A), one observes the interference pattern, i.e., the intensities which depend on the phase difference δ.

This experiment can be described completely within the framework of the two-dimensional Hilbert space $\mathcal{H}_2$. The observables are then given by projection operators $P(B)$ with eigenstates $|B\rangle$ and $|\neg B\rangle$ (for the path) and $P(A)$ with eigenstates $|A\rangle$ and $|\neg A\rangle$ (for the interference pattern). The state $|\varphi\rangle$ of the incoming photon can be decomposed in the B-basis as $|\varphi\rangle = 1/\sqrt{2}(|B\rangle + \exp(i\delta)|\neg B\rangle)$. Hence, the probabilities for the ways B and $\neg B$ are given by $p(\varphi, B) = |\langle\varphi|B\rangle|^2 = 1/2$ and $p(\varphi, \neg B) = |\langle\varphi|\neg B\rangle|^2 = 1/2$, respectively. The interference observable is given by $P(A)$ with eigenstates $|A\rangle = 1/\sqrt{2}(|B\rangle + |\neg B\rangle)$ and $|\neg A\rangle = 1/\sqrt{2}(|B\rangle - |\neg B\rangle)$. Hence, the probabilities for A and $\neg A$ are given by $p(\varphi, A) = 1/2(1 + \cos\delta)$ and $p(\varphi, \neg A) = 1/2(1 - \cos\delta)$, respectively. Obviously the observables $P(B)$ and $P(A)$ have a nonvanishing commutator $[P(B), P(A]_- = 1/2\{|B\rangle\langle\neg B| - |\neg B\rangle\langle B|\} \neq 0$ and can thus not be measured simultaneously.

Pure and mixed state operators of $\mathcal{H}_2$ can be represented by means of a three-dimensional unit sphere, the Poincaré sphere $\mathcal{P}$ (Figure 2). By using the Pauli operators $\{\mathbb{1}, \sigma_1, \sigma_2, \sigma_3\}$ with $[\sigma_j, \sigma_k] = 2\epsilon_{jkl}\sigma_l$ and $\{\sigma_j, \sigma_k\} = 2\delta_{jk}\mathbb{1}$, projection operators $P(X)$ are given by

$$P(X = 1/2(X_i\sigma_i + \mathbb{1}), \quad X_i \in \mathbb{R}^+, \sum X_i^2 = 1$$

and mixtures by

$$W(Y) = 1/2(Y_i\sigma_i + \mathbb{1}), \; Y_i \in \mathbb{R}^+, \sum Y_i^2 < 1$$

Here projection operators $P(X)$ are given by unit vectors $\vec{X}$ or points on the surface of $\mathcal{P}$ and mixed states by vectors $\vec{Y}$ with $|\vec{Y}| < 1$ or points in the interior of $\mathcal{P}$. Projection operators $P(X) = |X\rangle\langle X|$ and $\mathbb{1} - P(X) = P(\neg X) = |\neg X\rangle\langle\neg X|$ which project onto orthogonal states correspond to diametrical points of the Poincaré sphere. If the axes of $\mathcal{P}$ are chosen such that the operator $P(B)$ reads

$$P(B) = 1/2(B_i\sigma_i + \mathbb{1}) = 1/2(\sigma_3 + \mathbb{1})$$

the operator $P(A)$ of the interference pattern is given by

$$P(A) = 1/2(A_i\sigma_i + \mathbb{1}) = 1/2(\sigma_1 + \mathbb{1})$$

and thus $P(A)$ and $P(B)$ correspond to orthogonal vectors $\vec{A}$ and $\vec{B}$.

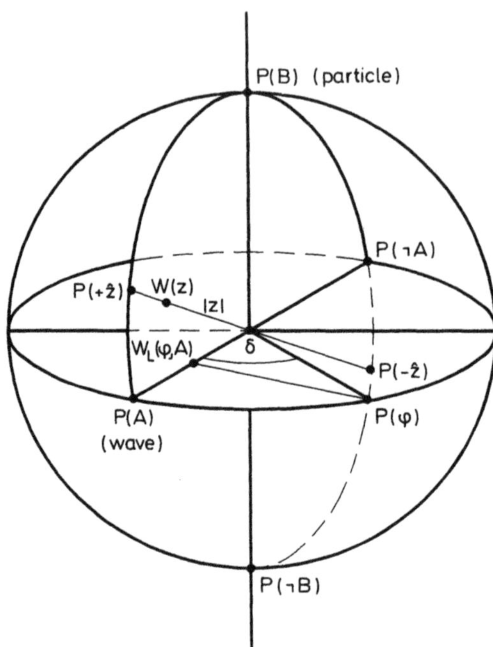

FIGURE 2. States and operators of the two-dimensional Hilbert space represented by the Poincaré sphere. Initial preparation $P[\varphi]$, phase shift δ, path observable $P(B)$, interference observable $P(A)$. Mixed states $W(Z)$ and $W_L(\varphi, A)$ lie in the interior of the sphere.

Mixed states are described by (positive trace one) operators $W(Z)$ and are given by $W(Z) = 1/2(Z_i\sigma_i + \mathbb{1})$ with vectors $|\vec{Z}| < 1$. The spectral decomposition then reads

$$W(Z) = W^+(Z)\cdot 1/2(\mathbb{1} + \hat{Z}_i\sigma_i) + W^-(Z)\cdot 1/2(\mathbb{1} - \hat{Z}_i\sigma_i)$$

with $\hat{Z}_i = Z_i/|\vec{Z}|$ and eigenvalues

$$W^+(Z) = 1/2(1 + |\vec{Z}|),\ W^-(Z) = 1/2(1 - |\vec{Z}|)$$

which correspond to the probabilities $p(W, Z^\pm)$, i.e., the expectation values of the projection operators $P(\pm\hat{Z}) = 1/2(\mathbb{1} \pm \hat{Z}_i\sigma_i)$. In the Poincaré sphere $\mathcal{P}$ the operator $W(Z)$ lies on the diameter given by $P(+\hat{Z})$ and $P(-\hat{Z})$ and has the distance $|\vec{Z}|$ from the origin of the sphere (Figure 2). Mixed states are of particular importance for the description of the measuring process.

For a given preparation $P[\varphi] = 1/2(\varphi_i\sigma_i + \mathbb{1})$ this sharp measurement of an observable $P(A) = 1/2(A_i\sigma_i + \mathbb{1})$ in the sense of von Neumann and Lüders consists of two steps (I) and (II)

$$P[\varphi] \overset{(I)}{\rightarrow} W_L(\varphi, A) \overset{(II)}{\rightarrow} P(A)$$

Step (I), the objectification, transforms the preparation $P[\varphi]$ into the Lüders mixture

$$W_L(\varphi, A) = p(\varphi, A)P(A) + p(\varphi, \neg A)P(\neg A)$$

whereas step (II) merely reduces the observer's subjective ignorance by reading the final result $P(A)$, say, of the measuring process. For the above-mentioned preparation $|\varphi\rangle = 1/2(|B\rangle + \exp(i\delta)|\neg B\rangle)$ one obtains either the mixture

$$W_L(\varphi, B) = 1/2P(B) + 1/2P(\neg B)$$

by measuring the path or the mixture

$$W_L(\varphi, A) = 1/2(1 + \cos\varphi)P(A) + 1/2(1 - \cos\varphi)P(\neg A)$$

by measuring the interference pattern.

In the Poincaré sphere the mixed states $W_L(\varphi, B)$ and $W_L(\varphi, A.)$ are obtained geometrically by orthogonal projections of the vector $\vec{\varphi}$ onto the radius vectors $\vec{B}$ and $\vec{A}$, respectively. For a given angle δ s.t. $\delta = \arccos \vec{\varphi}\cdot\vec{A}$ the probabilities for $P(A)$ and $P(\neg A)$ are $p(\varphi, A) = 1/2(1 + \cos\delta)$ and $p(\varphi, \neg A) = 1/2(1 - \cos\delta)$, respectively (Figure 2).

3. THE UNSHARP MEASURING PROCESS

In contrast to the sharp first kind measurement described above, the final result of an *unsharp measurement* is not a projection operator $P \in \mathcal{P}(\mathcal{H})$ but an effect operator $E \in \mathcal{E}(\mathcal{H})$. Effects are selfadjoint operators E, the spectrum E_λ of which lies within the interval [0,1]. In the two-dimensional Hilbert space $\mathcal{H}_2$ effects are given by operators of the kind

$$E_\alpha = \alpha/2(\lambda_i\sigma_i + \mathbb{1}), \sum\lambda_i^2 \leq 1, 0 \leq \alpha \leq 2/(1 + |\vec{\lambda}|)$$

For the sake of simplicity we will restrict the discussion here to the special case $\alpha = 1$. These special effect operators

$$E = 1/2(\lambda_i\sigma_i + \mathbb{1}), \sum\lambda_i^2 \leq 1$$

correspond to points in the interior of the Poincaré sphere, and are formally equivalent to mixture operators W. To any effect there exists a counter effect $\overline{E}$ which is given by

$$\overline{E} = \mathbb{1} - E = 1/2(\mathbb{1} - \lambda_i \sigma_i)$$

The spectral decomposition of an effect $E(\vec{\lambda})$ of this kind reads

$$E = 1/2(\vec{\lambda}\cdot\vec{\sigma} + \mathbb{1}) = 1/2(1 + |\vec{\lambda}|)P(\lambda) + 1/2(1 - |\vec{\lambda}|)P(-\lambda)$$

with projection operators

$$P(\pm\lambda) = 1/2(1 \pm \lambda_i \sigma_i / |\vec{\lambda}|)$$

The eigenvalues have the interpretation

$$E^+ = 1/2(1 + |\vec{\lambda}|) > 1/2\text{—reality degree of property } P(\vec{\lambda})$$
$$E^- = 1/2(1 - |\vec{\lambda}|) < 1/2\text{—unsharpness of property } P(\vec{\lambda})$$

where we have assumed that $E^+ > 1/2$ and $E^- < 1/2$.

By means of these considerations we are now prepared to describe the unsharp measuring process. An unsharp measurement of the observable $P(A) = 1/2(1 + A_i \sigma_i)$ leads to the effect

$$E(A,\lambda_A) = 1/2(\lambda_A A_i \sigma_i + \mathbb{1}), \ \Sigma\, (\lambda_A A_i)^2 \leqslant 1$$

as the final result of the measuring process. The eigenvalues $E^{\pm}(A,\lambda_A)$ of $E(A,\lambda_A)$, which follow from the spectral decomposition

$$E(A,\lambda_A) = 1/2(1 + \lambda_A)1/2(\mathbb{1} + A_i \sigma_i) + 1/2(1 - \lambda_A)1/2(\mathbb{1} - A_i \sigma_i)$$

are

$$E^+(A,\lambda_A) = 1/2(1 + \lambda_A) > 1/2\text{—reality degree of } P(A)$$
$$E^-(A,\lambda_A) = 1/2(1 - \lambda_A < 1/2\text{—unsharpness of } P(A)$$

and describe the degree of reality or the unsharpness of the observable $P(A)$.

In the two-dimensional Hilbert space $\mathcal{H}_2$ the special effects $E = 1/2(\lambda_i \sigma_i + \mathbb{1})$ with $\Sigma\, \lambda_i^2 \leqslant 1$ are equivalent to mixture operators, i.e., positive trace one operators. This accidental equivalence allows for a very simple illustration of the unsharp measuring process. Starting from the preparation $P[\varphi]$ an observable $P(C) = 1/2(C_i \sigma_i + \mathbb{1})$ with $C_i A_i = \lambda_A$ and $C_i(\vec{\varphi} \times \vec{A})_i = 0$ (Figure 3) is measured sharply with the result $P(C)$, say. The knowledge of $P(C)$ then provides the knowledge of the effect $E(A,\lambda_A)$ for the following reason: If starting from the preparation $P(C)$

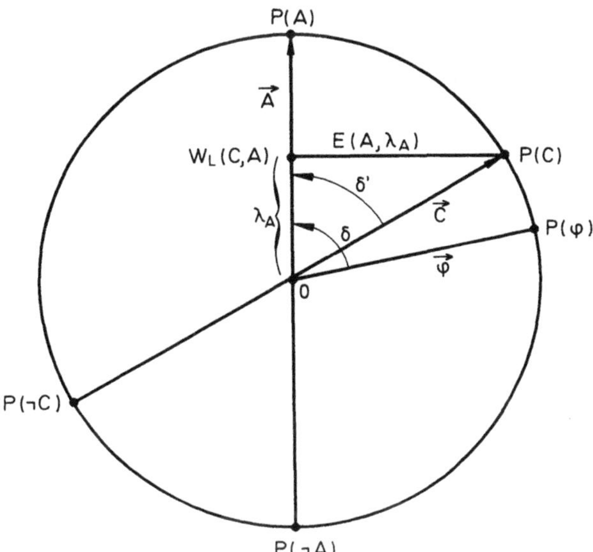

FIGURE 3. Unsharp measuring process $P(A)$ with preparation $P[\varphi]$. The vector $\vec{C}$ lies in the plane spanned by the vector $\vec{\varphi}$ and A.

one would measure $P(A)$ without reading, then one would obtain with *certainty* the Lüders mixture

$$W_L(C, A) = p(C, A)P(A) + p(C, \neg A)P(\neg A)$$

with $p(C, A) = 1/2(1 + \lambda_A)$ and $p(C, \neg A) = 1/2(1 - \lambda_A)$, which is equivalent to the effect $E(A,\lambda_A)$.

It is obvious that there are infinitely many projection operators $P(C')$ with $C_i'A_i = \lambda_A$ such that C' is not in the plane spanned by $\vec{\varphi}$ and $\vec{A}$. However, these projection operators $P(C')$ have a larger distance from $P[\varphi]$ than $P(C)$ and consequently the system with the preparation $P[\varphi]$ is stronger "demolished" by a $P(C')$-measurement than by a $P(C)$—measurement. Since the result, the effect $E(A,\lambda_A)$, is the same for all projection operators $P(C')$, the measurement of $P(C)$ with $C_i(\vec{\varphi} \times \vec{A})_i = 0$ corresponds to the "minimal demolishing measurement" of the effect $E(A,\lambda_A)$.

For the unsharp joint measurement of *two* observables (projection operators $P(A)$ and $P(B)$) we consider the *two* effects

$$E(A,\lambda_A) = 1/2(\lambda_A A_i \sigma_i + \mathbb{1}) \text{ with } \sum_i (\lambda_A A_i)^2 \leq 1$$

$$E(B,\lambda_B) = 1/2(\lambda_B B_i \sigma_i + \mathbb{1}) \text{ with } \sum_i (\lambda_B B_i)^2 \leq 1$$

and with eigenvalues $E(A,\lambda_A)$ and $E(B,\lambda_B)$, respectively. Two arbitrary observables M_1 and M_2 are called coexistent, if the ranges on the M_i are contained in the range of one joint observable M. For the unsharp observables $E(A,\lambda_A)$ and $E(B,\lambda_B)$ this means that these two effects are coexistent iff

$$1/2|\lambda_A\vec{A} + \lambda_B\vec{B}| + 1/2|\lambda_A\vec{A} - \lambda_B\vec{B}| \leqslant 1$$

If in particular $\vec{A}\cdot\vec{B} = 0$, this condition reads $\lambda_A^2 + \lambda_B^2 \leqslant 1$.

The special observables $P(A)$—(of the interference pattern) and $P(B)$—(of the path) fulfill the relation $\vec{A}\cdot\vec{B} = 0$. Moreover, we consider here the limiting case of maximal coexistence $\lambda_A^2 + \lambda_B^2 = 1$. The corresponding effects $E(A, \lambda_A)$ and $E(B, \lambda_B)$ can then be estimated jointly by the ideal Lüders measurements of an observable

$$P(C) = 1/2(C_i\sigma_i + \mathbb{1}) \text{ with } \sum C_i^2 = 1$$

which fulfills the two conditions $C_iA_i = \lambda_A$ and $C_iB_i = \lambda_B$. The observables $P(C)$ with $\vec{C}\cdot\vec{A} = \lambda_A$ lie on a circle C_A and those with $\vec{C}\cdot\vec{B} = \lambda_B$ on a circle C_B. From the geometry of the Poincaré sphere (see Figure 4) it is obvious that for values λ_A and λ_B with $\lambda_A^2 + \lambda_B^2 = 1$ there is exactly *one* observable $P(C)$ for which both conditions are fulfilled. However, since $\vec{C}$ is neither in the plane spanned by $\vec{\varphi}$ and $\vec{A}$ nor in the

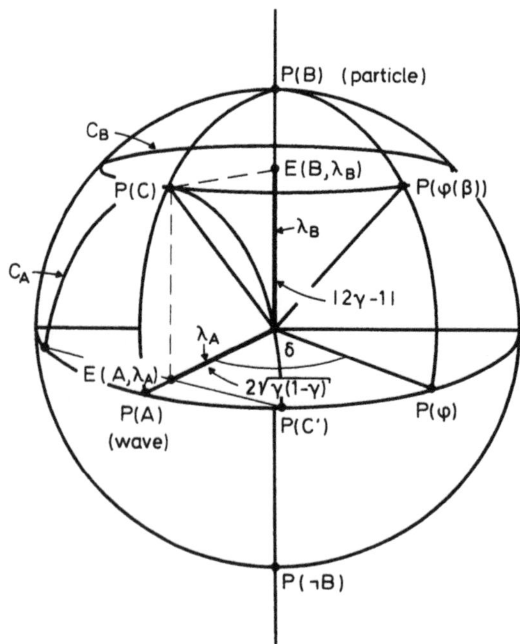

FIGURE 4. Poincaré sphere description of an unsharp joint measurement of the observables $P(A)$ and $P(A)$ by means of an intermediate observable $P(C)$. Effects $E(A, \lambda_A)$ and $E(B, \lambda_B)$; preparations $P[\varphi]$ and $P[\varphi(\beta)]$. Circles C_A and C_B of observables $P(C')$.

plane spanned by $\vec{\varphi}$ and $\vec{B}$, the joint measurement of A and B does not correspond to minimal demolishing measurement of both effects $E(A, \lambda_A)$ and $E(B, \lambda_B)$.

4. INFORMATION THEORETICAL CONSIDERATIONS

In the case of a sharp measurement the observer knows either the eigenvalue of the path observable or the eigenvalue of the interference observable. Unsharp measurements provide unsharp knowledge about the path and the interference pattern. It is useful to describe these different situations by means of information theory. For a preparation $W = P[\varphi]$ the deficiency of information about the result of a *sharp* measurement of the observable $P(A)$ can be expressed by the formula

$$H(\varphi, A) = -p(\varphi, A) \ln p)\varphi, A) - p(\varphi, \neg A) \ln p(\varphi, \neg A) \leq \ln 2$$

where $p(\varphi, A)$ and $p(\varphi, \neg A)$ are the probabilities of obtaining $P(A)$ and $P(\neg A)$ as measuring results. Conversely the (normalized) knowledge about the wave property $P(A)$ of a system with preparation $P[\varphi]$ is given by $W = 1 - H(\varphi, A)/\ln 2$.

Here we consider preparations $W = P[\varphi]$ which correspond to a pure state $|\varphi\rangle$ and observables which are not degenerate. Then the deficiency of information $H(\varphi, A)$ about the result of an ideal first kind measurement of $P(A)$ agrees with the von Neumann entropy of the Lüders mixture

$$W_L(\varphi, A) = p(\varphi, A)P(A) + p(\varphi, \neg A)P(\neg A)$$

which is obtained in step (I)—the objectification—of the measuring process, i.e., we have

$$H(\varphi, A) = Tr(W_L \ln W_L) =: H(W_L(\varphi, A))$$

where we have denoted the entropy of the mixture W_L by $H(W_L)$. Hence, the deficiency of information $H(\varphi, A)$ about the property A an be interpreted as the observer's subjective ignorance about A in the Lüders–Gemenge which is described by the mixture $W_L(\varphi, A)$.

The result of an unsharp measurement of $P(A)$ with the unsharpness $1/2(1 - \lambda_A)$ is given by the effect $E(A,\lambda_A)$. The gain of information, which is achieved by measuring the effect $E(A,\lambda_A)$ on a system with preparation φ should be determined by its minimal demolishing measurement, i.e., by the measurement of the projection operator $P(C)$, which in addition to $\vec{C}\cdot\vec{A} = \lambda_A$ also fulfills the condition $C_i\cdot(\vec{\varphi} \times \vec{A})_i = 0$. The deficiency of information $H(\varphi, C)$ about the property C in the state φ is then given by

$$H(\varphi, C) = -p(\varphi, C) \ln p(\varphi, C) - p(\varphi, \neg C) \ln p(\varphi, \neg C)$$

with (see Figure 3)

$$p(\varphi, C) = 1/2(1 + \cos(\delta - \delta')), \cos\delta' = \lambda_A$$
$$= 1/2(1 + \lambda_A \cos\delta + \sqrt{(1 - \lambda_A^2)}\sin\delta) =: p(\varphi; A, \lambda_A)$$

Hence, we write $H(\varphi, C) = H(\varphi; A, \lambda_A)$ for the gain of information which is achieved by measuring the effect $E(A, \lambda_A)$.

Since the deficiency of information $H(\varphi, C')$ for a projection operator $P(C')$ with $\vec{C}'\cdot\vec{A} = \lambda_A$ is a monotonically increasing function of the distance $d(\varphi, C') = \arccos(\vec{\varphi}\cdot\vec{C}')$ between $P[\varphi]$ and $P(C')$, and since the relation $d(\varphi, C) \leq d(\varphi, C')$ holds for any C' of this kind, we obtain $H(\varphi, C) \leq H(\varphi, C')$. Hence, the deficiency of information $H(\varphi; A, \lambda_A) = H(\varphi, C)$ is smaller than the value of any other $P(C')$, which provides the same knowledge about the effect $E(A, \lambda_A)$.

The remaining deficiency of information about the eigenvalue of $P(A)$ in the state $P(C)$ is then given by

$$H(C, A) = -p(C, A)\ln p(C, A) - p(C, \neg A)\ln p(C, \neg A)$$

with $p(C, A) = 1/2(1 + \cos\delta') = 1/2(1 + \lambda_A)$. This deficiency of information $H(C, A)$ is equal to the entropy of the Lüders mixture

$$W_L(C, A) = p(C, A)P(A) + p(C, \neg A)P(\neg A)$$

which is generated by the first step of the measuring process, i.e., we have $H(C, A) = H(W_L(C, A))$.

Since in the present case the probabilities read

$$p(C, A) = 1/2(1 + \lambda_A) \text{ and } p(C, \neg A) = 1/2(1 - \lambda_A)$$

the mixture $W_L(C, A)$ agrees with the effect $E(A, \lambda_A)$. Hence, the remaining deficiency of information about the value of $P(A)$ after the measurement of $P(C)$ is given by the entropy $H(E(A, \lambda_A))$ of the effect, which is considered to be the result of the $1/2(1 - \lambda_A)$—unsharp measuring process of $P(A)$. Hence, we obtain the result

$$H(C, A) = H(W_L(C, A)) = H(E(A, \lambda_A))$$

The special observables $P(A)$—(interference pattern) and $P(B)$—(path) fulfill the condition $\vec{A}\cdot\vec{B} = 0$. We will consider here again the case of maximal coexistence $\lambda_A^2 + \lambda_B^2 = 1$. The gain of information which is achieved by a joint measurement of the effects $E(A, \lambda_A)$ and $E(B, \lambda_B)$ is then given by the deficiency of information about the observable $P(C)$, the eigenstate of which provides the simultaneous knowledge of both effects, i.e., by

$$H(\varphi, C) = -p(\varphi, C)\ln p(\varphi, C) - p(\varphi, \neg C)\ln p(\varphi, \neg C)$$

where on account of $2\sqrt{\gamma(1-\gamma)} = \lambda_A$ (Figure 4) we have

$$p(\varphi, C) = 1/2(1 + \lambda_A \cos\delta) \text{ and } p(\varphi, \neg C) = 1/2(1 - \lambda_A \cos\delta)$$

5. REALIZATION OF THE UNSHARP JOINT MEASUREMENT OF $P(A)$ AND $P(B)$

According to the previous discussion for an unsharp joint measurement of the particle property—given by $P(B)$ and the wave property—given by $P(A)$, one has to perform a sharp measurement of an intermediate observable $P(C)$. In the photon split-beam experiment of Figure 1 this can be achieved by the following modification of the experimental setup: Whereas the transparency β of the first beam splitter $(BS)_1$ still has the value $\beta = \frac{1}{2}$, the transparency γ of the beam recombiner $(BS)_2$ is chosen such that γ is variable, i.e., $0 \leqslant \gamma \leqslant 1$. Each value of γ determines an observable $P(C(\gamma))$, which is measured sharply with the apparatus. In particular, for the value $\gamma = 0$ one would measure which way the photon came, i.e., one would measure the observable $P(C(0)) = P(B)$. If $\gamma = \frac{1}{2}$, one measures the interference pattern, i.e., the observable $P(C(1/2)) = P(A)$. In the general case $(0 \leqslant \gamma \leqslant 1)$, one measures the observable

$$P(C(\gamma)) = (\gamma - \sqrt{\gamma(1-\gamma)})\,\mathbb{1} + 2\sqrt{(\gamma(1-\gamma)}P(A) + |1 - 2\gamma|P(B)$$

which—in the Poincaré sphere representation of Figure 4—lies on the meridian circle between $P(A)$ and $P(B)$. The components of $P(C)$ in the direction of $P(A)$ and $P(A)$ are $2\sqrt{\gamma(1-\gamma)}$ and $|1 - 2\gamma|$, respectively. If one identifies these components with the parameters λ_A and λ_B, i.e., if one puts $\lambda_A = 2\sqrt{\gamma(1-\gamma)}$ and $\lambda_B = |1 - 2\gamma|$, then they determine the eigenvalues

$$E^{\pm}(A, \lambda_A) = 1/2(1 \pm \lambda_A) \text{ and } E^{\pm}(B, \lambda_B) = 1/2(1 \pm \lambda_B)$$

of the effects $E(A, \lambda_A)$ and $E(B, \lambda_B)$, respectively (Figure 4). Obviously the condition $\lambda_A^2 + \lambda_B^2 = 1$ of maximal coexistence is fulfilled here.

The visibility of the interference pattern depends on γ and is given here by $S(\gamma) = 2\sqrt{\gamma(1-\gamma)}$. It reaches its maximum for $\gamma = 1/2$ (when we do not know anything about the photon's path) and vanishes for $\gamma = 0$ and $\gamma = 1$ (when we know which way the photon came). On account of $S(\gamma) = \lambda_A$ the visibility is related to the reality degree $E^+(A, \lambda_A)$ of the interference observable $P(A)$ by

$$E^+(A, \lambda_A) = 1/2(1 + \lambda_A) = 1/2(1 + S(\gamma)) = 1/2(1 + 2\sqrt{\gamma(1-\gamma)})$$

In a similar way for the reality degree $E^+(B, \lambda_B)$ of the path observable $P(B)$ one obtains

$$E^+(B, \lambda_B) = 1/2(1 + \lambda_B) = 1/2(1 + |1 - 2\lambda|)$$

An experimental arrangement with the transparency γ of the second mirror and the visibility $S(\gamma)$ provides some information about the photon's particle and wave character. The corresponding formula for the deficiencies of information $H(\gamma, A)$ and $H(\gamma, B)$ about further measurements of $P(A)$ and $P(B)$ in an eigenstate $|C(\gamma)\rangle$ of $P(C(\gamma))$ are then given[2,4] by

$$H(\gamma, A) = -p(\gamma, A)\ln p(\gamma, A) - p(\gamma, \neg A)\ln p(\gamma, \neg A)$$
$$H(\gamma, B) = -p(\gamma, B)\ln p(\gamma, B) - p(\gamma, \neg B)\ln p(\gamma, \neg B)$$

where $p(\gamma, A)$ and $p(\gamma, B)$ are the probabilities of measuring $P(A)$ and $P(B)$, respectively, with the preparation $P(C(\gamma))$. These probabilities read

$$p(\gamma, A) = 1/2(1 + 2\sqrt{\gamma(1 - \gamma)}) = 1/2(1 + \lambda_A) = E^+(A, \lambda_A)$$
$$p(\gamma, B) = 1/2(1 + |1 - 2\gamma|) = 1/2(1 + \lambda_B) = E^+(B, \lambda_B)$$

and agree with the reality degrees of $P(A)$ and $P(B)$. For the deficiencies of information one thus obtains

$$H(\gamma, A) = -1/2(1 + S(\gamma))\ln 1/2(1 + S(\gamma)) - 1/2(1 - S(\gamma))\ln 1/2(1 - S(\gamma))$$
$$H(\gamma, B) = -\gamma\ln\gamma - (1 - \gamma)\ln(1 - \gamma)$$

Conversely, one can express the wave character and the particle character of a system in the state $|C(\gamma)\rangle$ by the (normalized) knowledge $W(\gamma) = 1 - H(\gamma, A)/\ln 2$ about the wave property and the knowledge $P(\gamma) = 1 - H(\gamma, B)/\ln 2$ about the particle property. The particle property ($P(\gamma)$ and the wave property $W(\gamma)$ are plotted against the visibility $S(\gamma)$ in Figure 5.

For the realization of the described experiment, 18 different transparency values γ for the beam recombiner $(BS)_2$ were used. The preparation $|\varphi(\delta)\rangle = 1/2(|B\rangle + \exp(i\delta)|\neg B\rangle)$ depends on the phase δ which was varied continuously. The intensities

$$I^+(\delta, \gamma) = |\langle\varphi(\delta)|C(\gamma)\rangle|^2, \; I^-(\delta, \gamma) = |\langle\varphi(\delta)|\neg C(\gamma)\rangle|^2$$

which are measured in the counters C_1 and C_2, respectively, are then given by

$$I^{\pm}(\delta, \gamma) = 1/2(1 \pm 2\sqrt{\gamma(1 - \gamma)}\cos\delta)$$

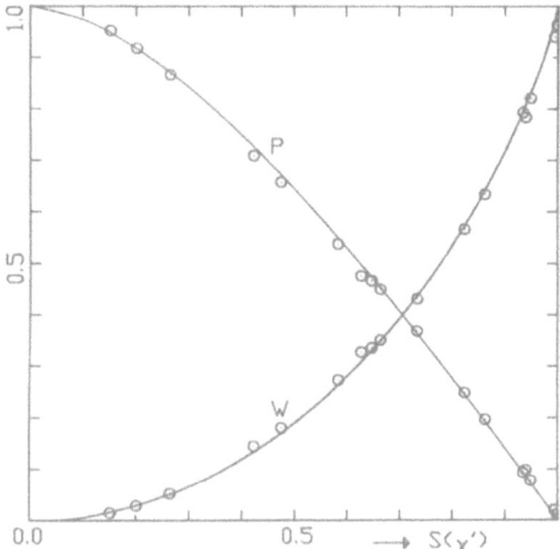

FIGURE 5. Particle property $P(\gamma)$ and wave property $W(\gamma)$ are plotted against the visibility $S(\gamma)$.

As an illustration of the experimental results, we consider two alternative situations which correspond to almost sharp measurements of the particle and the wave property (Figure 6; for further details see Mittelstaedt *et al.*[(4)]).

1. $\gamma_1 = 0.492$: here we have $W(\gamma_1) = 99.98\%$ wave property and $P(\gamma_1) = 0.02\%$ particle property. The measured visibility agrees with the theoretical value $S(\gamma_1) = 0.992$ and corresponds to a very significant interference pattern.
2. $\gamma_2 = 0.994$: here we have only $W(\gamma_2) = 1.8\%$ wave property and $P(\gamma_2) = 98.2\%$ particle property. Even in this extreme case the measured visibility $S(\gamma_2) = 0.145$ which agrees with the theoretical value is still sufficient for a significant interference pattern.

These results demonstrate in a very convincing way that the measurement of the observable $C(\gamma)$ provides an unsharp simultaneous knowledge of the photon's particle property as well as of its wave property. In accordance with previous theoretical expectations,[(2)] it follows that even if the path of the photon is known to a degree of 98.2% its wave character still manifests itself in a surprisingly significant interference pattern.

It should be mentioned that the present photon split-beam experiment is not yet a proper realization of the theory of the unsharp measuring process. It has been pointed out by Busch[(5)] that only the extended experiment of Figure 7 with four outcomes (in the counters C_1, C_2 and D_1, D_2) would provide a complete description

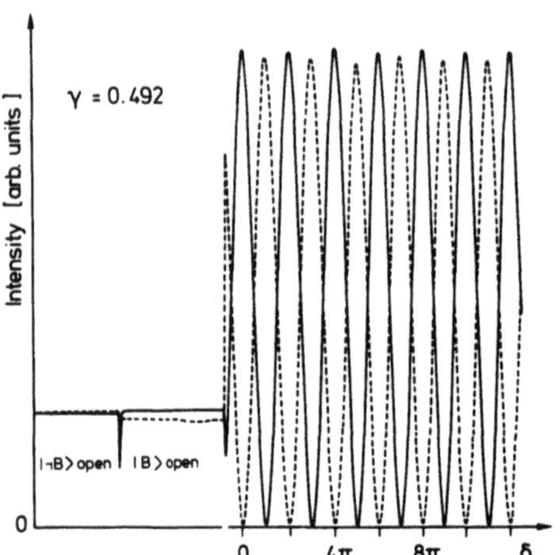

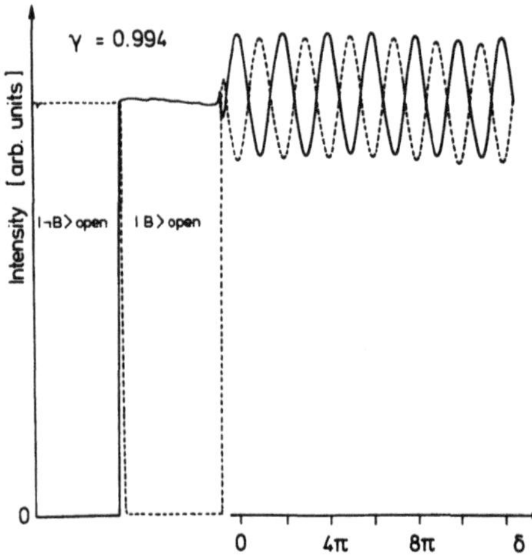

FIGURE 6. Experimental interference pattern for two transparencies. Upper part: $\gamma = 0.492$, i.e., 0.02% particle property, 99.98% wave property, and a visibility $S = 0.992$. Lower part: $\gamma = 0.994$, i.e., 98.2% particle property, 1.8% wave property, and a visibility $S = 0.145$.

of the unsharp measuring process. If the transparencies κ and λ of the additional mirrors are different, the counters C_1, C_2) and (D_1, D_2) can be considered as two distinct instruments for measuring unsharply the path $P(B)$ and the interference observable $P(A)$, respectively. Moreover, for $\kappa \neq \lambda$ the entire measuring instrument is statistically complete, i.e., the statistics of the four outcomes also provide complete information about the initial preparation $|\varphi(\beta, \delta)\rangle$ of the system which depends on the transparency β of the first beam splitter and of the phase δ.

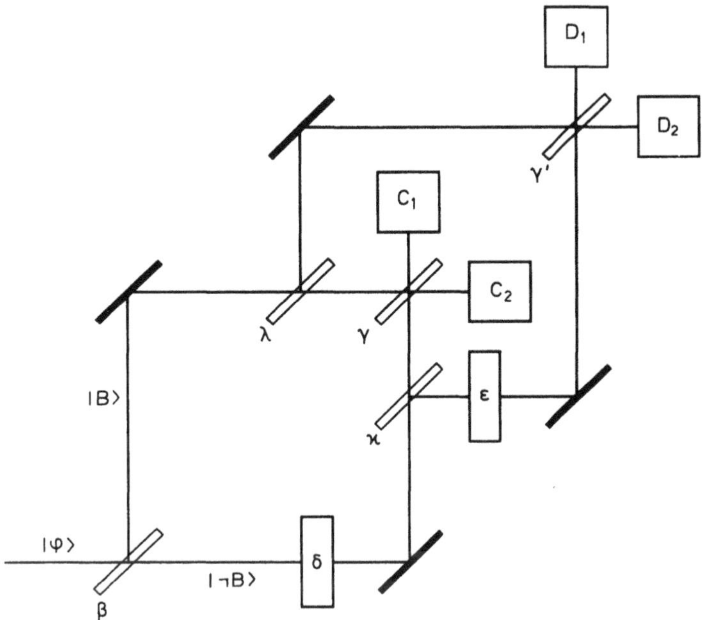

FIGURE 7. Extended version of the photon split-beam experiment of Figure 1 with two additional beam splitters (transparencies κ and λ) and two additional counters D_1 and D_2.

6. NEUTRON INTERFERENCE EXPERIMENTS

Another attempt to realize unsharp joint measurements of wave and particle properties are neutron interference experiments which were performed and discussed by Rauch and Summhammer,[6] Zeilinger,[7] Greenberger and YaSin,[8] and Rauch.[9] These experiments use a slightly different experimental setup (Figure 8). The incident beam is again split into beams B and $\neg B$ but in one of the beams, say B, an absorber is introduced which partially absorbs this beam. As an absorber of this kind, one could use another partly transparent mirror with transparency $\alpha < 1$. In this way the intensity of the path B can be arbitrarily reduced, whereas the $\neg B$ intensity remains unchanged.

Since in this experimental setup the normalization of the incident beam is destroyed by absorption, for the theoretical discussion we replace the absorbing mirror by a beam splitter with variable transparency β. The incoming beam can then be decomposed as

$$|\varphi(\beta, \delta)\rangle = \sqrt{1-\beta}|B\rangle + \sqrt{\beta}\exp(i\delta)|\neg B\rangle$$

If with this preparation the observable $P(A)$ of the interference pattern is measured, one obtains the probabilities

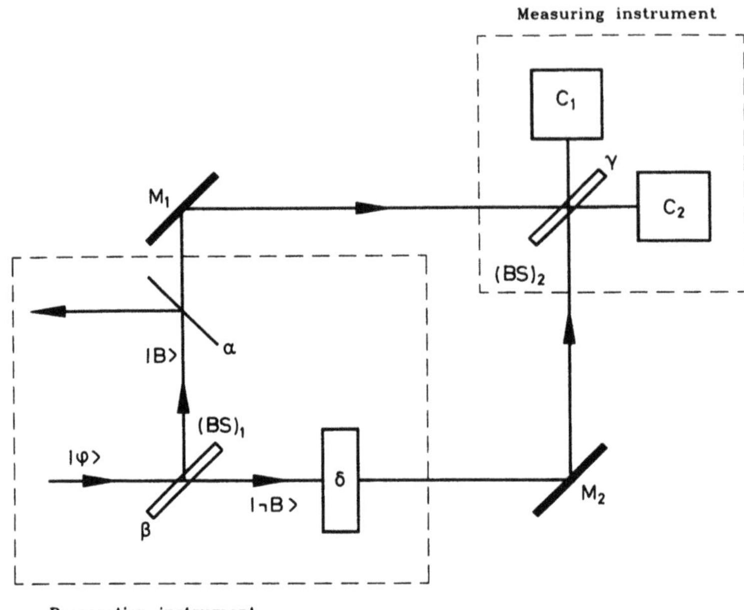

FIGURE 8. Modified photon split-beam experiment for unsharp joint measurements of the path and the interference pattern. Beam splitters $(BS)_1$ and $(BS)_2$ with transparencies β and γ. Absorber mirror with transparency α occurs at left.

$$p(\varphi, A) = 1/2(1 + \sqrt{\beta(1-\beta)}\cos\delta),$$
$$p(\varphi, \neg A) = 1/2(1 - 2\sqrt{\beta(1-\beta)}\cos\delta)$$

These expressions describe the probabilities for A and $\neg A$ if the system has the preparation $|\varphi(\beta)\rangle$. The ignorance about the path $P(B)$ is then given by

$$H(\varphi(\beta), B) = -(1-\beta)\ln(1-\beta) - \beta\ln\beta$$

In a situation with a preparation $|\varphi(\beta)\rangle$ which corresponds to 98.0% particle property $P(B) = 1 - H(\varphi(\beta), B)/\ln 2$, say, the remaining 2% wave property can be shown to still provide a significant interference pattern.[(8)] These intuitively surprising experiments confirm the theoretical predictions. However, they cannot be interpreted as unsharp joint measurements of the wave and particle properties A and B.

In order to discus the difference between the neutron interference experiment with ($\beta \neq \frac{1}{2}$, $\gamma = \frac{1}{2}$) and the unsharp joint measuring process with ($\beta = \frac{1}{2}$, $\gamma \neq \frac{1}{2}$) described above we must distinguish two different parts in the experimental arrangement of Figure 8: the preparation and the measuring instrument. In the neutron experiments the initial state $|\varphi(\beta, \delta)\rangle$ is prepared with $\beta \neq \frac{1}{2}$ (denoted here as "unsharp preparation") and lies on the meridian circle of the Poincaré sphere (Figure 4) between $P[\varphi]$ and $P(B)$. In the measuring part of the experimental setup one uses a beam recombiner with $\gamma = \frac{1}{2}$ and thus measures $P(A)$. The mixture which is obtained by this measuring process is given by

$$W_L(\varphi(\beta), A) = p(\varphi(\beta), A)P(A) + p(\varphi(\beta), \neg A)P(\neg A)$$

with eigenvalues $p(\varphi(\beta), A)$ and $p(\varphi(\beta), \neg A)$ which describe the relative frequencies of events in the counters C_1 and C_2. Hence, the final result of this measuring process is either in the state $P(A)$ or in the state $P(\neg A)$, but nothing is known about the particle property $P(B)$.

In contrast to this experiment, the unsharp joint measuring process starts with preparation $P[\varphi]$ and measures the observable $P(C(\gamma))$ with $\gamma \neq \frac{1}{2}$. The mixture which is obtained in this measuring process reads

$$W_L(\varphi, C(\gamma)) = p(\varphi, C(\gamma))P(C) + p(\gamma, \neg C(\gamma))P(\neg C(\gamma))$$

with eigenvalues $p(\varphi, C(\gamma))$ and $p(\varphi, \neg C(\gamma))$ which describe again the relative frequencies of events in the counters C_1 and C_2. However, the final result of the measuring process is the state $P(C)$ or the state $P(\neg C)$, both of which provide a simultaneous knowledge about the wave property $P(A)$ and the particle property $P(B)$.

It is obvious that for a preparation $P(\varphi(\beta))$ with $\beta = \gamma$ the probabilities

$$p(\varphi(\beta), A) = 1/2(1 + 2\sqrt{\beta(1 - \beta)}\cos\delta)$$

and

$$p(\varphi, C(\gamma)) = 1/2(1 + 2\sqrt{\gamma(1 - \gamma)}\cos\delta)$$

are equal. This means that the mixtures $W_L(\varphi(\beta), A)$ and $W_L(\varphi, C(\gamma))$ have the same eigenvalues and lead to the same counting rates in the counters C_1 and C_2. Hence, the two conceptually different situations of *unsharp preparation* ($\beta \neq 1/2$, $\gamma = 1/2$) and *unsharp measurement* ($\beta = 1/2$, $\gamma \neq 1/2$) seem to be indistinguishable from an experimental point of view. This is, however, not the case. The two mixed states $W_L(\varphi(\beta), A)$ and $W_L(\varphi, C(\gamma))$ can be distinguished by the expectation values of any observable which is not commensurable with $P(A)$ and $P(C(\gamma))$, e.g., by the observable $P(B)$. If we write for the probabilities

$$p^+ := p(\varphi, C(\gamma)) = p(\varphi(\beta), A)$$

$$p^{-} := p(\varphi, \neg C(\gamma)) = p(\varphi(\beta), \neg A)$$

one obtains

$$\mathrm{Tr}(W_L(\varphi(\beta), A)P(B)) = p^{+}1/2(1 + \cos\delta) + p^{-}1/2(1 - \cos\delta)$$

$$\mathrm{Tr}(W_L(\varphi, C(\gamma))P(B)) = p^{+}(1 - \gamma) + p^{-}\gamma$$

and these expectation values are obviously different. The two cases of unsharp preparation and unsharp measurement could also be distinguished by the extended split-beam experiment proposed by Busch.[(5)]

REFERENCES

1. N. BOHR, in: *Albert Einstein: Philosopher-Scientist* (P. A. SCHILPP, ed.), pp. 119–155, Open Court, La Salle, Ill. (1949).
2. K. WOOTTERS and W. H. ZUREK, *Phys. Rev. D* **19**, 473–484 (1979).
3. P. BUSCH, *Phys. Rev. D* **33**, 2253–2261 (1986).
4. P. MITTELSTAEDT, A. PRIEUR, and R. SCHIEDER, *Found. Phys.* **17**, 891–903 (1987).
5. P. BUSCH, *Found. Phys.* **17**, 905–937 (1987).
6. H. RAUCH and K. SUMMHAMMER, *Phys. Lett. A* **104**, 44–46 (1984).
7. A. ZEILINGER, *Physica B* **137**, 235–244 (1986).
8. D. M. GREENBERGER and A. YASIN, *Phys. Lett. A* **128**, 391–394 (1988).
9. H. RAUCH, in: *Proceedings of the 3rd International Symposium "Foundations of Quantum Mechanics,"* Physical Society of Japan, Tokyo (1989).

CHAPTER 11

Some Arguments against the Existence of de Broglie Waves

Wolfgang Mückenheim

1. DISTINGUISHING SCHRÖDINGER'S FROM DE BROGLIE'S WAVES

The degree of physical reality to be attributed to Schrödinger's wave function depends strongly on the degree of philosophical realism the "observer" is endowed with. The corresponding opinions span a wide range. The pure realist considers an electron as just a small particle, with the wave function being only a mathematical construct in his mind. The pure instrumentalist, on the other hand, refuses to talk about such notions as "reality" but accepts the wave function as the only entity corresponding to his idea of an electron (because a small particle would possess definite position and momentum). Of course, there are many epistemological niches between these extreme positions. However, if we dare to talk about an observer-independent reality (as we wish to do in this chapter), then we cannot accept Schrödinger's wave function as a part of this reality because its shape depends on our knowledge, changing (i.e., collapsing instantaneously) with every additional bit of information we accumulate.*

This feature of the wave function, which makes realists reluctant to add it to the inventory of their world, is just supplying that flexibility which is required to prevent quantum theory from internal contradictions and paradoxes.

*To give an example, the only wave function appropriate to describe a neutron emitted by a nucleus is a spherical wave. The additional knowledge (gained by a nondemolition measurement) that the neutron propagates close to a fixed direction leads to a plane wave. If further the decay time can be approximated (e.g., by detecting the recoiling nucleus) we have to construct a Gaussian wave packet. All these constructs describe one and the same neutron which has not suffered any physical interaction and which, after all, can be detected as a tiny particle with fixed mass and spin.

Wolfgang Mückenheim • Landshuter Allee 1, D-8903 Bobingen, Germany.

Wave–Particle Duality, edited by Franco Selleri. Plenum Press, New York, 1992.

This advantage is not shared by a different entity which, as a special structure of reality that can be described in terms of a formalism developed for waves, is a genuine part of reality. This structure has been known for a long time in the case of light, and it was predicted by de Broglie to apply to material particles too. Various experiments, starting with Young's double slit via Davisson–Germer's electron interferences to recent interference experiments involving neutrons and heavier particles, have proven the reality of these wave structures by presenting evidence in the form of distinct interference structures which could never be generated by ambiguous mathematical constructs existing merely in the brains of the experimenters.

2. DE BROGLIE'S INITIAL CONCEPT OF DE BROGLIE WAVES

For more than 50 years, de Broglie's initial ideas have been discussed and modified, with the unfortunate result that, today, it is not at all clear what we should understand by de Broglie waves, also called matter waves, empty waves, pilot waves, or phase waves. It is not necessary for our purposes to exhaustively discuss all features contained in the original concept* but we need to clarify some essential points to supply a common basis for the subsequent discussion. In order to remain as authentic as possible, we quote a few original statements.

After pointing out that there is a significant difference to Schrödinger's statistical wave function, de Broglie states that his wave "has a very low amplitude and does not carry energy, at least not in a noticeable manner. The particle is a very small zone of highly concentrated energy incorporated in the wave, in which it constitutes a sort of generally mobile singularity."[(2)] "Only its *phase*, related directly to the motion of the particle, seemed to me of fundamental significance. . . ."[(3)] ". . . we think that in wave mechanics one must introduce precise hidden parameters, which are the positions and velocities of the corpuscles. . . ."[(4)] "On conçoit alors l'onde continue comme guidant le mouvement de la particule. C'est une onde pilote."†[(5)] The reality attached by de Broglie to his wave becomes particularly evident by the following gedankenexperiment, set up to derive Bohr's postulate. "Passons maintenant au cas d'un electron décrivant d'une vitesse uniforme sensiblement inférieur à c une trajectoire fermée. Au temps $t = 0$, le mobile est en un point O. L'onde fictive associée, partant alors de O et décrivant toute la trajectoire avec la vitesse c/β, rattrape l'électron au temps τ en un point O′ tel que $OO' = \beta c\tau$."‡[(6)] There is scarcely any statement which more

*The interested reader is referred to a brief but competent review by de Broglie's former assistant G. Lochak.[(1)]

†The continuous wave is considered as guiding the motion of the particle. It is a pilot wave.

‡Let us consider now the case of an electron which describes a closed loop at a constant velocity significantly lower than c. At time $t = 0$ the electron is at point O. The fictitious wave associated with the electron starts at O and describes the same loop with the velocity c/β, overtaking the electron at time τ at a point O′ with $OO' = \beta c\tau$.

eloquently can underline the reality of this wave, propagating with phase velocity c/β, overtaking the particle and interacting with it.

Hence, we have to deal with a real wave without (measurable) energy or momentum, which is connected to the particle by

$$E = mc^2 = h\nu \tag{1}$$

$$p = mv = h/\lambda \tag{2}$$

with E and p energy and momentum of the particle, respectively. The particle's velocity (in free space)

$$v = dE/dp = d\sqrt{(m_0c^2)^2 + (pc)^2}/dp = p/m \tag{3}$$

is equal to the wave's group velocity

$$v = d\nu/d(1/\lambda) \tag{4}$$

The phase velocity, given by

$$V = \lambda\nu = c^2/v \tag{5}$$

is always larger than the speed of light for particles with nonvanishing rest mass m_0.

As mentioned above, there is overwhelming experimental evidence showing that certain parts of reality behave in a way that can be described by assuming the existence of de Broglie waves. The big question is, does it really make sense to assume the existence of such waves? And, if so, what is the oscillating medium? In particular, is this medium present everywhere and always (similar to the pre-relativistic ether) or does it emerge and vanish out of nothing into nothing (like the visual impression of a vapor trail), being created continuously by every particle?

3. DISCUSSION OF MODELS

If we prefer to stick to the opinion that reality consists of small particles which are guided by waves, we seem to have only the two models mentioned above. (This is dictated by the tertium non datur too: Either the undulatory medium is a permanent one or it is not.)

Let us for a moment assume the latter case. Then the medium of de Broglie waves has to be void of energy, because it is continuously created by particles which do not lose energy. The idea of something possessing nothing but "reality" but nevertheless influencing a particle by "guiding" it, is very hard to accept. Moreover, applying it to de Broglie's derivation of Bohr's postulate, we are forced to assume a wave-front propagation (or propagation of any part of the wave) with superluminal velocity V. If, further, the empty wave created by this process can influence a particle's motion or act in any other physically detectable way, this entails the possibility of sending signals with superluminal velocity. Hence, there would be a contradiction to special relativity. (If the wave cannot act in any physically detectable way, we need not further talk about it at all.) This verdict is strong enough to drop the "vapor trail" model. Consequently, the negative result of an experiment to detect the empty wave of a bunch of photons,[7] though not conclusive,[8,9] agrees with our conclusion.

The alternative to the vapor trail model is that of a permanent medium, i.e., the ether. But regardless of the problems which that notion raises in connection with questions concerning the isotropy of space independent of the state of the particle's motion, the same argument as given above holds also in this case. If a wave can "overtake" a particle in circular motion by propagating with superluminal velocity, and if this wave can act in a physical way, then we have the same contradiction with special relativity.

Hence, we can conclude that de Broglie's example,[6] fails. In order to save the existence of de Broglie waves, one can take the view[8] that the Fourier components form a wave packet such that outside a small bounded region the components cancel to zero. But it is easy to show that this view is in contradiction with de Broglie's initial concept. Let us consider a single particle. As quoted above, this is, for de Broglie, *incorporated* in the wave[2] and has *precise* position and velocity.[4] Hence, from Eqs. (1) and (2) we can deduce *precise* wave parameters λ and ν, leading to only *one* Fourier component. No superposition is possible. If, in order to remove this problem, the quantum mechanical uncertainty of energy and momentum is introduced, we end with Schrödinger's wave packet and de Broglie's initial concept is destroyed.

Although a single particle can be prepared and handled experimentally, we need not restrict our argument to this case but we can extend it to a bunch of several particles. Due to the independent existence of the de Broglie wave of each particle, it can by no means be expected that outside of the bunch these waves cancel each other. Should this be the case for a special bunch of particles (due to their energies, velocities, and relative locations), this must be considered an accidental and most improbable event. Likewise, there could be a bunch with parameters such that the amplitude of the common wave has the maximum possible amplitude outside of the narrow location containing the particles.

Only the Fourier components of a Schrödinger wave packet always will cancel outside of this region because they are *chosen* according to this requirement.

4. CONCLUSION

According to the discussion above it will be very difficult if not impossible to maintain de Broglie's picture of the independent existence of wave and particle.

On the other hand, there is so much experimental evidence for some wavelike structure of reality (as well as for its particlelike structure) that we cannot avoid to accept the "dualistic" view: Reality consists of entities inaccessible to our present means of reception which, according to our questions, answer in a way that can be interpreted as and be dealt with mathematical tools developed for waves or particles. But we cannot expect to grasp the complete reality by either of these notions.

REFERENCES

1. G. Lochak, in: *The Wave–Particle Dualism* (S. Diner, D. Fargue, G. Lochak, and F. Selleri, eds.), pp. 1–25, Reidel, Dordrecht (1984).
2. L. de Broglie and J. Andrade e Silva, *Phys. Rev.* **172**, 1284–1285 (1968).
3. L. de Broglie, *Non-linear Wave Mechanics*, Elsevier, Amsterdam (1960), Preface.
4. L. de Broglie, G. Lochak, J. A. Beswick, and J. Vassalo-Pereira, *Found. Phys.* **6**, 3–14 (1976).
5. L. de Broglie, *J. Phys. Radium* **8**, 225–241 (1927).
6. L. de Broglie, *C. R.* **177**, 507–510 (1923).
7. W. Mückenheim, P. Lokai, and B. Burghardt, *Phys. Lett. A* **127**, 387–390 (1988).
8. F. Selleri, *Phys. Lett. A* **132**, 72–74 (1988).
9. W. Mückenheim, *Phys. Lett. A* **132**, 75–76 (1988).

CHAPTER 12

On the "Completeness" of Quantum Mechanics

Thomas E. Phipps, Jr.

1. BACKGROUND

No study of the wave–particle dualism would be complete without examination of both the necessity and sufficiency of the mathematical-descriptive formalism that gives rise to it. Concerning sufficiency of the existing formalism, the issue of "completeness" of quantum mechanics as a physical theory was raised most poignantly by Einstein.[1] This matter is usually treated in connection with the Einstein–Podolsky–Rosen (EPR) paradox,[2,3] in the context of proposed "hidden variable" modifications or enhancements. The customary exposition then proceeds to Bell's theorem[4] and its modern developments, both theoretical and experimental[5]—the impression being created that there is a sort of championship-of-the-world fight in progress between clearly identified opponents: in one corner the recognized title-holder, "quantum mechanics," in the other a sequence of all possible (in general more generously parametrized) challengers to quantum mechanics.

As it happens, the assumption of the existence of such well-defined battle lines is a possibly fatal oversimplification. The conceptual fallacy in pitting quantum mechanics against all comers is simply that "quantum mechanics"—or any other product of theoretical physics—is not *in principle* a uniquely defined conceptual entity. Here we have in mind not the many possible versions or "interpretations" of accepted formalism, but substantive variants of that formalism. No theory is defined for purposes of physical description—meaning for purposes of acquiring observational support—except within a penumbra or con-

Thomas E. Phipps, Jr. • 908 South Busey Avenue, Urbana, Illinois 61801, USA.

Wave–Particle Duality, edited by Franco Selleri. Plenum Press, New York, 1992.

gruence of its class of what is termed "covering theories." Every experiment that upholds quantum mechanics upholds also all covering theories of quantum mechanics; for the definition of a covering theory is *any more richly parametrized theory that reduces identically to the "covered" theory for particular fixed values of the extra parameters.*

Any generalization of quantum mechanics that is a covering theory of quantum mechanics must by this definition inherit all the problems of quantum mechanics. Thus, it must encounter difficulties in resolving the EPR paradox to just the extent that ordinary quantum mechanics encounters such difficulties. No matter how richly parametrized they may be, such theories include quantum mechanics and thus include all of that theory's problems of mating to however-defined "physical reality." If we confine attention to covering theories, then in the battle between quantum mechanics and its opponents . . . which is which? (In the words of Pogo, "We have met the enemy and he is us.")

Since one seems thus to be "licked at the start" in any attempt to generalize quantum mechanics via the covering-theory approach, why bring up the subject? The answer is that one is licked only in respect to the answering of certain questions, such as those raised by EPR, and it may be in some sense (knowable only by hindsight from the vantage point of future history of science) that these are the *wrong questions*. In science the questions asked are all-important and nature provides no signposts pointing to the "right questions" for any particular era. The questions to which EPR leads—including metaphysical or ontological ones tending toward a definition of "reality"—may well be the wrong ones.

Perhaps the right ones for our time are more along the line of: *Why does our cherished mechanics of quanta fail to provide a specific dynamics of "nuclear" or "elementary" sub-atomic-scale quanta? Why is it parametrically incapable of describing Einsteinian "point events," not to mention "quantum jumps?" Why does it offer no distinction between the unique facts of history and the ensemble of possible futures? Why should the principle of relativity of physical size (which holds throughout the vast Newtonian range of sizes) suddenly fail at the threshold of subatomic scales?* Such questions lie clearly within the province of the physicist; whereas EPR (at least in its original Einsteinian form) can rapidly lead into territory whose ownership the philosopher may legitimately contest. The plainly "physical" questions just mentioned typify those upon which a study of covering theories of quantum mechanics casts a starkly revealing light.

There is a further, manifestly crucial, aspect of covering theories: invariant covering theories provide a royal road to "new physics," insofar as their extra parameters offer fresh descriptive possibilities. As the reader may have inferred, this chapter will be addressed mainly to covering theories of quantum mechanics—in particular to one most attractive candidate, a "top contender" that seems distinguished from the rest by its simplicity. Some of the implications of this covering theory for physical description will be sketched, and a potentiality will be demonstrated for vastly extending the mechanical descriptive purview. But it must be recognized at its outset that such a study can lead to no "resolution" of the wave–particle dualism. For that dualism, in precise analogy with EPR, poses questions that—though they inspire significant new experiments—may point less toward new physics than toward the clarification and mental integration of old physics.

In fact it is the writer's prejudice that too much attention to dualism distracts from perceiving and enhancing the *unity of mechanics*, which deserves a prominent place among the goals of physical theory. Concerning the *new physics* just mentioned, all experiments are in themselves ambiguous and offer physical insight only in conjunction with theory—so that any experiments, whether or not so intended, may in hindsight be perceived as the precursors of new physics. Humility is in order, for we still play among the pebbles bordering Newton's ocean, and despite much pebble-polishing and knowledge-squirreling have advanced farther in hubris than in wisdom.

2. COVERING THEORIES: AN EXAMPLE FROM ELECTROMAGNETISM

The subject of "completeness" of physical theory in general being intimately bound up with the status of covering theories, it may be instructive to digress briefly from our main subject of mechanics into the neighboring field of electromagnetism—which affords a singularly elegant, though little-known, example of the significance of the invariant covering theory.

In brief, Maxwell's electromagnetism rests upon field equations that are *not invariant* under any known coordinate transformation. This simple statement of fact invites the rebuttal that Maxwell's equations are covariant under Lorentz transformations and that *covariance is "just as good" as invariance*. To this latter contention the obvious response is, "How do you know until you've tried?" That is, the majority of today's physicists have imbibed covariant formalism from their cradle, and have never so much as sampled the flavor of a truly invariant electromagnetic formalism. Having been raised on *ersatz*, how can they render a judgment on *echt*?

The question of what electromagnetic "invariance" means cannot be separated from that of precisely identifying the invariants of kinematics. The subject of higher-order or "exact" description—which has been treated elsewhere[6,7]—lies outside the present purview. Fortunately, at first order in (v/c), Einsteinian and Newtonian identifications of the kinematic invariants are in close enough accord that we can proceed to illustrate here the invariant covering theory idea and physical role without concern about the description of very high-speed motions.

Confining ourselves then to first-order considerations, we invoke history by noting that Heinrich Hertz, the experimentalist who assured Maxwell's fame by confirming the existence of electromagnetic waves, was also a powerful theorist who published[8] a form of Maxwell's equations that was rigorously invariant under

Galilean (inertial) coordinate transformations. To accomplish this, Hertz banished "spacetime symmetry" by replacing Maxwell's partial time derivatives $\partial/\partial t$, wherever they occurred in the free-space field equations, with total derivatives,

$$d/dt = \partial/\partial t + \mathbf{v}\cdot\mathbf{\nabla}$$

The resulting modified field equations can be shown(7) to be Galilean invariant; in fact, they are readily expressed in manifestly invariant form,(7) such that each symbol in the field equations transforms invariantly under inertial transformations. Thus, there is no occasion to invoke covariance—which in any case could not obtain because of loss of spacetime symmetry (the partial space derivatives appearing in the Hertz equations being not symmetrical with the total time derivatives).

The formal maneuver just described introduces into the field equations a new velocity-dimensioned parameter "**v**" not present in Maxwell's version of electromagnetism. Hence, the Hertz theory is more richly parametrized than the Maxwell theory. In fact, we see that Hertz's is a covering theory of Maxwell's—because, on assigning to $\mathbf{v} = (v_x, v_y, v_z)$ the fixed numerical values (0,0,0), Hertz's equations reduce identically to Maxwell's (in view of $d/dt \to \partial/\partial t$). The fact of noninvariance of Maxwell's theory under any coordinate transformation, taken with the fact of Galilean invariance of Hertz's theory, means that Hertz's equations constitute an *invariant covering theory* of Maxwell's noninvariant theory.

Why has the reader never heard about this? Since there is a lesson about physics (sociology of) to be learned, it is worth a brief further historical detour to answer this. First, Hertz used an archaic (nonvector) notation that concealed the simplicity of his total time derivative modification. Second, he gave no proof of "invariance" but simply asserted it. (A modern commentator,(9) on beholding this assertion, failed to check the mathematics but blandly remarked that Hertz must have meant "covariance." Not so; he meant what he said, but took too much for granted about his readers' intelligence. In fact, he could not have meant "covariance" because his equations lack spacetime symmetry.)

Third, Hertz made a genuine mistake on the side of physical modeling or interpretation (the soft underbelly or Achilles' heel of all mathematical physics, as Hertz himself implied in his famous putdown, "Maxwell's theory is Maxwell's equations"): Hertz assumed that "**v**" measured an ether velocity, and further borrowed an old assumption due to G. Stokes that ether was 100% convected by material bodies. So he interpreted "**v**" as the observable velocity of such bodies in the laboratory, and consequently termed his theory an "electrodynamics of moving bodies."(8) His equations were thus interpreted(9) as predicting the production of a magnetic field by motion of a dielectric in the laboratory. This effect was looked for and not found.(10) Hence, Hertz's theory (viewed *not* as Hertz's equations but as these plus Hertz's fanciful interpretation) was discredited and discarded in favor of Maxwell–Lorentz theory.

The latter, being spacetime symmetrical, was used by Einstein–Minkowski as the basis for the famous hypothesis of the "metric nature of spacetime." Once that went into the curriculum of the global village there was no turning back from covariance to invariance, and Hertz's important mathematical discovery of an invariant covering theory was lost to history . . . until recently, when several investigators (including S. Kosowski of Poland, F. D. Tombe of Northern Ireland, and C. I. Mocanu of Romania, as well as the present author) independently rediscovered Hertz's invariant mathematics. Naturally, it has been necessary to find a better physical interpretation than Hertz's, and here opinions differ to this day.

This is not the place to go into the subject, which has been treated elsewhere,[7] but it may be remarked that by interpreting "**v**" as the velocity in the laboratory of a particular tangible object, the *field detector*, it is possible to avoid both the taint of metaphysics and the trap of false predictions into which Hertz unhappily stumbled. There is—as must be true of all covering theories—predictive agreement of the covering theory (Hertz's) with all observational evidence that supports the covered theory (Maxwell's). The special case in which covered and covering theories become identical is that in which field detector velocity vanishes, $\mathbf{v} = (0,0,0)$, which is precisely Maxwell's case of the *field detector at rest in the laboratory*. In summary: Maxwell got the mathematical physics "right in one laboratory," but not in all variously moving laboratories. That remained for Hertz, who yet tripped at the final step of converting mathematical physics into physics.

The issue of covariant versus invariant description may appear academic, pedantic, or even metaphysical. On the contrary, it relates directly and decidably to observable facts of experience: It is an issue of real physics. According to Einstein–Minkowski *all forces in nature must be expressible in covariant form*, whereas according to Hertz no such requirement can hold for electromagnetism. The discovery of a single example of a noncovariant force in nature would settle the issue in favor of Hertz's invariant covering theory and would disprove spacetime symmetry.

As it happens, there is growing evidence for the existence of noncovariant electromagnetic forces. The original Ampere law[11] of force between current elements, for example, obeyed Newton's third law and thus was noncovariant. Ampere's law, though never known to be violated,[11] was replaced in the favor of physicists by the Lorentz force law. The two laws differ by a quantity Q that is an exact differential.[12] Thus, their predictive differences (together with the Lorentz violation of Newton's third law) vanish when Q is integrated around any closed circuit external to the test current element. Yet if the circuit containing the test element is itself considered, the nontest portion of that circuit forms an open loop. In quantifying action-reaction it is the action upon the test element of the nontest portion—not of the total circuit—that must be considered. The integral of Q around a partial circuit need not vanish and the distinction between the two laws should be measurable. (In this theoretical conclusion we venture to differ from Maxwell.[(13)])

In fact, the experiment has been done[(14)] and has indicated that nature votes for Ampere—consequently for Hertz, for invariance, for Newton's third law as acting between current elements, and against homogeneous "spacetime" and its alleged metric nature or symmetry. Earlier evidence of Graneau[(11)] and others showing support for Ampere's law by exploding wire and railgun-buckling observations at very high pulsed currents is thus confirmed by low-current evidence valid under conditions precluding alternative explanations such as conductor melting.

It is clear from this electromagnetic example that issues of real physics, decidable by crucial experiment, devolve from such seemingly moot questions as covered versus covering theory, invariance versus covariance, etc. We now return to our main topic of mechanics.

3. A SIMPLE COVERING THEORY OF QUANTUM MECHANICS

The foregoing introduction via covering theories to the "completeness" of physical theory was concerned with what might be termed the "sufficiency" of physical description. Let us now address "necessity." In order to reason about this topic it is essential to proceed from *a priori* principles of some sort. Fortunately in the case of mechanics we do not start from a *tabula rasa*: A takeoff point for all other kinds of mechanics is provided by classical mechanics, which is well understood in all its aspects and may lay some claim to being the most broadly successful of all physical theories.

The most highly evolved form of classical particle mechanics is (arguably) the Hamilton–Jacobi form,

$$H = -\frac{\partial S}{\partial t}, \qquad H = H(q_j, p_j, t) \tag{1a}$$

$$p_j = \frac{\partial S}{\partial q_j} \tag{1b}$$

$$-P_j = \frac{\partial S}{\partial Q_j}, \qquad S = S(q_j, Q_j, t) \tag{1c}$$

which we seize on here because of its marked formal resemblance to the Schrödinger and Dirac equations. Two features of Eq. (1) are noteworthy: (1) The huge invariance group of the "contact" (or canonical) transformations under which these equations of motion remain unchanged. This group includes but far exceeds both linear and nonlinear coordinate transformations. (2) The tremendous range of sizes of physical systems to which the equations and their concomitant "point particle" idealization apply.

The second item suggests a physical principle capable of guiding the development of a covering theory of ordinary quantum mechanics—namely,

Principle of Relativity of Physical Size[7]*: The equations of motion of point-particle mechanics are expressible in a form that does not connote absolute largeness or smallness of the physical system described.*

The vital issue is: Over what range of sizes is the Newtonian idealization of the *mathematical point particle* physically permissible? We know that for many purposes it is quite acceptable to treat our sun's planets as mathematical points, and to do the same for baseballs, buckshot, and smoke particles. If Dirac[15] is to be believed, however, all this changes dramatically at the threshold of the atomic world; for, he asserts, *quantum mechanics is the discipline that sets an absolute size scale to the world*. In other words, the grand cavalcade of size relativity comes to a jarring halt right in the province of the chemist. That makes chemists very important people . . . absolutely.

So, anyway, *says* Dirac. Now let us see what he *does*[15]: Heedless of the doctrine of size absolutivity, he applies the idealization of the point particle to the smallest and lightest of the known massive particles, the electron. (Newton could hardly have done more . . . nor less.) Specifically, he cooks up a felicitous point operator form of the function H appearing in Eq. (1a), together with a formal operand Ψ, and thus extends into the smallest physical size range—far below the atomic—the point particle idealization embodied in the size relativity principle. In other words, he ignores size absolutivity and applies Newtonian size relativity . . . and this self-refutation is attended with astounding success! Since we ourselves must humbly decline to succeed better than success, let us follow this great man in doing as he does, not as he says—by applying without stint the size relativity principle.

Note that the principle, as stated above, speaks to "a form" of the mechanical equations of motion. It does not imply that *any* form will do . . . we have to look for a particular form. Fortunately, our task is such an easy one that it practically performs itself. Proceeding from Eq. (1)—since we wish to share Dirac's success in describing electrons—we know that we shall have to supply a formal operand Ψ_f at least to Eq. (1a). This means looking on the symbols of Eq. (1a) in general as operators acting toward the right; and since classically (1b) and (1c) are on an equal footing with (1a) it is natural to think of these additional equations as containing rightward-acting operators equally in need of operands. Being parsimonious, we do not part with operands easily, and so propose sharing the same operand:

$$H\Psi_f = -\frac{\partial}{\partial t}S\Psi_f \tag{2a}$$

$$p_j\Psi_f = \frac{\partial}{\partial q_j} S\Psi_f \tag{2b}$$

$$-P_j\Psi_f = \frac{\partial}{\partial Q_j} S\Psi_f \tag{2c}$$

Although this hypothesis has been arrived at here somewhat in the manner of doodling, it turns out to be quite a satisfactory "form" to represent *equations of motion for all mechanics*, invariant on all size scales in the sense of the size relativity principle. Indeed, we shall now show that not only is it invariant on all size scales of likely interest to physics, but it is a covering theory of all known forms of mechanics. Thus, we have again to deal with an invariant covering theory. Equation (2) is seen to possess three distinct classes of solution:

Class I. Ψ_f = constant. In this case the constant can be canceled from Eq. (2) and what remains is identically the Hamilton–Jacobi equations. Thus, Eq. (2) is a covering theory of classical mechanics, Eq. (1). That classical motions are included among the exact solutions of our postulated generalized equations of motion, Eq. (2), is a fact of profound significance for measurement theory. It alters the relationship of classical and quantum physics, since ordinary quantum mechanics treats classical motion states always as mere approximations to "exact" superpositions of quantum states, never as best-available descriptors in their own right. In dealing with Eq. (2) we have to get used to the idea that *all mathematical solutions are for physical descriptive purposes approximations*. What physical theory offers in any given problem is merely a modest choice between poor and less poor approximation, not a choice between drab *approximate* and gorgeous *exact*. (We describe *things*, and descriptions are not things. Ergo descriptions are never exact, for only things can be exactly things.)

Class II. S = constant = $\hbar/i$. In this case, Eq. (2) reduces to

$$H\Psi_f = -\frac{\hbar}{i}\frac{\partial}{\partial t}\Psi_f \tag{3a}$$

$$p_j\Psi_f = \frac{\hbar}{i}\frac{\partial}{\partial q_j}\Psi_f \tag{3b}$$

$$-P_j\Psi_f = \frac{\hbar}{i}\frac{\partial}{\partial Q_j}\Psi_f \tag{3c}$$

The value $\hbar/i$ of the constant is chosen to agree with experiment. Equations (3a), (3b) are of the form of the Schrödinger–Dirac equations. Equation (3c) is a stranger involving extra parameters (Q_j, P_j) that classically are constants of the motion and that retain the character of constants regardless of the class of solution considered. Obviously, Class II solutions describe quantum (atomic) states of motion. By inspection we see that Eq. (3c) has the solution

$$\Psi_f = \Phi(q_i, t) \exp\left[-\frac{i}{\hbar} \Sigma_j P_j Q_j \right] . \tag{4}$$

The exponential multiplier appearing here is just a constant phase factor, in general absorbed into the wave function normalization factor. Thus, $|\Psi_f|^2 = |\Phi|^2$, Φ being just the Schrödinger or Dirac wave function. Since all the physical predictions of quantum mechanics depend on sums or integrals of mean-value products such as $\Psi_f^* A \Psi_f = \Phi^* A \Phi$, from which the constant phase factor cancels, it is clear that the class of observational agreements of Eq. (3) coincides with that of ordinary quantum mechanics, viz.,

$$H\Phi = -\frac{\hbar}{i}\frac{\partial}{\partial t}\Phi \tag{5a}$$

$$p_j\Phi = \frac{\hbar}{i}\frac{\partial}{\partial q_j}\Phi \tag{5b}$$

This results from canceling the constant phase factors from Eqs. (2a), (2b).

So, Eq. (2), which we saw above is a covering theory of classical mechanics, is now shown to be a covering theory of ordinary quantum mechanics as well. This puts us on familiar ground. Nevertheless, there is something new: The "constant" phase factor of Eq. (4) contains not universal constants but *dynamical constants* (Q,P), which in general "jump" to new values when the descriptive problem (system Hamiltonian) changes . . . thus furnishing an entirely new mechanism of discontinuity and a way of severing that "von Neumann chain" of phase connections which in the Copenhagen view joins all physical descriptive problems into one endless, seamless whole. In short, we acquire a *post facto* way of describing happenings, point events, or what used to be called "quantum jumps"—*without*, however, acquiring any new predictive capabilities. Thus, the transition from covered to covering theory has profound implications for quantum measurement theory. These have been examined elsewhere[7,16] and need not detain us here.

In sum: As a scheme for calculating observable quantities, quantum mechanics is altered not a bit by substitution of the covering theory, Eq. (2). But the richer parametrization of the latter has a great impact on measurement theory. For example: (1) It permits us to view *history as a fact*, not as the sort of statistical ("class of facts") ensemble appropriate to prediction. (2) Through this nontrivial distinction between prediction and retrodiction, it gives substance at the quantum level to "time's arrow." (3) It secures the logical sufficiency of mechanics without need or call for extra axiomatics (e.g., a projection postulate).

Class III. $S \neq$ constant, $\Psi_f \neq$ constant. In this case the mathematical character of the problem changes. Equation (2c) is no longer a "fifth wheel," but becomes a "second equation in the second unknown." That is, in each of the other classes of solution there is only one unknown function, S or Ψ_f. But here both of these functions are unknown and have to be solved for simultaneously. An example of such simultaneous solution has been given.(7,17) It appears to describe nuclear-scale stationary bound states of electron–positrons in terms of states of *imaginary particle momentum* but real mass–energy. Thus, the ability of a covering theory to lead to "new physics" (right or wrong) is reaffirmed. Further evolutionary developments of the theory are needed. Here we confine attention to noting some attributes of the Class III formalism.

The most notable formal features of the Class III solutions are that (a) the classical-analog (CA) operators become in general non-Hermitean and (b) the Heisenberg postulate is violated. (It is for the latter reason that electron–positrons, as noted above, can exist on the nuclear size scale.) The Heisenberg postulate is generalized to

$$(p_k q_j - q_j p_k)\Psi_f = \left[\left(\frac{\partial}{\partial q_k}\right) S q_j - q_j \left(\frac{\partial}{\partial q_k}\right) S\right]\Psi_f = S\partial_{jk}\Psi_f \tag{6}$$

This leads to a tripartite interpretation of the quantity S: On the classical (Class I) scale S is Hamilton's principal function; on the atomic (Class II) scale S is Heisenberg's constant $(\hbar/i)$; and on the nuclear (Class III) scale S measures the degree of departure of the commutator of (q,p) mechanical variables from the Heisenberg value. That is, the commutator is no longer universally constant but becomes a space-time variable function subject to the boundary condition that quantum mechanics be recovered far from the scene of nuclear (subatomic) description; i.e., $S(r) \to \hbar/i$ as $r \to \infty$.

If we split off a real function s by the definition $S = (\hbar/i)s$, we see from a well-known theorem (viz., that the product of two noncommuting Hermitean operators is non-Hermitean) that a CA operator such as H is non-Hermitean; for from Eq. (2a) we have

$$H\Psi_f = \frac{\hbar}{i}\frac{\partial}{\partial t} s\Psi_f \tag{7}$$

which exhibits H as the product of a Hermitean operator $-(\hbar/i)\partial/\partial t$ and a Hermitean (real) operator s. The real property is imposed upon s by the physical requirement that the transformations

$$\begin{aligned} \mathcal{H} &= Hs^{-1} \\ \Psi &= s\Psi_f \end{aligned} \tag{8}$$

render the resulting time-conjugate operator $\mathcal{H}$ Hermitean—for these transformations, applied to eq. (2a), reduce it to

$$\mathcal{H}\Psi = -\frac{\hbar}{i}\frac{\partial}{\partial t}\Psi \tag{9}$$

which identifies $\mathcal{H}$ as the operator conjugate to time.

It would in any formalism be disastrous to have the time-conjugate operator turn out to be non-Hermitean . . . and that misfortune is avoided by the simple but crucial formal transformations (8). To confirm the Hermitean property of $\mathcal{H}$, for example in the case of a nonrelativistic one-body problem, we observe that

$$\begin{aligned} \mathcal{H} &= Hs^{-1} = [(1/2m_0)\mathbf{p}\cdot\mathbf{p} + V]s^{-1} \\ &= -(\hbar^2/2m_0)\nabla s\cdot\nabla + Vs^{-1} \end{aligned} \tag{10}$$

which is readily seen to be Hermitean if and only if s is real. (A similar demonstration for the Dirac Hamiltonian is even more immediate.) Here the CA momentum operator $\mathbf{p}$ is seen from Eq. (2b) to be the non-Hermitean product of the Hermitean operator $(\hbar/i)\nabla$ and the real function s. A transformation analogous to Eq. (8) yields a Hermitean momentum,

$$\mathbf{P} = \mathbf{p}s^{-1} = \frac{\hbar}{i}\nabla \tag{11}$$

The transformations (8), by producing Hermitean operators, reduce the Class III formalism to mathematically familiar terms. But it must not be overlooked that the specific form of the Hamiltonian is affected by the transformation, and it is this *specific form* that contains all the physics. Thus, a new theory that will not in the least interest mathematicians may be of considerable interest to physicists.

4. ALTERNATIVE "NECESSITATIONS" OF THE COVERING THEORY, EQUATION (2)

An alternative general principle from which Eq. (2) may equally well be inferred is the following:

Principle of Correspondence Reversibility: The formal correspondence between classical and atomic-scale mechanics shall proceed with equal facility in either direction and shall in either case yield a complete mechanics.

Early in the history of quantum mechanics, Pauli[18] advanced a claim that one could start with the ordinary quantum mechanical equations of motion, Eq. (5), and recover the equations of motion of classical mechanics as a limiting case.

This claim was important at the time for confirming the legitimacy of the new form of mechanics, as it improved its connection with known successful mechanics by making correspondence a two-way street. Unfortunately, Pauli's claim is spurious, as the only links he established were between Eqs. (5a), (5b) on the one hand and Eqs. (1a), (1b) on the other. No mention was made of Eq. (1c), without which no Newtonian mechanics is possible. Recently it was recognized[19] that what one gets by Pauli's route is not Newtonian particle mechanics but Liouville-type statistical mechanics. That is, the formal absence of the constant parameters (Q_j,P_j)—known classically as the "new canonical variables"—from ordinary quantum mechanics deprives that discipline of the specificity needed to describe point events and leaves it with only the capacity to describe (statistical) classes of events. This is a shared disability of quantum mechanics and classical statistical mechanics, as contrasted with classical point particle mechanics. The cause, a parametric deficiency, is likewise shared.

If the correspondence reversibility principle is imposed as mandatory, it is apparent that ordinary quantum mechanics is in violation and must be replaced by some other theory. The need to enrich parametrization on the quantum side in order to improve specificity for *post facto* point event description—and to *prevent any change in number of parameters* during the correspondence transition in either direction—then suggests the covering theory approach, and one quickly gets to Eq. (2) by fairly obvious inferences. There is no need to elaborate here. Equation (2) offers a true point particle descriptive mechanics for all physical size scales.

Finally, setting aside all "principles," there is a direct empirical route to something like Eq. (2) via the reader's personal knowledge. Most physicists have had the experience of observing in a darkened room the scintillations, e.g., of a zinc sulfide phosphor. These are flashes of light that have the appearance of originating at definite times from pinpoint locations. It may be supposed that these attributes of localization would persist if the phosphor were examined under varying magnifications up to the most powerful. We seem thus to experience (and to retain in "historical" memory) personal detection of a specific space-time constellation of point events. Yet since the equations of motion of ordinary quantum mechanics, Eq. (5), lack parameters capable even of after-the-fact description of such a particular event constellation, the Copenhagen interpretation assures us that quantum-level "historical" retrodiction is as futile as prediction and that such an experience, as well as the memory of it, is consequently an illusion.

Physics was chartered to describe human experience, not to denigrate that experience as illusion. The illusion, according to the argument of this chapter, is that Eq. (5) forms the basis for a "complete" mechanical description of nature. The conservative approach to formal completion involves exploiting the wealth of possibilities for parametric enrichment—while "holding fast to the good"—offered by invariant covering theories.

REFERENCES

1. A. Einstein, in: *Albert Einstein: Philosopher-Scientist* (P. A. Schilpp, ed.), Library of Living Philosophers, Evanston, Ill. (1949).
2. A. Einstein, B. Podolsky, and N. Rosen, *Phys. Rev.* **47**, 777–780 (1935).
3. F. Selleri (ed.), *Quantum Mechanics versus Local Realism: The Einstein–Podolsky–Rosen Paradox*, Plenum Press, New York (1989).
4. J. S. Bell, *Physics* **1**, 195 (1964).
5. A. Aspect, *Phys. Rev. Lett.* **47**, 460 (1981); **49**, 1804 (1982).
6. T. E. Phipps, Jr., *Ann. Fond. Louis de Broglie* **8**, 325–344 (1983); **9**, 41–64 (1984).
7. T. E. Phipps, Jr., *Heretical Verities: Mathematical Themes in Physical Description*, Classic Non-fiction Library, Urbana, Ill. (1987).
8. H. R. Hertz, *Electric Waves*, Teubner, Leipzig (1892); Dover, New York (1962), last chapter.
9. A. I. Miller, *Albert Einstein's Special Theory of Relativity Emergence (1905) and Early Interpretation (1905–1911)*, Addison–Wesley, Reading, Mass. (1981), pp. 11–14.
10. A. Eichenwald, *Ann. Phys.* **11**, 1, 421 (1903).
11. P. Graneau, *Ampere–Neumann Electrodynamics of Metals*, Hadronic Press, Nonantum, Mass. (1985).
12. C. Christodoulides, *Am. J. Phys.* **56**, 357–362 (1988).
13. J. C. Maxwell, *A Treatise on Electricity and Magnetism*, Clarendon, Oxford (1891); Dover, New York (1954), p. 319.
14. T. E. Phipps and T. E. Phipps, Jr., *Phys. Lett. A* **146**, 6–14 (1990); also T. E. Phipps, Jr., *Phys. Essays* **3**, 198–206 (1990).
15. P. A. M. Dirac, *The Principles of Quantum Mechanics*, 3rd ed., Clarendon, Oxford (1947), pp. 3–4.
16. T. E. Phipps, Jr., *Phys. Essays* **1**, 20–23 (1988).
17. T. E. Phipps, Jr., *Phys. Rev.* **118**, 1653 (1960).
18. W. Pauli, *Handbuch der Physik*, 2nd ed., Springer, Berlin (1933), p. 241.
19. P. Ajanapon, *Am. J. Phys.* **55**, 159–163 (1987).

CHAPTER 13

Neutron Interferometric Tests of Quantum Mechanics

Helmut Rauch

Neutron interferometers based on wave-front and amplitude division have been developed in the past. Most experiments have been performed with the perfect crystal neutron interferometer which provides widely separated coherent beams enabling new experiments in the field of fundamental nuclear and solid-state physics. A nondispersive sample arrangement and the difference of stochastic and deterministic absorption have been investigated. The verification of the 4π-symmetry of spinors and of the quantum mechanical spin-superposition experiment on a macroscopic scale are typical examples of interferometry in spin space. These experiments were continued with two resonance coils in the beams, and the results showed that coherence persists, even if an energy exchange between the neutron and the resonator system occurs with certainty. A quantum beat effect was observed when slightly different resonance frequencies were applied to both beams. In this case, an extremely high energy sensitivity of 2.7×10^{-19} eV was achieved. This effect can be interpreted as a magnetic Josephson-effect analog. Phase echo systems, experiments with chopped beams and multiplate interferometry are discussed as examples for forthcoming experiments. All the results obtained to date are in agreement with the formalism of quantum mechanics but stimulate the discussion about the interpretation of this basic theory.

1. INTRODUCTION

Three different kinds of neutron interferometers have been tested in the past. The slit interferometer is based on wave-front division and provides long beam

Helmut Rauch • Atominstitut der Oesterreichischen Universitäten, A-1020 Wien, Austria.

Wave–Particle Duality, edited by Franco Selleri. Plenum Press, New York, 1992.

paths but only a very small beam separation.[1,2] The perfect crystal interferometer is based on amplitude division and is now most frequently used due to its wide beam separation and its universal availability for (1) fundamental, (2) nuclear, and solid-state physics research.[3,4] The interferometer based on grating diffraction is a recent development and has its main application for very slow neutrons.[5] A schematical comparison is shown in Figure 1. The perfect crystal interferometer provides the highest intensity and highest flexibility for beam handling.

In this chapter the development and the application of the perfect crystal interferometer are reviewed. The first successful test of such an interferometer was in 1974 at our small 250-kW TRIGA reactor in Vienna[3] (Figure 2).

The perfect crystal interferometer represents a macroscopic quantum device with characteristic dimensions of several centimeters. The basis for this kind of neutron interferometry is provided by the undisturbed arrangement of atoms in a monolithic perfect silicon crystal.[3,6] An incident beam is split coherently at the first crystal plate, reflected at the middle plate, and coherently superposed at the third plate (Figure 1). It follows immediately from general symmetry considerations that the wave functions in both beam paths, which compose the beam in the forward direction behind the interferometer, are equal ($\psi_0^{\mathrm{I}} = \psi_0^{\mathrm{II}}$), because they are transmitted–reflected–reflected (TRR) and reflected–reflected–transmitted (RRT), respectively. The system is based on Bragg diffraction from perfect crystals; therefore, the de Broglie wavelength of the neutrons is about 1.8 Å and their energy is about 0.025 eV.

The whole theoretical treatment of the diffraction process is based on the dynamical diffraction theory, which can also be found in the literature for the neutron case.[7–10] Inside the perfect crystal, two wave fields are excited when the incident beam fulfills the Bragg condition, one of them having its nodes at the position of the atoms and the other in between them. Therefore, their vectors are slightly different ($k_1 - k_2 \simeq 10^{-5}k_0$) and due to mutual interference processes, a rather complicated interference pattern is built up, which changes substantially over a characteristic length Δ_0—the so-called Pendellösung length, which is on the order of 50 μm for an ordinary silicon reflection. To preserve the interference properties over the length of the interferometer, the dimensions of the monolithic system have to be accurate on a scale comparable to this quantity. Therefore, the whole interferometer crystal has to be placed on a stable goniometer table under conditions avoiding temperature gradients and vibrations.

A phase shift between the two coherent beams can be produced by nuclear, magnetic, or gravitational interactions. In the first case, the phase shift is most easily calculated using the index of refraction[11,12]:

$$n = \frac{k_{\mathrm{in}}}{k_0} = 1 - \frac{\lambda^2 N}{2\pi}\sqrt{b_c^2 - \left(\frac{\sigma_r}{2\lambda}\right)^2} + i\frac{\sigma_r N\lambda}{4\pi} \tag{1}$$

(1) simplifies for weakly absorbing materials ($\sigma_r \to 0$) to

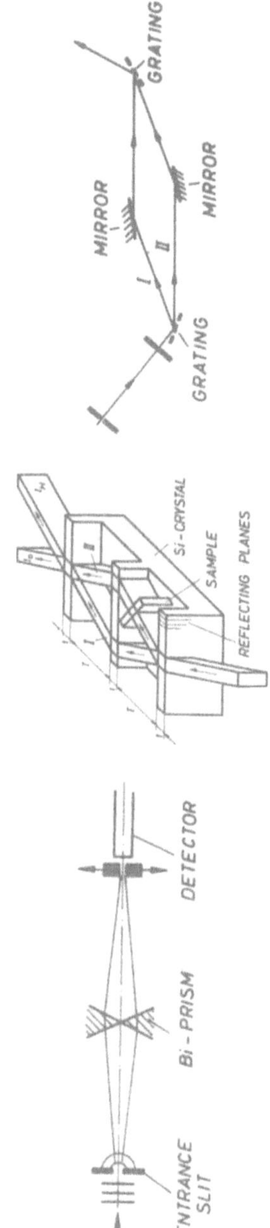

FIGURE 1. Scheme of a slit, a perfect crystal, and a grating interferometer.

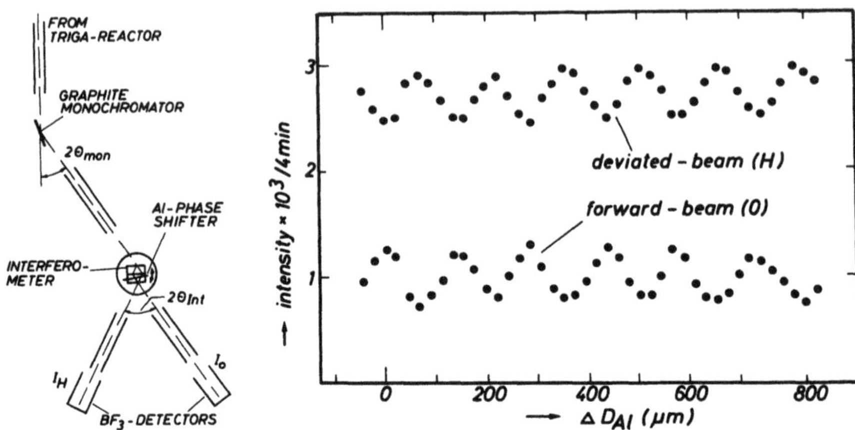

FIGURE 2. First observation of interference fringes with a perfect crystal interferometer.[3]

$$n = 1 - \lambda^2 \frac{Nb_c}{2\pi} \tag{2}$$

where b_c is the coherent scattering length and N is the particle density of the phase shifting material. As in ordinary light optics the change of the wave function is obtained as follows:

$$\psi \rightarrow \psi_0 e^{i(n-1)kD} = \psi_0 e^{-iNb_c\lambda D} = \psi_0 e^{i\chi} \tag{3}$$

Therefore, the intensity behind the interferometer is given by

$$I_0 \propto |\psi_0^{\mathrm{I}} + \psi_0^{\mathrm{II}}|^2 \propto (1 + \cos\chi) \tag{4}$$

The intensity of the beam in the deviated direction follows from particle conservation:

$$I_0 + I_{\mathrm{H}} = \text{const} \tag{5}$$

Thus, the intensities behind the interferometer vary as a function of the thickness D of the phase shifter, the particle density N, or the neutron wavelength λ.

Any experimental device deviates from the idealized assumptions made by the theory: the perfect crystal can have slight deviations from its perfectness, and its dimensions may vary slightly; the phase shifter contributes to imperfections by variations in its thickness and inhomogeneities; and even the neutron beam itself contributes to a deviation from the idealized situation because of its wavelength spread $\delta\lambda$. Therefore, the experimental interference patterns have to be described by a generalized relation

$$I \propto A + B\cos(\chi + \phi_0) \tag{6}$$

where A, B, and ϕ_0 are characteristic parameters of a certain setup. It should be mentioned, however, that the idealized behavior described by Eq. (4) can nearly be approached by a well-balanced setup.[13] The reduction of the contrast at high order results from the longitudinal coherence length which is determined by the wavelength spread of the neutron beam ($\Delta\chi_L = \lambda^2/\Delta\lambda$). This causes a change in the amplitude factor of Eq. (6) as ($B \rightarrow B\exp[-(\Delta\lambda/\lambda_0)^2\chi_0^2/2]$). The wavelength dependence of χ in Eq. (3) disappears when the surface of the sample is oriented parallel to the reflecting planes and the path length through the interferometer becomes $D_0/\sin\theta_B$ and, therefore, the phase shift $\chi = -2d_{hkl}Nb_cD_0$ becomes independent of the wavelength. In this case the damping at high interference orders due to the wavelength spread does not appear as in the standard position. Related results of a recent experiment where the interference pattern in the 256th interference order have been measured in the dispersive and the nondispersive sample position are shown in Figure 3.[14] The much higher visibility of the interferences in the nondispersive sample arrangement is clearly seen.

All of the results of interferometric measurements obtained to date can be explained well in terms of the wave picture of quantum mechanics and the complementarity principle of standard quantum mechanics. Nevertheless, one should bear in mind hat the neutron also carries well-defined particle properties, which have to be transferred through the interferometer. These properties are summarized in Table 1 together with a formulation in the wave picture. Both particle and wave properties are well established and therefore, neutrons seem to be a proper tool for testing quantum mechanics with massive particles, where the wave–particle dualism becomes obvious.

All neutron interferometric experiments pertain to the case of self-interference, where during a certain time interval, only one neutron is inside the interferometer, if at all. Usually, at that time the next neutron has not yet been born and is still contained in the uranium nuclei of the reactor fuel. Although there is no interaction between different neutrons, they have a certain common history within predetermined limits which are defined, e.g., by the neutron moderation process, by their movement along the neutron guide tubes, by the monochromator crystal, and by the special interferometer setup. Therefore, any real interferometer pattern contains single particle and ensemble properties together. In the following sections, typical experiments performed mainly by our group within the last 15 years will be presented.

2. STOCHASTIC VERSUS DETERMINISTIC ABSORPTION

A certain beam attenuation can be achieved either by a semitransparent material or by a proper chopper system. The transmission probability in the first

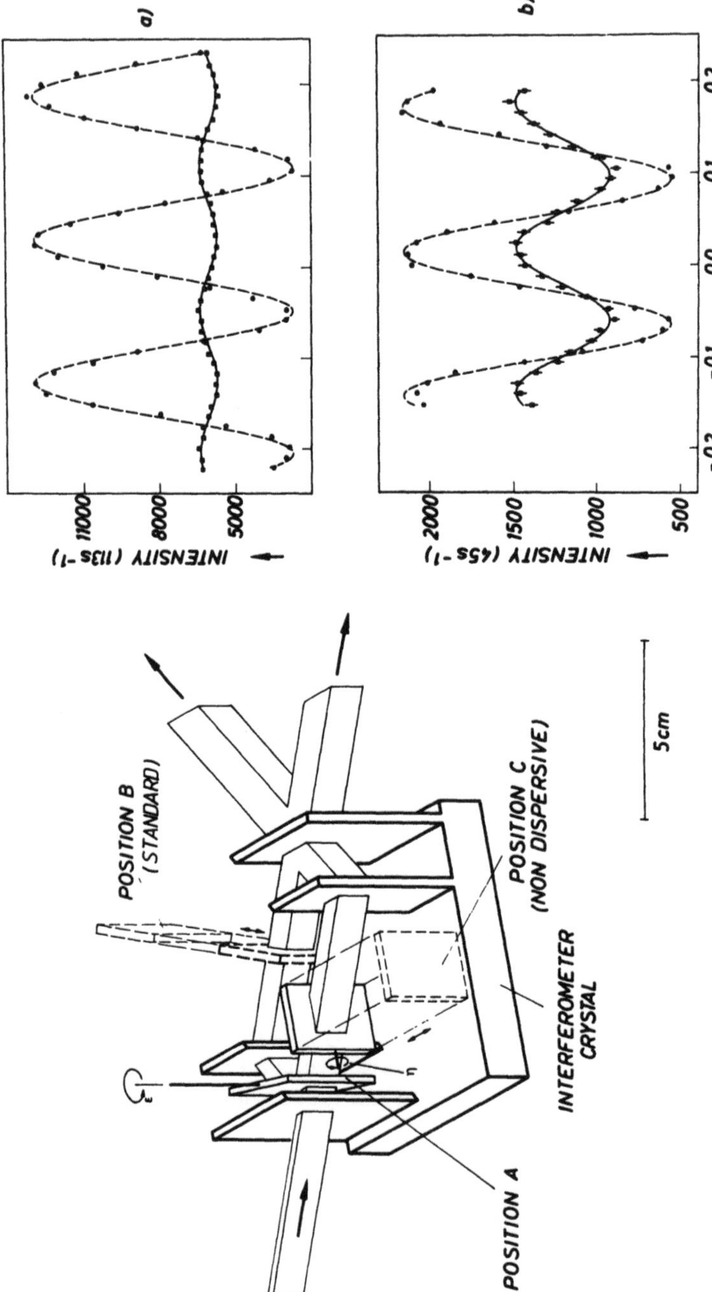

FIGURE 3. Interference pattern observed at high order (m = 256) with a dispersively (above) and a nondispersively arranged sample[14] (dashed lines correspond to measurements at low order).

TABLE I
Neutron Properties

Particle properties	
Mass	$m_0 = 1.6749286(10) \times 10^{-24}$g
Spin	$s = \frac{1}{2}\hbar$
Magnetic moment	$\mu = -1.91304275(45)\ \mu_K$
Lifetime	$\tau = 888.6(3.5)$ s
Electric charge	$q < 2.2 \times 10^{-20}\ e$
Electric dipole moment	$d < 4.8 < 10^{-25}\ e{\cdot}cm$
Electrical polarizability	$\alpha_e = 1.1^{+0.4}_{-0.6} \times 10^{-3}$ fm^{-3}
Confinement radius	$R = 0.7$ fm
Quark structure	$n = u{-}d{-}d$
Wave properties	
Compton wavelength	$\lambda_c = \dfrac{h}{mc} = 1.32 \times 10^{-13}$ cm
de Broglie wavelength	$\lambda_B = \dfrac{h}{mv} = 1.8 \times 10^{-8}$ cm*
Coherence length	$\lambda_c = \lambda^2/\Delta\lambda = 1 \times 10^{-6}$ cm*
Packet length	$\lambda_p = v{\cdot}\Delta t = 1 \times 10^{0}$ cm*
Decay length	$\lambda_d = v{\cdot}T_{1/2} = 2 \times 10^{8}$ cm*
Phase difference	$0 \leq \chi \leq 2\pi$

*Values belong to thermal neutrons (λ_B = 1.8 A, v = 2200 m/s).

case is defined by the absorption cross section σ_a of the material [$a = I/I_0 = \exp(-\sigma_a ND)$] and the change of the wave function is obtained directly from the complex index of refraction [Eq. (1)]:

$$\psi \rightarrow \psi_0 e^{i(n-1)kD} = \psi_0 e^{i\chi} e^{-\sigma_a ND/2} = e^{i\chi}\sqrt{a}\psi_0 \tag{7}$$

Therefore, the beam modulation behind the interferometer is obtained in the following form:

$$I_0 \propto |\psi_0^{\rm I} + \psi_0^{\rm II}|^2 \propto [(1 + {\rm a}) + 2\sqrt{a}\cos\chi] \tag{8}$$

On the other hand, the transmission probability of a chopper wheel or another shutter system is given by the open to closed ratio, $a = t_{\rm open}/(t_{\rm open} + t_{\rm closed})$, and one obtains after straightforward calculations

$$\begin{aligned} I &\propto [(1 - a)|\psi_0^{\rm II}|^2 + a|\psi_0^{\rm I} + \psi_0^{\rm II}|^2] \\ &\propto [(1 + a) + 2a\cos\chi] \end{aligned} \tag{9}$$

i.e., the contrast of the interference pattern is proportional to $\sqrt{a}$, in the first case, and proportional to a in the second case, although the same number of neutrons have been observed in both cases. The absorption represents a measuring process in both cases because a compound nucleus is produced with an excitation energy of several MeV, which is usually deexcited by captured gamma rays. These can easily be detected by different means.

Figure 4 shows a typical result for the transmission probabilities near $a = 0.25$ as well as the dependence of the normalized interference amplitude on the transmission probability.[15,17] The different contrast becomes especially obvious for low transmission probabilities where the interfering part of the interference pattern is distinctly larger than the transmission probability through the semitransparent absorber sheet. The difference diverges for $a \rightarrow 0$ but it has been shown that in this regime the variations of the transmission due to variations of the thickness or of the density of the absorber plate have to be taken into account which shifts the points below the $\sqrt{a}$ curve.[18]

The region between the linear and the square root behavior can be achieved by very fast-rotating chopper slits or by a narrow transmission lattice, where one starts to lose information of through which individual slit the neutron went. The critical slit width is connected to the Pendellösung length and to the fact that certain neutrons become "labeled" neutrons due to slit diffraction which makes a separate detection possible in principle.

3. WAVE PARTICLE MEASURE

The results of the previous section show that the degree of contrast is a sensitive measure for the particle and wave character of the physical system. Therefore, an attempt is made to get a more general formulation, which accounts for stochastic and deterministic processes and for situations where a preparatory stage is created or where a real measuring process with a collapse of the wave field occurs.

For this more general discussion, the formulas can be used in the idealized form

$$I = T[1 + V\cos\phi'] \tag{10}$$

where we use the transparency of the system $T = (a + 1)/2$ and the visibility of the interference fringes $V = V_s = 2a/(a+1)$ for the statistical and $V = V_d = 2a^2/(a + 1)$ for the deterministic case. Both quantities can be obtained from the interference pattern or by separate measurements of the beam attenuation factor a. The quantity TV denotes the amplitudes of the interference fringes as they are shown in Figure 4. If the reduction of the contrast is also caused by high-order coherence phenomena or due to the roughness of the sample, the visibility factor in the equations becomes a product of the different visibility factors.

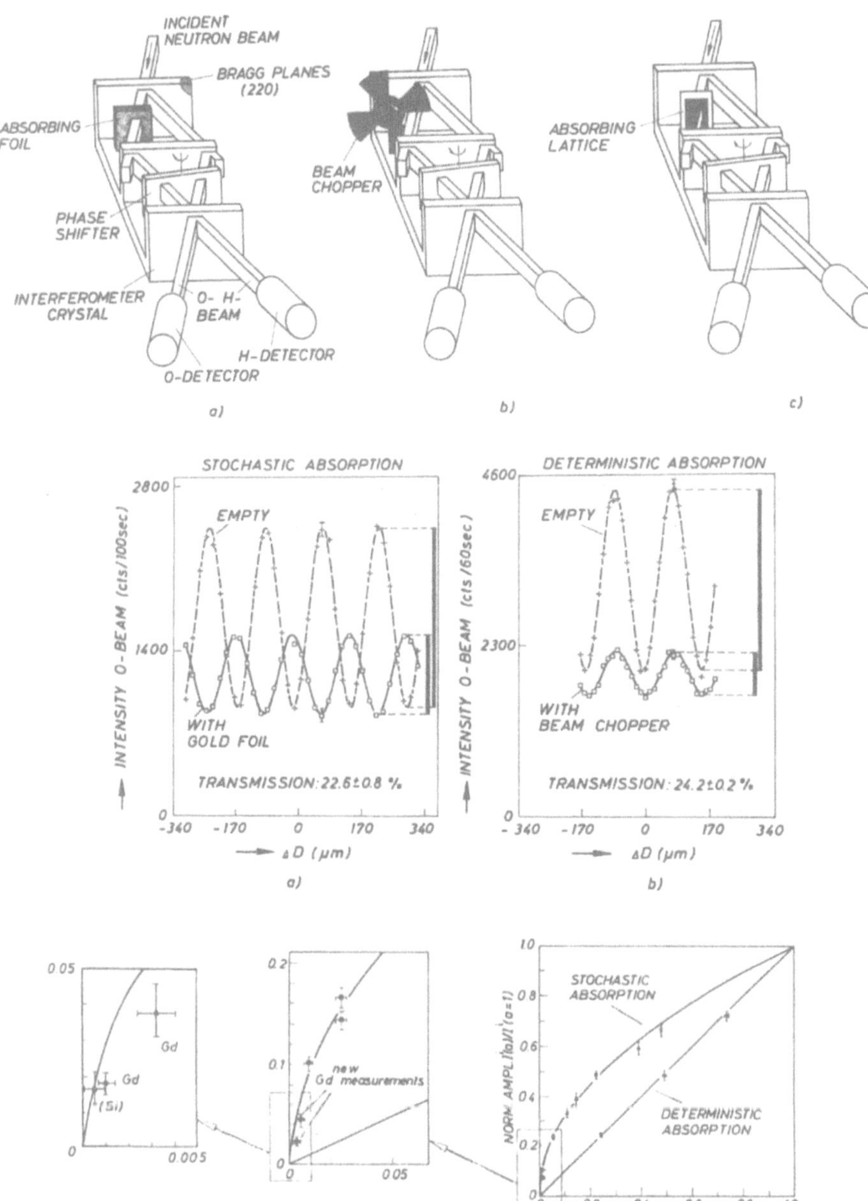

FIGURE 4. Sketch of the experimental arrangement for absorber measurements (above): (a) stochastic absorption; (b) deterministic absorption; (c) attenuation by a transmission grating. Typical results for stochastic and deterministic absorption (middle). Reduction of the contrast as a function of beam attenuation for different absorption methods (below).[16]

A measure for the particle nature can be found, if a quantity is taken which is the sum of the noninterfering intensity plus the probability of the neutrons for being absorbed in beam I or II, respectively[19]:

$$P(TV) = T - TV + (1-T) = 1 - TV = 1 - \langle|\psi^{I}\psi^{II}|\rangle \tag{11}$$

and the wave nature from the amplitude of the interference pattern as

$$W(TV) = TV = \langle|\psi^{I}\psi^{II}|\rangle \tag{12}$$

This rather simple formulation fulfills the relation

$$P + W = 1 \tag{13}$$

and can also be used for cases where the contrast attenuation is caused by large phase shifts which are on the order of the coherence length. In this case, the visibility varies but T remains 1. Obviously, this formulation is in complete agreement with the experimental results concerning the beam attenuation and the loss of contrast measurements discussed before. The whole results can be summarized in a single figure (Figure 5). Measurements with equal beam attenuation in both beam paths belong to the line $V=1$, high-order loss of contrast and other deterministic phase mixing measurements are described by the curve $T=1$, and many incoherent phase mixing processes lie near the origin.

This kind of formulation may facilitate a combination with a Shannon information-theoretic entropy approach, where various attempts have been made in the past.[20,21]

4. 4π-SYMMETRY OF SPINORS

The magnetic interaction is caused by the dipole coupling of the magnetic moment of the neutron $\vec{\mu}$ to a magnetic field $\vec{B}$ ($H = -\vec{\mu}\vec{B}$). Therefore, the propagation of the wave function is given by

$$\psi \rightarrow \psi_0 e^{-i(Ht/h)} = \psi_0 e^{-i(\vec{\mu}\vec{B}t/h)} = \psi_0 e^{-i\vec{\sigma}\vec{\alpha}/2} = \psi(\alpha) \tag{14}$$

where $\vec{\alpha}$ represents a formal description of the Larmor rotation angle around the field $\vec{B}$ ($\alpha = (2\mu/h)\int B\,dt \simeq (2\mu/hv)\int B\,ds$). This wave function shows the typical 4π-symmetry of a spinor

$$\begin{aligned} \psi(2\pi) &= -\psi(0) \\ \psi(4\pi) &= \psi(0) \end{aligned} \tag{15}$$

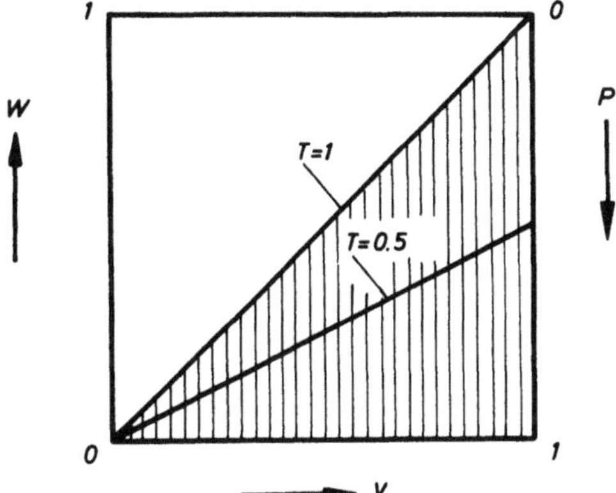

FIGURE 5. Synopsis of the wave–particle character in different interference experiments.(19)

whereas 2π-symmetry exists only for the expectation values

$$|\psi(2\pi)|^2 = |\psi(0)|^2 \tag{16}$$

The 4π-periodicity becomes visible in interferometer experiments, as predicted theoretically,(22–24) and has been verified experimentally in early neutron interferometric experiments,(25,26) where the intensity for unpolarized incident neutrons was found to be

$$I_0 \propto |\psi(0) + \psi(\alpha)|^2 \propto \left(1 + \cos\frac{\alpha}{2}\right) \tag{17}$$

The results of the first related experiment are shown in Figure 6. These results are widely debated in the literature. It should be mentioned that this 4π-symmetry can always be attributed to real rotations in the case of fermions.(27,28) Today, the most precise value for the periodicity factor is $\alpha_0 = 715.87 \pm 3.8$ degrees.(29) This value provides only a small margin for speculation about SU(2)-symmetry breaking, but a new and more precise determination of α_0 is recommended. The 4π-periodicity effect has been observed for unpolarized as well as polarized neutrons, which demonstrates the intrinsic feature of this phenomenon and the self-interference properties involved in these kind of experiments. New attention should be drawn to an interferometric observation of the Berry phase(30) which represents a topological phenomenon and is therefore of central interest.

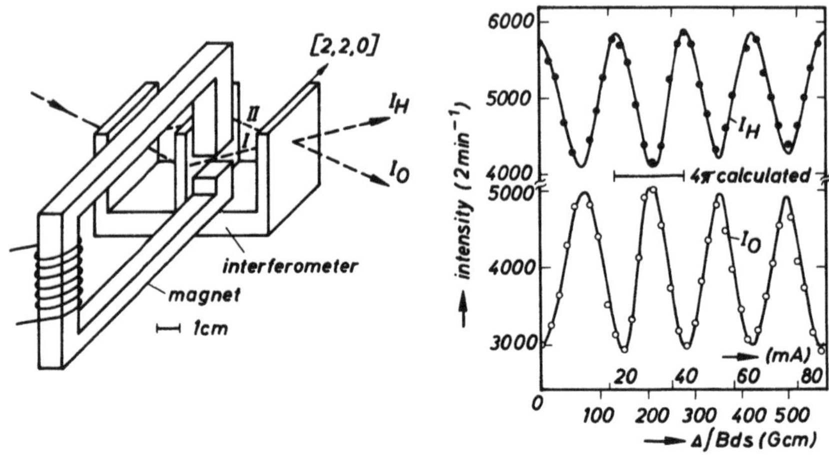

FIGURE 6. First observation of the 4π-symmetry factor of spinors.[25]

5. SPIN STATE INTERFEROMETRY

In this case the polarization vector can be influenced differently in the two coherent beam paths and these beams can be superposed at the end of the interferometer. The principles of these experiments and the most important results are summarized in Figure 7[31,32] More experimental details can be found in these references. There is a marked difference between the action of a static flipper and a resonance flipper, which has to be discussed in more detail.

In the first case (static flipper) the wave function is changed by the flipper according to Eq. (14), which has to be applied for polarized incident neutrons:

$$\psi \rightarrow e^{i\chi}e^{-i\vec{\sigma}\vec{\alpha}/2}|z\rangle = e^{i\chi}e^{-i\sigma_y\pi/2}|z\rangle = -i\sigma_y e^{i\chi}|z\rangle = e^{i\chi}|-z\rangle \tag{18}$$

The rotation of the polarization vector around the y-axis has been postulated to be π.[33] Thus, two wave functions with opposite spin directions are superposed at the third plate

$$\psi \propto (|z\rangle + e^{i\chi}|-z\rangle) \tag{19}$$

which corresponds to the situation proposed by Wigner[34] in 1963 to verify the quantum mechanical spin superposition law. In this case the intensities in the O and H beams are equal, and the beams are polarized in the (x,y)-plane, i.e., perpendicular to both the initial spin directions. The angle of the polarization in the (x,y)-plane is given by the nuclear phase shift:

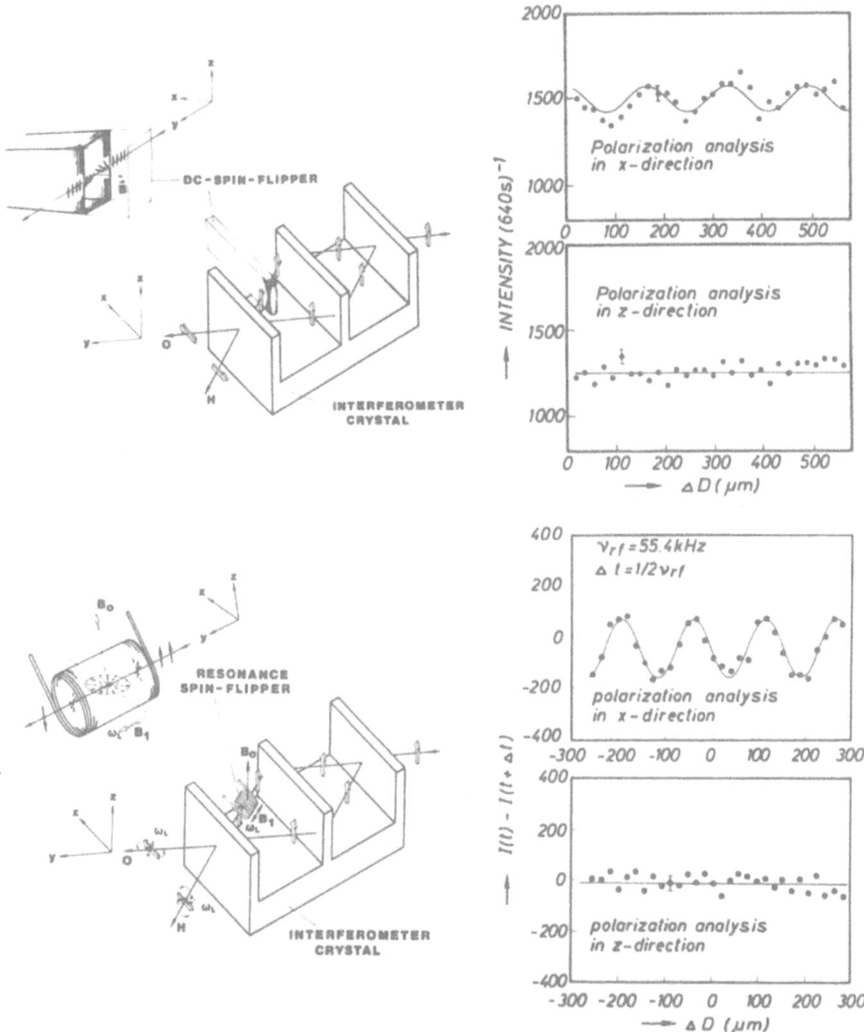

FIGURE 7. Sketch of the static (above) and the time-dependent (below) spin superposition experiment with characteristic results.[31,32]

$$\vec{P}_0 = \frac{\langle\chi|\vec{\sigma}|\psi\rangle}{I_0} = \begin{bmatrix} \cos\chi \\ \sin\chi \\ 0 \end{bmatrix} \tag{20}$$

Thus, a pure initial state is transferred to a pure final state which is different to both states existing before superposition. The interference pattern appears only, if a polarization analysis is performed in the $|x\rangle$ or in the $|y\rangle$ direction. If the analyzer is set in the $|z\rangle$ (or $|-z\rangle$) direction, no intensity modulation is observed.

The second version of the spin superposition experiment was performed with a Rabi-type resonance flipper which is also commonly used in polarized neutron physics. This kind of interaction is time-dependent and, in addition to the spin-inversion, an exchange of the resonance energy $E_{HF} = h\omega_r$ occurs between the neutron and the resonator system, which has to be considered in the interferometric experiment. This energy exchange was observed in a separate experiment, where the energy resolution ΔE of the apparatus was better than the Zeeman energy splitting ($\Delta E < E_{HF}$).[35] This experiment was performed according to a proposal of Drabkin and Zhitnikov.[36] For a complete spin reversal the frequency of the field has to match the resonance condition and the amplitude B_1 has to fulfill the relation $|\mu|B_1 l/hv = \pi$, where l is the length of the coil. Oscillating fields are used instead of purely rotational fields and, therefore, only one component contributes to the resonance which causes a slight shift of the resonance frequency from the Larmor frequency $\omega_L = 2|\mu|B_0/h$ due to the Bloch–Siegert effect ($\omega_r = \omega_L[1 + (B_1{}^2/16B_0^2)]$).[37,38] Thus, the wave function of the beam with the flipper changes according to

$$\psi \rightarrow e^{i\chi} e^{i(\omega - \omega_r)t} |-z\rangle \tag{21}$$

Therefore, a spin-up and a spin-down state are superposed at the position of the third plate. The final polarization of the beam in the forward direction is given by

$$\vec{P} = \begin{bmatrix} \cos(\chi + \omega_r t) \\ \sin(\chi + \omega_r t) \\ 0 \end{bmatrix} \tag{22}$$

and lies again in the (x,y)-plane but now rotates within this plane with the resonance (Larmor) frequency without being driven by a magnetic field. A stroboscopic method was needed for the observation of this effect. The direction of the polarization in the (x,y)-plane depends on the status (phase) of the resonance field and, therefore, has to be measured synchronously with this phase.

The observed interference pattern (Figure 7) demonstrates that coherence persists, although a well-defined energy exchange between the neutron and the apparatus exists. Thus, an energy exchange is not automatically a measuring process. As we will see later, the exchanged photon cannot be used for a quantum nondemolition measurement. In our experiment, the following argument based on different uncertainty relations can be used: First, a single absorbed or emitted photon of the resonator cannot be detected because of the photon number–phase uncertainty relation, which can be written in the form[39,40]

$$(\Delta N)^2 \frac{(\Delta S)^2 + (\Delta C)^2}{\langle S\rangle^2 + \langle C\rangle^2} \geq \tfrac{1}{4} \tag{23}$$

where S and C can be expressed by the creation and annihilation operators, $C = (a_- + a_+)/2$ and $S = (a_- - a_+)/2i$, whose matrix elements couple coherent Glauber states. For our purpose this relation can be used in its simpler form.

$$\langle \Delta N^2\rangle\langle \Delta\theta^2\rangle \geq \tfrac{1}{4} \tag{24}$$

The uncertainty of the photon number of the resonator is minimized for a coherent state resonator by $\Delta N = \sqrt{\langle N\rangle}$[41] and, therefore, the lower limit for the phase uncertainty becomes $\Delta\theta \approx 1/(\sqrt{\langle N\rangle})$. Because in this kind of spin-superposition experiment the phase determination of the flipper field is required to be better than $\theta \leq \frac{1}{2}$ for the stroboscopic method, it is therefore impossible to observe a single absorbed or emitted photon ($\Delta N \geq 1$).

A second version of the beam path detection may be based on the observation of the energy change of the neutron. This can only be achieved if the energy resolution of the instrument fulfills the relation $\Delta E \leq 2\mu B_0$. On the other hand, the stroboscopic measuring method requires time channels $\Delta t \leq 1/2\nu_{HF} = h/4|\mu|B_0$, which provides another constraint on the experiment. Both conditions cannot be fulfilled with respect to the energy–time uncertainty relation concerning the beam parameters $\Delta E\Delta t \leq h/2$. Therefore, we conclude that a simultaneous detection of the beam path through the interferometer and of the interference pattern remains impossible. This kind of experiment has also been analyzed in terms of a coherent state or of a number state resonator.[42] These authors came to the same conclusion, that interference becomes destroyed if a signal is extracted from the resonators.

It has been argued by Vigier's group[43,44] that new information about the particle–wave duality can be obtained with resonator coils in both coherent beams. The corresponding experiments will be discussed in the next section.

6. DOUBLE COIL EXPERIMENTS

The experimental arrangement for the double coil experiment is shown in Figure 8.[45,46] The final polarization lies in the $|-z\rangle$ direction and the energy transfer $\hbar\omega_r$ can be smaller or larger than the energy resolution ΔE because this information cannot be in any way associated with a beam path detection. The layout of the experiment followed the proposal of Vigier's group.[43,44] According to our previous considerations, the change of the wave functions with the resonance flippers turned to the resonance frequency can be written for polarized incident neutrons ($|z\rangle$) and for different modes of operation as follows.

a. Both flippers are operated synchronously without a phase shift between the slipper fields:

$$\psi \rightarrow e^{i(\omega-\omega_r)t}|-z\rangle + e^{i\chi}e^{i(\omega-\omega_r)t}|-z\rangle \tag{25}$$

This results in an intensity modulation

$$I_0 \propto 1 + \cos\chi \tag{26}$$

which is independent of the flipper fields.

b. Both flippers are operated synchronously with a distinct phase relation Δ:

$$\psi_0 \rightarrow e^{-i(\omega-\omega_r)t}|-z\rangle + e^{i\chi}e^{i\Delta}e^{i(\omega-\omega_r)t}|-z\rangle \tag{27}$$

In this case, the intensity modulation is given by

$$I_0 \propto 1 + \cos(\chi + \Delta) \tag{28}$$

c. Both flippers are operated asynchronously with statistically fluctuating phase differences $\Delta(t)$ which average out during the measuring interval. Then.

$$I_0 \propto \text{const} \tag{29}$$

It should be mentioned in this context that even in this case, coherence phenomena can be observed if a stroboscopic investigation is performed ($I_0 = I_0(\Delta)$).

The results of these related experiments are shown in Figure 8. Complete agreement with the theoretical predictions is found. The interference properties are preserved, although an energy exchange $\hbar\omega_r$ certainly takes place. Only quanta within a narrow energy band around $\hbar\omega_r$ and no others are excited inside the flipper resonator. Therefore, one could believe that the spin flip and the energy transfer process to the neutron occurred inside one of the two coils, which would demonstrate that the neutron has chosen one of the two possible paths. But even this rather weak statement would require the concept of pilot waves, quantum potentials, etc., leading immediately to questions about the interpretation of quantum mechanics, which are not the subject of this chapter. In this connection, an experiment is proposed which shows this situation even more clearly[(47)] (Figure 9). In this case, the flippers are operated only if the neutron burst is inside the coils and a strong magnetic field causes an energy shift $\Delta E_{HF} = 2\mu B_0$ which is larger than the energy width of the beam ($\Delta E_{HF} > \Delta E$) and, therefore, measurable with a high-resolution time-of-flight system. In this situation, it looks even more likely that the neutron has chosen one of both possible beam paths although we do not know which one. In this case the field (phase) variation becomes faster than the coherence time of the beam ($\Delta t_c = \Delta x/v$) which causes a phase-chopping effect, reducing the interference contrast.

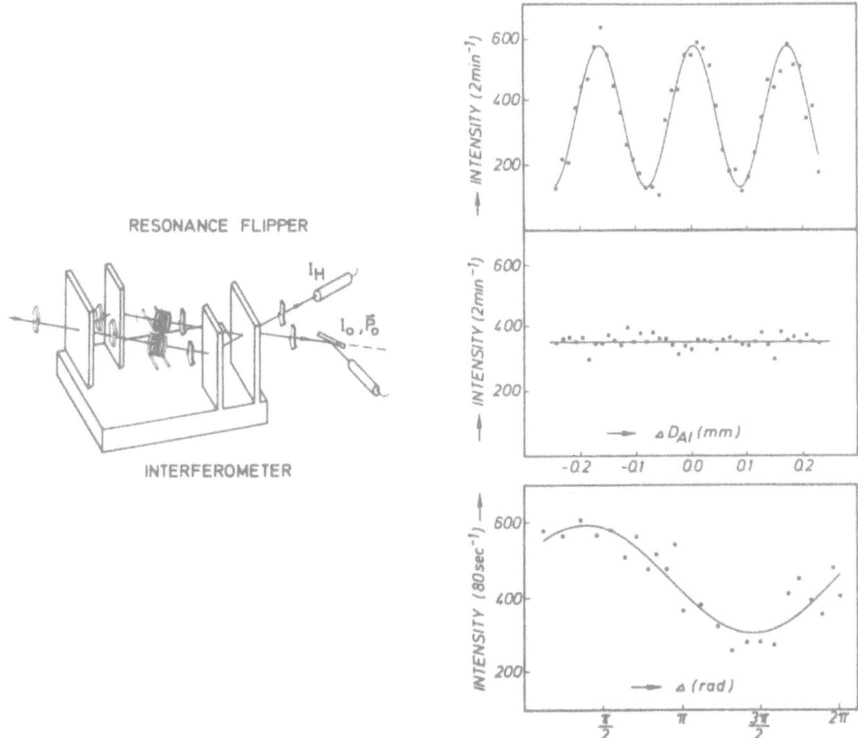

FIGURE 8. Sketch of the double resonance coil experiment (left) and results of the double coil experiments (right)[45,46] Above: Synchronous flipper fields with ν_r = 71.90 kHz [Eq. (26)]. Middle: Two slightly fluctuating independent flipper fields with ν = 71.92 ± 0.02 kHz [Eq. (29)]. Below: Interference pattern as a function of the phase shift Δ between both flipper fields at ν_r = 71.90 kHz [Eq. (28)].

7. MACROSCOPIC QUANTITIES IN UNCERTAINTY RELATIONS

By means of perfect crystal neutron optics the resolution in momentum or energy space can be increased to such an extent, that the conjugate quantities each macroscopic dimensions.

First, the perfect crystal can be envisaged as a very narrow collimator, whose angular divergence is roughly given by the ratio of the lattice constant divided by the thickness of the crystal (d_{hkl}/t). This feature becomes visible in multiple Laue-rocking curves, where the convolution of the individual reflection curves exhibits a very narrow central peak[48,49] which has similarities to the diffraction focusing effect first treated for X rays.[50] The individual reflection curves are well known from dynamical diffraction theory and can be written as[7,8]

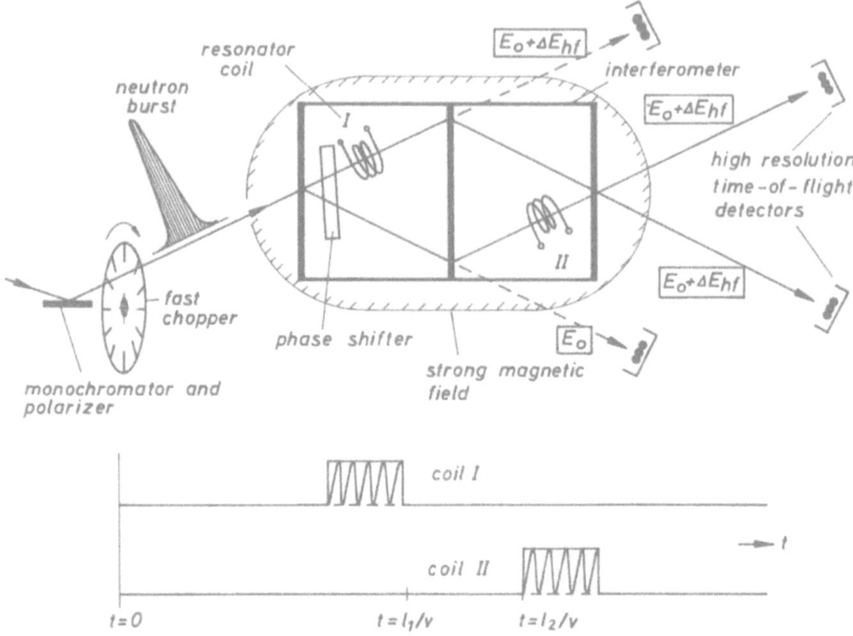

FIGURE 9. Proposed double coil experiment where pulsed neutron beams and pulsed neutron flippers are used and where interference and energy change of the neutrons can be observed simultaneously.[47]

$$P(y) = \frac{\sin^2 A\sqrt{1 + y^2}}{1 + y^2} \tag{30}$$

where A is a reduced quantity related to the thickness of the perfect crystal and y describes the deviation from the exact Bragg angle. At the same time, a narrow incident beam is spread out across the whole Borrmann fan, whose dimension at the exit is given by the thickness (t) of the crystal plates as $2t{\cdot}tg\theta_B$. The rapid variation of the intensity across the reflection curve or the Borrmann fan (Pendellösung fringes) is caused by the rapid variation of the phases of the excited waves inside the crystal. The corresponding rocking curves are given by the convolution of such fine structured curves. They can also be interpreted as self-correlation functions, which can be connected with the uncertainty relations.[51] The analytical form of the central peak of these multiple Laue rocking curves can be written as[52]

$$I_p \propto \frac{J_1(2Ay)}{Ay} \tag{31}$$

The width at half maximum is given by

$$\delta\theta = 0.7\frac{d_{\text{hkl}}}{t} \tag{32}$$

which is on the order of several thousandths of a second of arc. The related momentum uncertainty ($\Delta k \simeq k\delta\theta$) is on the order of cm^{-1} and, therefore, Δx in the space–momentum uncertainty relation becomes on the order of cm. The formulas for the triple case Laue rocking curves and for various other contributions to the rocking curves can be found in the literature.[52]

The experiments have been performed with monolithic multiplate systems by rotating a wedge-shaped material around the beam axis; this provides a proper control of small beam deflections in the horizontal plane, which is the only sensitive plane for perfect crystal reflections. This deflection is given by the properties of the material, by the angle β of the wedge, and by the rotation angle α around the beam axis

$$\delta = \frac{Nb_c\lambda^2}{\pi} tg\left(\frac{\beta}{2}\right)\sin\alpha \tag{33}$$

The broadening of the central peak due to the insertion of a macroscopic slit is shown in Figure 10. More recently, measurements with macroscopic transmission

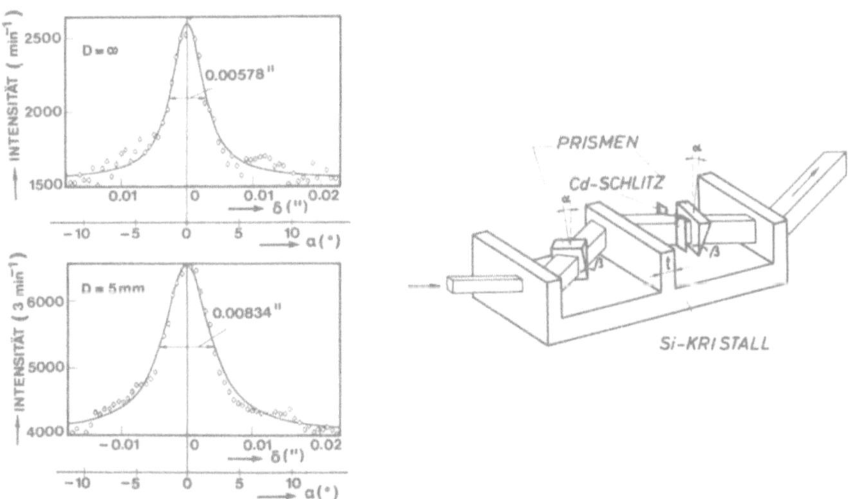

FIGURE 10. Experimental arrangement for the observation of triple Laue rocking curves and broadening of the narrow central peak due to diffraction from a slit with a width of 5 mm.[53]

lattices have yielded similar results. Although the wavelength of the neutrons is smaller than the width of the slit, by a factor of about 10^8 the broadening becomes visible due to the high angular resolution of such systems. This can also be understood by the large transversal coherence length of the beam across the whole Borrmann fan ($2t \cdot rg\theta_B$).

As an alternative to the high-momentum resolution just discussed, an extremely high-energy resolution can be achieved by a slight modification of the double coil experiment described in Section 6. If the frequencies of the two coils are chosen to be slightly different, the energy transfer becomes different too ($\Delta E = h(\omega_{r1} - \omega_{r2})$). The frequency difference can be made very small, if high-quality synthesizers are used for the field generation. The flipping efficiencies for both coils are always very close to 1 (better than 0.99). Now, the wave functions change according to

$$\chi \rightarrow e^{i(\omega-\omega_{r1})t}|-z\rangle + e^{i\chi}e^{i(\omega-\omega_{r2})t}|-z\rangle \tag{34}$$

Therefore, the intensity behind the interferometer exhibits a typical quantum beat effect, given by

$$I \propto 1 + \cos[\chi + (\omega_{r1} - \omega_{r2})t] \tag{35}$$

Thus, the intensity behind the interferometer oscillates between the forward and deviated beam without any apparent change inside the interferometer.[45,46] The time constant of this modulation can reach a macroscopic scale which is again correlated to an uncertainty relation $\Delta E \Delta t \leq h/2$. Figure 11 shows the result of an experiment, where the periodicity of the intensity modulation, $T = 2\pi/(\omega_{r1} - \omega_{r2})$, amounts to $T = (47.90 \pm 0.15)$s caused by a frequency difference of about 0.02 Hz. This corresponds to a mean difference of ΔE energy transfer between the two beams, $E = 8.6 \times 10^{-17}$ eV, and to an energy sensitivity of 2.7×10^{-19} eV,

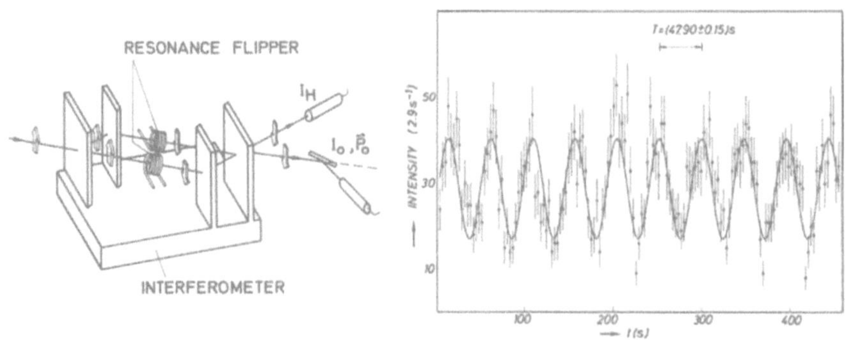

FIGURE 11. Quantum beat effect observed when the frequencies of the two flipper coils differ by about 0.02 Hz around 71.89979 kHz.[46]

which is better by many orders of magnitude than that of other advanced spectroscopic methods. This high resolution is strongly decoupled from the monochromaticity of the neutron beam, which was $\Delta E_B \simeq 5.5 \times 10^{-4}$ eV around a mean energy of the beam $E_B = 0.023$ eV in this case. It should be mentioned that the result can also be interpreted as being the effect of a slowly varying phase $\Delta(t)$ between the two flipper fields [see Eq. (28)], but the more physical description is based on the argument of a different energy transfer. The extremely high resolution may be used for fundamental, nuclear, and solid-state physics applications.

The quantum beat effect can also be interpreted as a magnetic Josephson effect analog.[54] In this case, the phase difference is driven by the magnetic energy

$$\frac{\delta}{\delta t}(\Delta_2 - \Delta_1) = \omega_{r2} - \omega_{r1} = \frac{1}{\hbar} 2\mu B_0 \tag{36}$$

which yields the observed intensity modulation [compare eq. (35)]

$$I \propto (1 + \cos \Delta(t)) \tag{37}$$

where $\Delta(t) = 2\mu B_0 t/h$. This is analogous to the well-known Josephson effect in superconducting tunnel junctions,[55] where the phase of the Cooper pairs in both superconductors is related according to

$$\frac{\delta}{\partial t}(\phi_2 - \phi_1) = \frac{1}{\hbar}(E_2 - E_1) = \frac{1}{\hbar} 2 \text{ eV} \tag{38}$$

which is driven by the electrical potential V between both superconductors.

An intrinsic lower limit for the energy width ΔE_i of the neutron beam which is caused by its lifetime, $\tau = 925$ s, which yields according to $\Delta E_i \tau > h/2$, $\Delta E_i \simeq 3.5 \times 10^{-24}$ eV. The decay appears in both beam paths and contributes an attenuation factor $\exp(-t/\tau) = \exp(-l\tau/v)$which is similar to those discussed in Section 2.

8. EXPERIMENTS IN PROGRESS OR IN PREPARATION

We feel that most of the fundamental experiments for testing quantum mechanics by neutron interferometry have already been performed but there will be the need to demonstrate the features of quantum mechanics in more detail. The proposal shown in Figure 9 is such an example. The outcome of such an experiment can be calculated in advance but it suggests that the neutron has a trajectory inside the interferometer although which one remains unknown. The transition to the quantum limit has to be considered if an uncertainty limit is touched (lattice diffraction, diffraction in time, phase chopping). There also remain possibilities for more accurate repetitions of experiments under even better conditions which could produce spectacular results at any time. In addition, there may be new experiments, which could push the development of advanced neutron optics further. Some examples of such experiments are discussed below.

a. Phase echo system. Such systems are similar to spin echo systems known in advanced neutron spectroscopy,[(56)] but use the phase of the wave function instead of the Larmor precession angle as the measurable quantity.[(57)] The interference pattern disappears, if the longitudinal shift of the wave packets due to a phase shifter becomes larger than the longitudinal coherence length of the beam ($\chi \cdot \lambda / 2\pi > \lambda^2/\Delta\lambda$, see also Figure 3). This behavior has been observed experimentally.[(58,59)] By applying an opposite phase shift in the same beam, or the same phase shift in the second beam of the interferometer, the smeared interference pattern can be restored to full contrast, shown schematically in Figure 12. Such experiments demonstrate that phase information can exist although the measured signal looks like a statistical mixture. The coherence properties can be recovered if a proper measuring method is applied. Recovery of a smeared-out interference pattern has been demonstrated recently by using combinations of thick Bi- and Ti-phase shifters.[(60)] This method will establish a new horizon if combined with multiplate interferometry (part c), where interference properties of a dephased beam can be recovered in the following interferometer loop.

b. Pulsed beam interferometry. It could be argued that there is always an overlap of wave functions in a stationary situation (or at least of plane wave components of the wave packet from both beam paths) at the position of the beam splitter and at the place of superposition. This can be avoided by using a chopper which produces bursts, whose lengths are smaller than the dimension of the interferometer[(61)] (Figure 13). The known spatial spreading of the wave function of matter waves

$$[\Delta x(t)]^2 = [\Delta x(0)]^2 + \left[\frac{(h/2m)t}{\Delta x(0)}\right]^2 \tag{39}$$

has no influence on the interference properties for all practical situations, where the length of the bursts Δx is much larger than the coherence length, $\lambda^2/\Delta\lambda$.

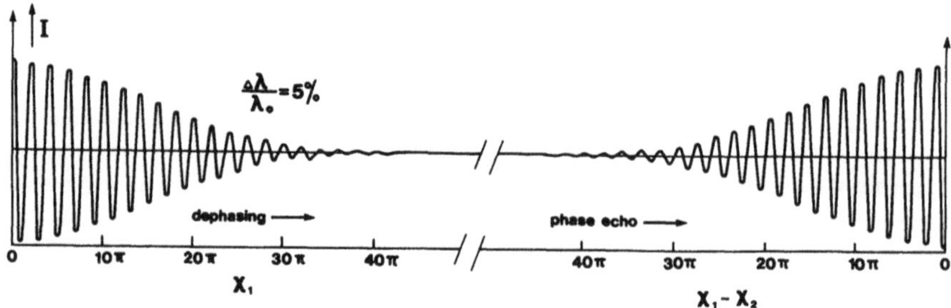

FIGURE 12. Principle of a phase echo system.[(57)]

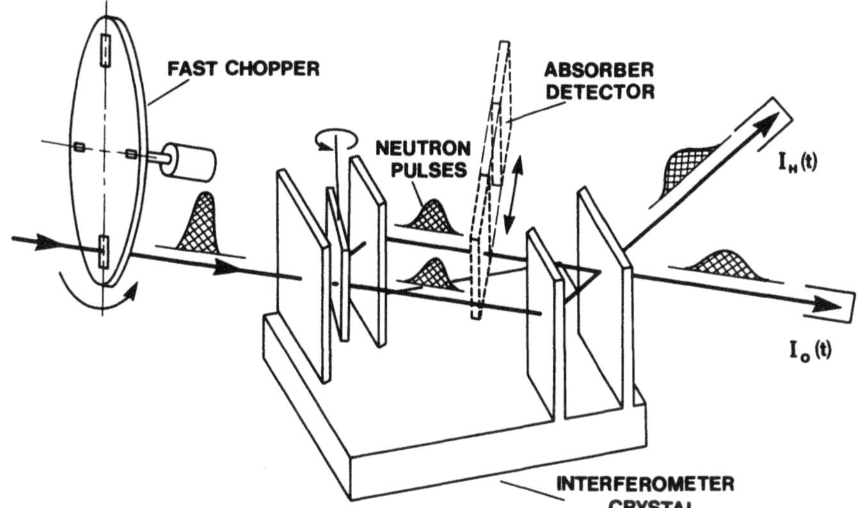

FIGURE 13. Sketch of the apparatus for interference experiments with pulsed beams.[61]

Nevertheless, such experiments will make the discussion about the collapse of the wave field in the case of an absorber (see Section 2) more profound and new types of delayed choice experiments will become feasible where the decision about interference or beam path detection can be made after the burst has passed the beam splitter. Very interesting heat effects have been observed when the interference pattern of the slower neutrons from a chopper slit in front overlaps with the interference pattern of the faster neutrons from the following slit.[61,62]

c. Multiplate interferometry. A five-plate interferometer is shown in Figure 14. In this setup, different interferometer loops are linked together by common beam paths. The theoretical description follows the formalism developed for the standard triple-plate case,[9,10] but the expected interference properties show some new features, which do not exist for the standard interferometer. The whole theoretical description follows the treatment for the X-ray case[63]; here we refer to a four-plate interferometer and specify averaged intensities of the interfering beams behind this device:

$$
\begin{aligned}
I_3 &= K_2[3 + 2\cos(\chi_A + \chi_B) + 2\cos\chi_A + 2\cos\chi_B] \\
I_4 &= K_1 + 2K_3 - 2K_2[\cos(\chi_A + \chi_B) + \cos\chi_B] + 2K_3\cos\chi_A \\
I_5 &= 2K_2 + K_4 + 2K_2\cos(\chi_A + \chi_C) - 2K_3[\cos\chi_A + \cos\chi_C] \\
I_6 &= K_1 + 2K_3 - 2K_2[\cos(\chi_A + \chi_C) + \cos\chi_A] + 2K_3\cos\chi_C
\end{aligned}
\tag{40}
$$

$$
\text{with } K_1 = \frac{417\pi}{1048},\ K_2 = \frac{79\pi}{1048},\ K_3 = \frac{65\pi}{1048},\ K_4 = \frac{175\pi}{1048}
$$

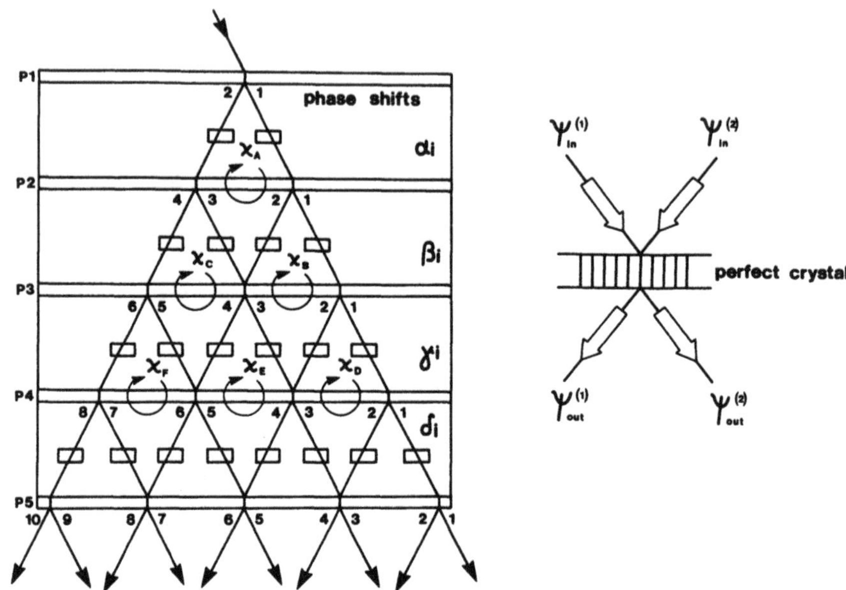

FIGURE 14. Sketch of a multiplate interferometer with an indication of different interferometer loops (left) and principle of coherent beam mixing (right).

Each intensity depends on two net phase shifts of two interferometer loops. First experimental investigations agree with these predictions.[(64)] The intensities and interference properties of the loops B and C can partly be controlled by the first interferometer loop. There exist additional positions of four-wave mixing, which provide new aspects for coherent neutron optics. The intensities inside these interferometer loops can be coherently influenced not only at the position of the splitter but also at the position of the mirror. This may be useful for the achievement of squeezed neutron states[(65)] and new bunching systems. The formulas have been checked as to whether a phase shifter can have an influence, if it is placed within a beam with zero intensity (e.g., γ_3 if $\chi_A = \pi$). They predict no such influence. Absorbing and thick phase shifters producing phase shifts on the order of the coherence length lead to additional effects.

9. DISCUSSION

All the results of the neutron interferometric experiments are well described by the formalism of quantum mechanics. According to the complementarity principle of the Copenhagen interpretation, the wave picture has to be used to describe the observed phenomena. The question as to how the well-defined particle properties of the neutron are transferred through the interferometer, is not meaningful within this interpretation, but it should be an allowed one from the physical point of view. Therefore, other interpretations should also be included in the discussion of such experiments. The particle picture can be preserved if pilot waves are postulated or if a quantum potential guides the particle to the predicted position. Related calculations have been performed for a simplified interferometer system.[66,67] Unfortunately, the results of these calculations are identical with the results of ordinary quantum mechanics and, therefore, to decide between both points of view remains an epistemological problem. The nonlocal quantum potential and the beam trajectories are shown in Figure 15. The alternative view according to the wave picture is visualized in Figure 16 where the position of the nodes of the superposed wave fields relative to the lattice points determine where the waves proceed behind the interferometer.

We have always tried to perform unbiased experiments and do not wish to interfere with any epistemological interpretation of quantum mechanics. Perhaps in the future new proposals for experiments will be formulated that permit a unique decision between different interpretations. As an experimentalist, one appreciates the pioneering work of the founders of quantum mechanics, who created this basic theory with so little experimental evidence. Now we have much more direct evidence, even on a macroscopic scale, but, nevertheless, one notices that the

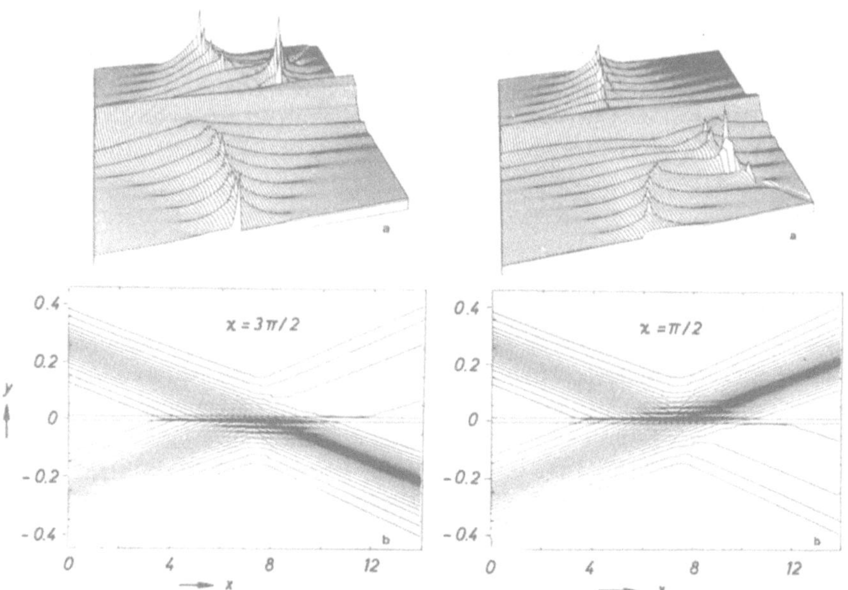

FIGURE 15. Quantum potential and beam trajectories at the place of beam superposition for a phase shift of $\chi = 3\pi/2$ (left) and $\chi = \pi/2$ (right).[39]

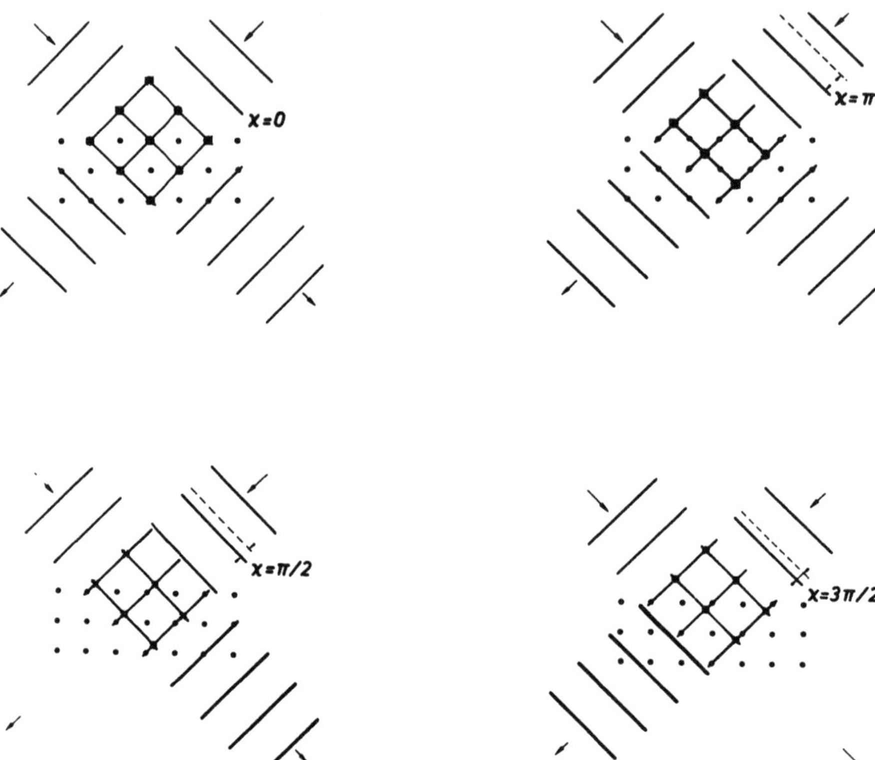

FIGURE 16. Nodes of the wave field and lattice points at the third interferometer plate. The relative position between the nodes of the wave field and the lattice points depends on the phase shift and determines the beams behind the interferometer.

interpretation of quantum mechanics goes beyond human intuition in certain cases. Only two aspects of the experiments discussed before should be repeated: How can every neutron in the spin-superposition experiment be transferred from an initial pure state in the $|z\rangle$ direction into a pure state in the $|x\rangle$ direction behind the interferometer, if no spin turn occurs in one beam and a complete spin reversal occurs in the other beam path? How can every neutron have information about which beam to join behind the interferometer, when a slightly different energy exchange occurs in both beams inside the interferometer and the time constant of the beat effect is many orders of magnitude larger than the time of flight through the system?

Experiments of our group and such which are related to fundamental physics problems have been discussed in this chapter. Several recent review articles can supplement a broader scope about the status of neutron interferometry.[68–72]

ACKNOWLEDGMENT. All the experimental results discussed in detail have been obtained by our Dortmund–Grenoble–Vienna interferometer group working at the high flux reactor in Grenoble. The cooperation within this group and especially the cooperation with colleagues from our Institute, which are cited in the references, are gratefully acknowledged.

A first version of this manuscript was published in *Helv. Phys. Acta* **61**, 589 (1988).

REFERENCES

1. H. MAIER-LEIBNITZ and T. SPRINGER, *Z. Phys.* **167** (1962).
2. R. GAEHLER, J. KALUS, and W. MAMPE, *J. Phys. E* **13**, 546 (1980).
3. H. RAUCH, W. TREIMER, and U. BONSE, *Phys. Lett. A* **47**, 369 (1974).
4. W. BAUSPIESS, U. BONSE, H. RAUCH, and W. TREIMER, *Z. Phys.* **271**, 177 (1974).
5. A. I. IOFFE, V. S. ZABIYANKAN, and G. M. DRABKIN, *Phys. Lett.* **111**, 373 (1985).
6. U. BONSE and M. HART, *Appl. Phys. Lett.* **6**, 155 (1965).
7. H. RAUCH and D. PETRASCHECK, in: *Neutron Diffraction* (H. DACHS, eds.), Springer-Verlag, Berlin (1978), Chapter 9.
8. V. F. SEARS, *Can. J. Phys.* **56**, 1261 (1978).
9. W. BAUSPIESS, U. BONSE, and W. GRAEFF, *J. Appl. Crystallogr.* **9**, 68 (1976).
10. D. PETRASCHECK, *Acta Phys. Aust.* **45**, 217 (1976).
11. M. L. GOLDBERGER and F. SEITZ, *Phys. Rev.* **71**, 294 (1947).
12. V. F. SEARS, *Phys. Rep.* **82**, 1 (1982).
13. U. BONSE and H. RAUCH (eds.), *Neutron Interferometry*, Clarendon Press, Oxford (1979).
14. H. RAUCH, E. SEIDL, D. TUPPINGER, D. PETRASCHECK, and R. SCHERM, *Z. Phys. B* **69**, 69 (1987).
15. H. RAUCH and J. SUMMHAMMER, *Phys. Lett. A* **104**, 44 (1984).
16. J. SUMMHAMMER, H. RAUCH, and D. TUPPINGER, *Phys. Rev. A* **36**, 4447 (1987).
17. H. RAUCH, J. SUMMHAMMER, and E. JERICHA, *Phys. Rev. A* **42**, 3726 (1990).
18. M. NAMIKI and S. PASCAZIO, *Phys. Lett.* **147**, 430 (1990).
19. H. RAUCH, *Proc. 3rd Int. Symp. Found. Quantum Mechanics* (S. KOBAYASHI *et al.*, eds.), Phys. Soc. Japan (1989), p. 3.
20. W. K. WOOTTERS and W. H. ZUREK, *Phys. Rev. D* **19**, 473 (1979).
21. P. BUSCH, *Found. Phys.* **17**, 905 (1987).
22. Y. AHARONOV and L. SUSSKIND, *Phys. Rev.* **158**, 1237 (1967).
23. H. J. BERNSTEIN, *Phys. Rev. Lett.* **18**, 1102 (1967).
24. G. EDER and A. ZEILINGER, *Nuovo Cimento B* **34**, 76 (1976).
25. H. RAUCH, A. ZEILINGER, G. BADUREK, A. WILFING, W. BAUSPIESS, and U. BONSE, *Phys. Lett. A* **54**, 425 (1975).
26. S. A. WERNER, R. COLELLA, A. W. OVERHAUSER, and C. F. EAGENIR, *Phys. Rev. Lett.* **35**, 1053 (1975).
27. A. ZEILINGER, *Nature* **294**, 544 (1981).
28. H. J. BERNSTEIN, *Nature* **315**, 42 (1985).
29. H. RAUCH, A. WILFING, W. BAUSPIESS, and U. BONSE, *Z. Phys. B* **29**, 281 (1978).
30. M. V. BERRY, *Proc. R. Soc. London Ser. A* **392**, 45 (1984).
31. J. SUMMHAMMER, G. BADUREK, H. RAUCH, U. KISCHKO, and A. ZEILINGER, *Phys. Rev. A* **27**, 2523 (1983).
32. G. BADUREK, H. RAUCH, and J. SUMMHAMMER, *Phys. Rev. Lett.* **51**, 1015 (1983).
33. A. ZEILINGER, Ref. 13, p. 241.

34. E. P. WIGNER, *Am. J. Phys.* **31**, 6 (1963).
35. B. ALEFELD, G. BADUREK, and H. RAUCH, *Z. Phys. B* **41**, 231 (1981).
36. G. M. DRABKIN and R. A. ZHITNIKOV, *Sov. Phys. JETP* **11**, 729 (1960).
37. F. BLOCH and A. SIEGERT, *Phys. Rev.* **57**, 522 (1940).
38. H. KENDRICK, J. S. KING, S. A. WERNER, and A. AROTT, *Nucl. Instrum. Methods* **79**, 82 (1970).
39. P. CARRUTHERS and M. M. NIETO, *Rev. Mod. Phys.* **40**, 411 (1968).
40. R. JACKIW, *J. Math. Phys.* **9**, 339 (1968).
41. R. J. GLAUBER, *Phys. Rev.* **131**, 2766 (1963).
42. M. O. SCULLY and H. WALTHER, *Phys. Rev. A* **39**, 5229 (1989).
43. C. DEWDNEY, P. GUERET, A. KYPRIANIDIS, and J. P. VIGIER, *Phys. Lett. A* **102**, 291 (1984).
44. J. P. VIGIER, *Pramana* **25**, 397 (1985).
45. G. BADUREK, H. RAUCH, and D. TUPPINGER, *Proc. Int. Conf. New Techniques and Ideas in Quantum Measurement Theory*, New York Academy of Science (1986), p. 133.
46. G. BADUREK, H. RAUCH, and D. TUPPINGER, *Phys. Rev. A* **34**, 2600 (1986).
47. H. RAUCH and J. P. VIGIER, *Phys. Lett. A* **151**, 269 (1990); **157**, 377 (1991).
48. U. BONSE, W. GRAEFF, R. TEWORTE, and R. RAUCH, *Phys. Status Solidi A* **43**, 487 (1977).
49. U. BONSE, W. GRAEFF, and H. RAUCH, *Phys. Lett. A* **69**, 420 (1979).
50. G. M. ALADZHADZHYAN, P. A. BEZIRGANYAN, O. S. SEMERDZHYAN, and D. M. VARDANYAN, *Phys. Status Solidi A* **43**, 399 (1977).
51. J. B. M. UFFINK and J. HILGEVOORD, *Phys. Lett. A* **105**, 176 (1984).
52. D. PETRASCHECK and H. RAUCH, *Acta Crystallogr. A* **40**, 445 (1984).
53. H. RAUCH, U. KISCHKO, D. PETRASCHECK, and U. BONSE, *Z. Phys.* **51**, 11 (1983).
54. H. RAUCH, *Proc. Symp. Found. Modern Physics*, Joensuu 1990, World Sci. Publ. (1991), p. 347.
55. B. D. JOSEPHSON, *Rev. Mod. Phys.* **46**, 251 (1974).
56. F. MEZEI (ed.), *Neutron Spin Echo*, Lecture Notes in Physics **128**, Springer-Verlag, Berlin (1980), p. 180.
57. G. BADUREK, H. RAUCH, and A. ZEILINGER, in Ref. 56, p. 136.
58. H. RAUCH, in Ref. 7, p. 161.
59. H. KAISER, S. A. WERNER, and E. A. GEORGE, *Phys. Rev. Lett.* **50**, 560 (1983).
60. R. CLOTHIER, H. KAISER, S. A. WERNER, H. RAUCH, and H. WÖLWITSCH, *Phys. Rev. A* **44**, 5357 (1991).
61. M. HEINRICH, H. RAUCH, and H. WÖLWITSCH, *Physica B* **156/157**, 588 (1989).
62. H. RAUCH, H. WÖLWITSCH, R. CLOTHIER, H. KAISER, and S. A. WERNER, *Phys. Rev. A* (July, 1992), in press.
63. P. A. BEZIRGANYAN, F. O. EIRAMDSHYAN, and K. G. TRUNI, *Phys. Status Solidi A* **20**, 611 (1973).
64. M. HEINRICH, D. PETRASCHECK, and H. RAUCH, *Z. Phys. B* **72** 357 (1988).
65. B. YURKE, *Phys. Rev. Lett.* **56**, 1515 (1986).
66. C. DEWDNEY, *Phys. Lett. A* **109**, 377 (1985).
67. C. DEWDNEY, P. R. HOLLAND, and A. KYPRIANIDIS, *Phys. Lett. A* **119**, 259 (1986).
68. A. G. KLEIN and S. A. WERNER, *Rep. Prog. Phys.* **46**, 259 (1983).
69. D. GREENBERGER, *Rev. Mod. Phys.* **55**, 875 (1983).
70. H. RAUCH, *Contemp. Phys.* **27**, 345 (1986).
71. S. A. WERNER and A. G. KLEIN, *Methods Exp. Phys.* **23**(A), 259 (1986).
72. V. F. SEARS, *Neutron Optics*, Oxford University Press, London (1989).

CHAPTER 14

GEDANKEN EXPERIMENTS ON DUALITY

LUIZ CARLOS RYFF

1. APPROACH

If we accept experimental results[1–3] as evidence of violations of Bell's inequalities,[1,4] we either have to abandon realism or introduce some faster-than-light (FTL) interaction.* This is indeed quite an amazing result, one which deserves thorough theoretical and experimental investigation. My approach is a first attempt in this direction. As I intend to show, nonlocal realism, just like local realism, can be investigated experimentally.

I have considered experiments that are combinations of experiments on wave–particle duality[6] and on nonlocality.[2] In an experiment to test Bell's inequalities using two-channel polarizers, a photon of a correlated pair has either to follow one or the other of two possible paths. On the other hand, in an experiment with a Mach–Zehnder interferometer the photon has to follow both paths at the same time. An appealing idea is to combine these experiments in order to find out what will then happen to the photon. From the standpoint of realism, what happens to the photon at the two-channel polarizer must be independent of the kind of experiment that is being performed.

The following question can be raised when a FTL interaction is assumed: when is this interaction triggered?† According to realism, and contrary to the usual interpretation, this triggering has to be caused by some physical process, whether

*Actually, some loopholes still remain, and we may try to explain the experimental results using a local realistic approach.[5]

†This question is obviously related to that of knowing just when the collapse of the state vector takes place, in the usual interpretation of quantum mechanics. Owing, however, to the essentially subjectivistic character of this interpretation, no satisfactory answer can be expected here.

LUIZ CARLOS RYFF • Universidade Federal do Rio de Janeiro, Instituto de Física, Cidade Universitária, 21945 Rio de Janeiro - RJ, Brazil.

Wave–Particle Duality, edited by Franco Selleri. Plenum Press, New York, 1992.

it is observed or not. In the cases to be examined, there are two evident possibilities: splitting of the photon at the polarizer or photon detection at the photomultiplier.

Even if we assume that quantum mechanical predictions for the experiments *so far* performed are correct, it seems natural to suppose that the assumption of realism may lead to experimental consequences in conflict with the usual interpretation of quantum mechanics. Naturally, there is no rigorous mathematical procedure to show this. Nonetheless, there are interesting consequences if we start with some simple, physically sound assumptions.*

Many of the results to be presented here have already been discussed elsewhere.[7–9] In the present chapter these conclusions will be made more rigorous and will be extended to new situations.

2. WAVE–PARTICLE DUALITY

In the first experiment to be considered,[6] a light wave packet containing a single photon impinges on a beam splitter (the argument would follow the same lines if it impinged on a two-channel polarizer instead). Two mutually exclusive phenomena may then be observed: photon anticorrelation (the photon follows one of two possible paths) or photon interference (using a Mach–Zehnder interferometer). Consistent with realism, I will assume that a photon that impinges on a beam splitter has no means of "guessing" the kind of experiment to which it will be subjected. Hence, taking into account the possibility of observing interference phenomena, I have to assume that the initial wave packet is split into two wave packets at the beam splitter. I will then tentatively assume that the wave packets are identical in every respect, both being capable of producing a detector click. Since, when an anticorrelation experiment is performed, only one detector can and must click each time, I also have to assume that some aleatory factors make one detector click before the other and that at the same time a signal is sent to prevent the other detector from clicking as well. This idea could easily be checked by removing one of the detectors; since there would then be no neutralizing signal, the remaining detector would detect twice the number of events, as if the photon always followed the same path—and this is not physically sound.

An alternative is to assume that the probability of a packet producing a click is $\frac{1}{2}$, and that the first packet impinging on a detector sends a message to the other packet informing it whether the first packet has been detected or *not*. If detected, the probability that the second packet will produce a click thus drops to zero; if not, it jumps to one.

*Although the terms *simple* and *physically sound* may be considered somewhat vague, I think their meaning will become quite clear in the present context. Something similar occurs in the interpretation of the experimental tests of Bell's inequalities. We can consider that local realism has been disproved *if* and *only if* the possibility of a conspiracy of nature is discarded. On the other hand, what is meant by *conspiracy of nature* is open to debate.

Naturally, more complicated situations, involving many polarizers and beam splitters, could also be devised. Whenever a detector is not triggered, this information will be sent to the other packets and probabilities then readjusted appropriately. Although this is a logical possibility, it seems quite unreasonable. (As will be shown later, it seems still more unreasonable when nonlocality is taken into account.)

I will then assume that the photon wave packet is split into two different wave packets upon passing through either a beam splitter or a two-channel polarizer: one incapable of producing a detector click and the other capable of doing so, that is, an *empty* and a *photonic* wave packet. From the standpoint of realism, this is an objective fact, independent of any observation or detection. This does not exclude the possibility of having the empty and photonic packets recombine after passing through a beam splitter or a two-channel polarizer, in accordance with the wave-like properties of light. I will then adopt the pilot-wave interpretation as introduced by de Broglie, according to which[(10)]

(a) Light is composed of photons *and* waves. The latter account for the wave-like properties of light, while the former, which are "guided" by the waves, account for its particle-like properties.*

3. NONLOCALITY

In the second experiment to be analyzed,[(2)] two correlated photons, ν and ν', are emitted in opposite directions and made to pass through two-channel polarizers, with different orientations, before being detected. As has been emphasized,[(1)] and experimentally corroborated,[(11)] in order to maintain a realistic viewpoint, we must assume some sort of FTL interaction to explain the observed correlations.† Consistent with realism, I will assume that this interaction is triggered by some physical process. In the present case, there are two possible candidates: (1) photon detection at the photomultiplier and (2) splitting of the wave packet at the polarizer.

I will then consider a slightly modified version of the experiment in Ref. 2, in which one of the detectors is removed (Figure 1).‡ Whenever only ν' is detected, I know (I am considering the ideal case) that ν has followed the path that does not have a detector, and this can also be taken as a kind of detection. The probabilities

*The term *photon* in (a) is used in the broad sense and does not necessarily imply the existence of a localized particle inside the photonic packet.

†Actually, local realistic approaches based on experimental limitations are possible.[(5)]

‡I will not use the usual ν_1 and ν_2 to refer to the photons, since here it does not matter which photon is emitted first, but rather which arrives at the polarizer or is detected first. In particular, the source of Figure 1 can either be the same as that of Ref. 2 or as that of Ref. 3, in which case the photons are emitted simultaneously. ν (ν') denotes the photon emitted toward the left (right) in the drawing.

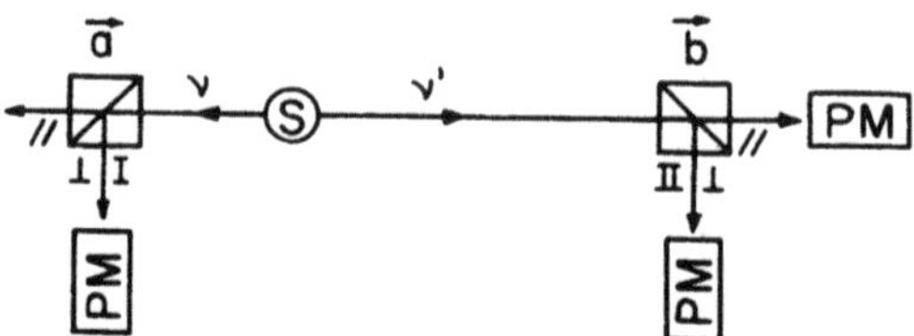

FIGURE 1. A source (S) emits two correlated photons (ν, ν'). ν impinges on pol. I and is detected before ν' impinges on pol. II. One of the detectors (which could register photon ν) has been removed.

of coincident detections may then be determined. The usual interpretation of quantum mechanics says we must obtain the same results we would obtain with the detector in place. Therefore, if realism agrees with quantum mechanics on this particular point, no detection is needed to trigger the FTL interaction.*

The experiment of Figure 1 deserves closer examination. According to quantum mechanics,

$$p(a, b) = p(a_\perp, b_\perp) = \tfrac{1}{2}\cos^2(a, b) \tag{1a}$$

and

$$p(a, b_\perp) = p(a_\perp, b) = \tfrac{1}{2}\sin^2(a, b) \tag{1b}$$

where $p(a, b_\perp)$ is the probability of ν' being detected in a polarization state perpendicular to $\vec{b}$ while ν is indirectly detected in a polarization state parallel to $\vec{a}$, and so on. Hence,

$$p(b) = p(a, b) + p(a_\perp, b) = p(a, b_\perp) + p(a_\perp, b_\perp) = p(b_\perp) = \tfrac{1}{2} \tag{2}$$

as expected. On the other hand, if the FTL interaction is triggered by photon detection, we must have

$$p(a, b) = \tfrac{1}{2}\gamma,\ p(a, \mathrm{b}_\perp) = \tfrac{1}{2}(1 - \gamma) \tag{3}$$

where $\gamma[(1 - \gamma)]$, the probability of ν' being transmitted (reflected) at pol. I, must be different from $\cos^2(a, b)$ $[\sin^2(a, b)]$, since the quantum mechanical result cannot be reproduced in this case. (In fact, if all the detectors are removed in Figure 1,† the other probabilities can also be written in terms of γ, and the

*It would be interesting to examine the consequences of this conclusion for existent nonlocal realistic theories.[12]

†Although the photons cannot be observed without the detectors, that they are transmitted or reflected at the polarizers can be an objective fact, from the standpoint of realism.

correlation function calculated. Naturally, Bell's inequalities would have to be satisfied, since we would then have local realism.) On the other hand, the other probabilities would still be expressed by (1a,b) and (2) would not be satisfied anymore. By simply determining the ratio $p(b)/p(b_\perp)$, we would thus be able to know if a distant detector had been removed or not. Superluminal communication would then be possible.

An alternative, if we insist on the idea of detection triggering but reject superluminal communication, is to reintroduce the supposition that the first packet impinging on a detector sends a message to the other packets informing whether the first packet has been detected or not. In the present context, however, this idea will be still more unreasonable than before. To see this, we need only place a beam splitter between the source and the polarizer on the path followed by photon ν. When the left-hand detector is not triggered, photon ν has either been transmitted at the polarizer or reflected at the beam splitter. Hence, to obtain the correct probabilities of detection, the FTL signal has to force ν' into a state such that the different possibilities are taken into account. In other words, ν' has to be informed about the probabilities of ν being transmitted at the polarizer or reflected at the beam splitter.

I will then assume that the FTL interaction is triggered whenever the first photon of a correlated pair is split at a polarizer into an empty and a photonic packet. In this case, to be in perfect agreement with quantum mechanics, we have to assume that the second photon is forced into the same state as the first photon. A variation of the experiment represented in Figure 2 serves in illustration. In this variation, the distance between the source and the polarizers is such that ν always arrives at pol. I and is detected before ν' can reach pol. II. Many different optical devices, such as retarders, polarizers, absorbers, and phase shifters, can be placed along the two arms of the interferometer. If ν is found in a polarization state perpendicular to $\vec{a}$, for example, and we want to know in which state ν' can

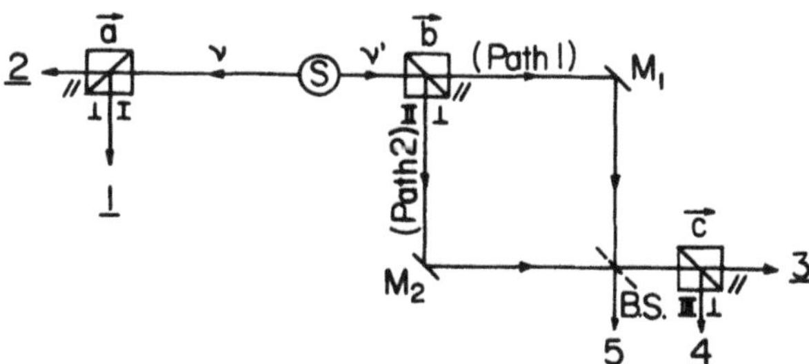

FIGURE 2. A source (S) emits a pair of correlated photons (ν, ν'). The right-hand part of the apparatus consists of a Mach–Zehnder interferometer which allows light reaching pol. III to be found in the same polarization state as when it reached pol. II. The distance from pol. I to the source can be varied, as can be the orientations of pol. I and III.

be found at $\underline{5}$, and with what probability, we need only assume that ν' impinges on pol. II in the same state in which ν was found (i.e., perpendicular to $\vec{a}$) and then apply our knowledge of wave optics.

In the usual approach, the previous conclusion is correct only when ν is in fact detected at $\underline{1}$ while according to my approach it is correct whenever splitting at the polarizer forces ν into a polarization state perpendicular to $\vec{a}$, *even if no detection occurs*. I will then assume that

(b) When the first photon wave packet of a correlated pair is split at a two-channel polarizer into an empty and a photonic packet, the second photon wave packet is forced into the same polarization state as that of the photonic packet.*

In other words, when the first photon of a correlated pair is forced into a certain polarization state upon passing through a polarizer, the second one is forced into the same state.

4. COMBINING EXPERIMENTS ON NONLOCALITY AND WAVE–PARTICLE DUALITY

To make predictions for the experiment represented in Figure 2, some extra assumptions are needed. We have to consider whether (and in which circumstances) the passage of the first photon of a correlated pair through a polarizer is a reversible or an irreversible process, insofar as after this passage whatever happens to either one of the distant photons may or may not influence the other. In order to clarify this point, I will initially discuss the simple experiment represented in Figure 3. The distances between the source and the polarizers are such that we are certain that ν' will reach pol. III before ν reaches pol. I. In this case, if ν' is detected at $\underline{3}$, ν impinges on pol. I in a polarization state parallel to $\vec{b}$, since ν' has passed through pol. II before reaching pol. III. Naturally, at least in the case in which the empty and photonic waves do not recombine, the following conclusion can be drawn:

(c) Whatever may happen to either one of the photonic packets following the process described in (b), this will have no effect on the other, distant photonic packet.

*Actually, assumption (b) applies only to entangled states of the same type, as that produced in the experiments of Ref. 2, for example. In the case of the decay of the positronium,[13] the second photon would be forced into a state orthogonal to that of the photonic packet. Naturally, (b) can be adapted for experiments using one-channel polarizers.

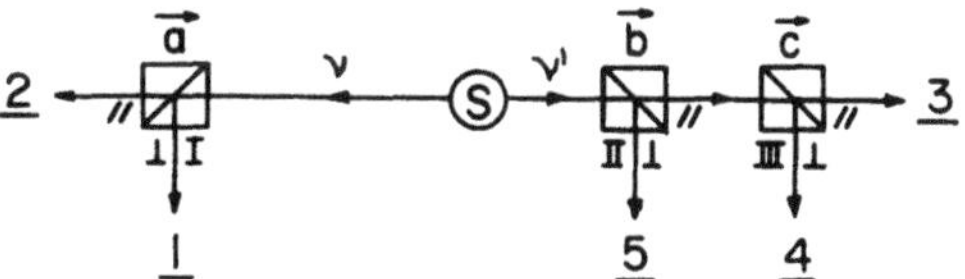

FIGURE 3. A source (S) emits a pair of correlated photons (ν, ν'). $L_1 > L_3$, where L_1 (L_3) is the distance from pol. I (III) to the source.

In an attempt to maintain complete agreement with the usual predictions of quantum mechanics, we could assume, in the case of the experiment represented in Figure 2, that (c) does not apply when the photonic and empty packets are recombined. However, this assumption is not in itself sufficient for achieving the aforementioned agreement, as can be seen from the experiment represented in Figure 4. According to the usual approach, when the removable mirror (RM) is not in place and path 3 is freed, $(P_{24})_Q$ (RM removed) = 0, where P_{24} is the probability of detecting ν at 2 and ν' at 4. This result, easily derived from quantum mechanical formalism, can be inferred from the fact that the correlation state is such that ν and ν' must be found in the same polarization state. Now I will consider the situation in which a photonic packet follows path 2 and an empty one, path 1. According to (b), ν' will reach pol. II in a polarization state perpendicular to $\vec{a}$, and as a consequence, no empty wave will propagate along path 3. Hence, when ν follows path 2, it is irrelevant whether path 3 is blocked or not, since there is no empty wave propagating along path 3 to recombine with the photonic wave propagating along path 4. But if path 3 is blocked—and we can see that things happen as if it were—the same conclusion (c), derived from the experiment in

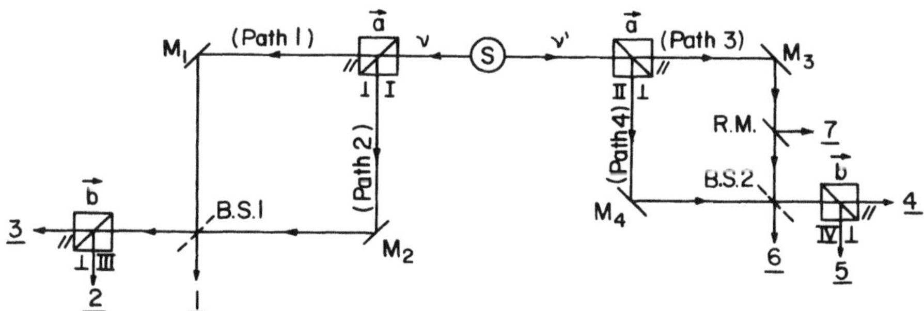

FIGURE 4. A source (S) emits a pair of correlated photons (ν,ν'). The apparatus consists of two Mach–Zehnder interferometers, which allow light reaching pol. III (IV) to be found in the same polarization state as when it reached pol. I (II). The distances are such that ν impinges on pol. I before ν' can reach pol. II, and ν' impinges on pol. IV before ν can reach pol. III. R.M. is a removable mirror, which allows us to perform two kinds of experiments: with the R.M. in place and with it removed.

Figure 3, would be valid. A similar line of reasoning is valid when ν follows path 1. Therefore, we would obtain $(P_{24})_R$ (RM removed) = $(P_{24})_R$ (path 3 blocked) $+(P_{24})_R$ (path 4 blocked) $\neq 0$, in strong disagreement with the quantum mechanical result.

Although in principle, since there is simply no empty wave to recombine in the above experiment, it could be used to test nonlocal realism, even if assumption (c) does not hold when the photonic and empty waves recombine, in reality some questions may be raised. I have been discussing an ideal situation but the photons actually are not perfectly correlated and the polarizers display some degree of leakage. How correlated will ν and ν' be in this case, after the waves recombine? Strictly speaking, from the standpoint of realism and considering our present knowledge, this kind of question can only be answered through experimentation. However, some inferences are possible. We can imagine different situations in the experiment of Figure 4, such that pol. I and II do not have the same orientation. The intensities of the empty and photonic waves on the arms of the right-hand interferometer can thus be varied. If the usual quantum mechanical results are still observed, we may conclude that relative intensity has no effect on these results. However, relative intensity can also be varied by placing an absorber along one of the arms of the interferometer and maintaining the initial orientations of the polarizers. The results obtained would then differ from the previous ones. For example, if ν' is detected in a polarization state parallel to $\vec{b}$, ν is not forced into the same state, thus contradicting the previous conclusion. We may then complicate the model still further by imaging that a FTL signal travels backward along the arms of the interferometer, and what these signals find on their way is what affects the results. For instance, one signal may or may not impinge on an absorber, depending on the experiment that is being performed, and this will change the experimental outcome. A little reflection shows that the signal would have to be endowed with properties similar in some respects to those of the usual electromagnetic waves. In principle, this possibility could be tested by placing Pockels cells on the paths of the photons, for example, so that the wave following one of the two possible paths (either 3 or 4) is blocked, while the signal traveling backward is allowed to pass.*

*An experiment on the same line, in which the photon paths are intercepted by strong laser beams, has been proposed.[14] Although not devised with this intention, the experiment of Ref. 11, using acousto-optical switches, might help to clarify this point. However, the experiment cannot be considered conclusive, since we can always imagine that the FTL signal follows backward along the path left by the photon, not being acted on by the switches. It might be interesting to perform an experiment in which a Faraday rotator is placed on the path followed by ν'. Since in this case the direction of rotation of the plane of polarization does not depend on the direction of propagation of the impinging light, different results from those predicted by quantum mechanics would be obtained whenever ν' is detected before ν has reached Pol. I, if the possible FTL signal indeed has properties similar to those of the usual electromagnetic waves. From a practical viewpoint, this experiment has the advantage of not requiring any kind of switch.

The above digression is intended to show that even more cumbersome hypotheses are in principle testable. However, from a conceptual standpoint, it would appear more consistent to assume that (c) is valid even when the empty and photonic waves recombine. In other words, after the process described in (b), the two-photon system is found to be in a disentangled state that can no longer be changed back into an entangled state. This possibility is akin to the concept of collapse of state vector. However, in a realistic approach this disentanglement is an objective fact, whether it is observed or not. This will be discussed in detail in the next section.

5. TESTING THE PROPOSED APPROACH

I will discuss the experiment represented in Figure 2, assuming perfect correlation and polarizers but considering that not all photons are collected and that available detectors are far from ideal. This will not place any essential restriction on conclusions, since available polarizers and the observed correlated states for the angle usually subtended by the collecting lenses are very near ideal. On the other hand, the realistic approach yields results that strongly disagree with the usual interpretation of quantum mechanics.

Two different situations will be considered. In the first, the distances between the source and the polarizers are such that photon ν impinges on pol I *before* ν' has reached pol. II. The interferometer is designed so that ν' impinges on pol. III in the same polarization state in which it impinges on pol. II. I will calculate the probabilities P_{23} and P_{24} of coincident detection. The quantum mechanical result can be obtained in a straightforward manner, which, as a consequence of assumption (b), is also valid for the realistic approach. Let f (f') be the probability of ν (ν') entering the left (right) collimating system; let η (η') be the efficiency of the left (right) detectors, and let g' (g) be the probability of ν' (ν) entering the right (left) collimating system if ν (ν') has entered the left (right) collimating system. The probability of ν being found in a polarization state parallel to $\vec{a}$ is $\frac{1}{2}$; ν' is then forced into the same polarization state. Since the probability of ν' reaching pol. III is $\frac{1}{2}$, using Malus's law it follows that

$$(P_{23})_Q = (P_{23})_R = \tfrac{1}{4}\eta\eta' fg' \cos^2(a, c) \tag{4a}$$

$$(P_{24})_Q = (P_{24})_R = \tfrac{1}{4}\eta\eta' fg' \sin^2(a,c) \tag{4b}$$

Hence,

$$\left(\frac{P_{24}}{P_{23}}\right)_Q = \left(\frac{P_{24}}{P_{23}}\right)_R = \left(\frac{R_{24}}{R_{23}}\right) = \tan^2(a, c) \tag{5}$$

where R_{23} and R_{24} are the measurable detection rates.

In the second situation to be considered, the distances between the source and the polarizers are such that photon ν impinges on pol. I *after* ν' has reached pol. II. Looked at from the usual standpoint, the situation must still be the same. Hence, according to (4a,b)

$$\left(\frac{P'_{24}}{P'_{23}}\right)_Q = \left(\frac{P_{24}}{P_{23}}\right)_Q = \left(\frac{R_{24}}{R_{23}}\right) = \tan^2(a,\, c) \tag{6}$$

On the other hand, if assumption (c) holds true, we are faced with a new situation. Let $A(b,\, c)$ $[B(b,\, c)]$ be the probability of ν' being transmitted at pol. III when it follows path 1 (2). Summing the probabilities arriving at using (b) and (c) (which represent two mutually exclusive possibilities), we easily obtain*

$$(P'_{23})_R = \eta\eta' f' g[A(b,\, c)\cos^2(a,\, b) + B(b,\, c)\sin^2(a,\, b)] \tag{7}$$

since ν is forced into a state parallel (perpendicular) to $\vec{b}$ when ν' is transmitted (reflected) at pol. II. Similarly, let $A'(b,\, c)$ $[B'(b,\, c)]$ be the probability of ν' being reflected at pol. III when it follows path 1 (2). Hence,

$$(P'_{24})_R = \eta\eta' f' g[A'(b,\, c)\cos^2(a, b) + B'(b,\, c)\sin^2(a,b)] \tag{8}$$

Since light impinging on pol. III is in a totally unpolarized state, then

$$A(b,\, c) + B(b,\, c) = A'(b,\, c) + B'(b,\, c) \tag{9}$$

Formulas (7) and (8) conflict with (4a,b). In particular, when angle $(a,\, b) = 45°$, using (9) we obtain

$$\left(\frac{P'_{24}}{P'_{23}}\right)_R = \left(\frac{R_{24}}{R_{23}}\right) = 1 \tag{10}$$

independent of angle (a,c) and in disagreement with (6).

6. ANOTHER POSSIBLE EXPERIMENT

Assumptions (b) and (c) cannot deal with all imaginable situations. In particular, the experiment represented in Figure 5 raises curious and interesting questions. The interferometer is designed in such a way that the parallel components of the incident beam following direction $\underline{4}$ ($\underline{5}$) via path 1 and 2 interact constructively (destructively). The perpendicular component reaching beam split-

*A calculation in terms of hidden variables can be found in Ref. 7.

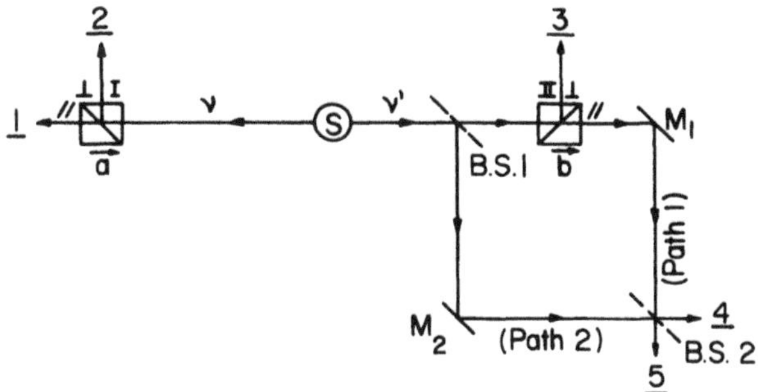

FIGURE 5. A source (S) emits a pair of correlated photons (ν, ν'). The right-hand part of the apparatus consists of a Mach–Zehnder interferometer with pol. II located along one of its arms, allowing light following direction 5 to be found perpendicularly polarized to direction $\vec{b}$.

ter 2 (BS 2) via path 2 is distributed equally in directions $\underline{4}$ and $\underline{5}$. As a consequence, photon ν' following direction $\underline{5}$ is always found in a polarization state perpendicular to direction $\vec{b}$.

I will only consider the following sequence of events, which suffices to clarify the kind of question a nonlocal realistic approach may raise: (1) ν' impinges on BS 1 and is split. A photonic wave packet follows path 2 and an empty one follows path 1 and is split at pol. II. (2) After that, ν is detected at $\underline{2}$. Assuming (and this can be experimentally verified) that neither the splitting of ν' at BS 1 nor the splitting of the empty wave packet at pol. II disentangles the two-photon state, ν' is then forced into a polarization state perpendicular to $\vec{a}$, according to (b).

What happens to the empty wave packet following path 1? To simplify the reasoning, I will consider that $\vec{a} = \vec{b}$. If nothing happens, then, according to (a), when the beams following paths 1 and 2 recombine at BS 2, there is a nonnull probability that ν' will be found at $\underline{5}$ not in a polarization state perpendicular to $\vec{b}$, since the beams are in mutually orthogonal polarization states. Hence, by simply observing the polarization state of the photon following direction $\underline{5}$, we are able to infer whether ν has been made to pass through a polarizer or not. A superluminal telegraph might be conceived in this manner.

An alternative, albeit a complicated one, is to assume that the splitting of ν also changes the state of the empty wave packet in such a way that things happen as if ν' had already been in a state of polarization perpendicular to $\vec{a}$ prior to impinging on BS 1. When $\vec{a} = \vec{b}$, the amplitude of the empty wave packet following path 1, which has already passed the polarizer, thus suddenly drops to zero. When $\vec{a} \neq \vec{b}$, the amplitude will have to be modified appropriately in order to ensure that ν' will always be found at $\underline{5}$ in a polarization state perpendicular to $\vec{b}$.

Therefore, the combination of a FTL interaction with the pilot-wave concept leads to a situation in which superluminal communication cannot be discarded *a priori*. This should be viewed as a natural conclusion since, if we accept a FTL interaction, superluminal communication becomes an open possibility. Moreover, to accept a FTL interaction we have only to accept quantum mechanical nonlocality and the existence of a real external world with definite properties—and these can hardly be considered bizarre ideas. Conversely, we may interpret the proposed experiment as an attempt to determine the consequences of assuming that superluminal communication is impossible, in a nonlocal realistic approach. As seems to become evident, from the standpoint of realism this kind of question can only be answered through experimentation.

7. COMPARISON WITH CONCRETE SITUATIONS

As was underscored in Section 5, the results then obtained would not have been essentially any different if imperfect correlation and polarizers had been assumed. Nevertheless, it is not immediately obvious how assumption (b) can be extended to make it consistent with quantum mechanical nonlocality in nonideal situations. Thus, it is important, at least for conceptual reasons, to try to clarify this point. This section shows how this can be done.

To indicate the degree of correlation between two photons, I will introduce the *pair correlation coefficient*, α, whose value can vary from zero to one. To make this point clear, I will initially consider the ideal case of perfect polarizers. I will assume that if the first photon wave packet of a correlated pair is split into an empty and a photonic packet upon passing through a polarizer, and if the pair correlation coefficient is α, then a fraction α of the second photon wave packet is forced into the same polarization state as the photonic packet, and a fraction $(1 - \alpha)$ remains unpolarized.* When $\alpha = 1$, there is a perfect correlation; when $\alpha = 0$ there is no correlation at all.

I will consider only the pairs of photons actually reaching the polarizers. Let $p_\theta(\alpha)\,d\alpha$ be the probability of the pair correlation coefficient falling between α and $\alpha + d\alpha$, where θ is the half-angle subtended by the collecting lenses. Then

$$\int_0^1 p_\theta(\alpha)\,d\alpha = 1 \tag{11}$$

When $\theta = 0$, there is a perfect correlation: $p_0(1) = 1$, $p_0(\alpha \neq 1) = 0$; but a null probability of detecting a photon.[15]

In concrete situations, the packet will emerge from the polarizer in a partially polarized state. Then, if a fraction ρ of the photonic packet is in a polarized state, a fraction $\rho\alpha$ of the second wave packet will be forced into the same polarization

*In a hidden-variables approach, we have to assume that a fraction α is forced into a polarization state and a fraction $(1 - \alpha)$ remains in the same initial hidden-variables state.

state as the fraction ρ of the photonic packet, and a fraction $1 - \rho\alpha$ will remain unpolarized. Thus, assumption (b) can be substituted by assumption:

(b′) When the first photon wave packet of a correlated pair whose pair correlation coefficient is α is split at a polarizer into an empty and a photonic wave packet such that a fraction ρ of the photonic packet is polarized, the second photon wave packet is forced into a partially polarized state such that a fraction $\rho\alpha$ is in the same polarization state as that of the photonic packet.

I will now show how the pair correlation coefficient α can be related to the usual correlation coefficient $F(\theta)$, present in the expression for the probability of coincident detections that are used in experimental tests of Bell's inequalities. In the experiment discussed in Refs. 1 and 15, two correlated photons, ν_1 and ν_2, impinge on one-channel polarizers oriented parallel to $\vec{a}$ (pol. I) and $\vec{b}$ (pol. II), respectively. Let ϵ_M^1 and ϵ_m^1 (ϵ_M^2 and ϵ_M^2) be the transmittances of the first (second) polarizer for light polarized parallel and perpendicular to the polarizer axis. Hence, $p_1(a)$ and $p_2(b)$, the probabilities of ν_1 and ν_2 being detected, are

$$p_1(a) = \tfrac{1}{2}\eta_1 f_1 \epsilon_+^1 \tag{12a}$$

and

$$p_2(b) = \tfrac{1}{2}\eta_2 f_2 \epsilon_+^2 \tag{12b}$$

where f_1 (f_2) is the probability of ν_1 (ν_2) reaching pol. I (II), η_1 (η_2) is the efficiency of the first (second) photomultiplier, and

$$\epsilon_\pm^1 = \epsilon_M^1 \pm \epsilon_m^1,\ \epsilon_\pm^2 = \epsilon_M^2 \pm \epsilon_m^2 \tag{13}$$

The incident beam can be decomposed into two incoherent components, one parallel and the other perpendicular to the axis of the polarizer. A fraction ϵ_M (ϵ_m) of the parallel (perpendicular) component is transmitted at the polarizer. Hence, formulas (12a,b) follow naturally from assumption (a) of Section 2, since the probability of ν_1 (ν_2) being transmitted at pol. I (II) is $\epsilon_+^1/2$ ($\epsilon_+^2/2$). Part of the parallel component combines with the perpendicular component, producing an unpolarized component. Thus, the fraction of the beam emerging from pol. I in a polarization state parallel to $\vec{a}$ is $\rho = \epsilon_-^1/\epsilon_+^1$. A fraction $\rho\alpha$ of the second beam is then forced into a state parallel to $\vec{a}$, and a fraction $(1 - \rho\alpha) = (\epsilon_+^1 - \epsilon_-^1\alpha)/\epsilon_+^1$ remains unpolarized. The component parallel to $\vec{a}$ can be decomposed into components parallel and perpendicular to $\vec{b}$. Using Malus's law plus the definition of ϵ_M and ϵ_m, we see that the fraction of the component parallel to $\vec{a}$ that is transmitted at pol. II is $(\epsilon_M^2 \cos^2 \Phi + \epsilon_m^2 \sin^2 \Phi)$, where Φ = angle (a, b). Since the fraction of the unpolarized component that is transmitted at pol. II is $\epsilon_+^2/2$, the probability of coincident detection as a function of α can be written as

$$p(a, b, \alpha) =$$

$$\eta_1\eta_2\frac{\epsilon_+^1}{2}f_1g\left[\frac{\epsilon_-^1}{\epsilon_+^1}\alpha(\epsilon_M^2\cos^2\Phi + \epsilon_m^2\sin^2\Phi) + \frac{\epsilon_+^1 - \epsilon_-^1\alpha}{\epsilon_+^1}\frac{\epsilon_+^2}{2}\right] \tag{14}$$

where g is the conditional probability. (14) can be rewritten:

$$p(a, b, \alpha) = \tfrac{1}{4}\eta_1\eta_2 f_1 g(\epsilon_+^1\epsilon_+^2 + \alpha\epsilon_-^1\epsilon_-^2\cos 2\Phi) \tag{15}$$

Multiplying (15) by $p_\theta(\alpha)\,d\alpha$ and integrating, using (11), we obtain the usual result[1,15]

$$p(a, b) = \tfrac{1}{4}\eta_1\eta_2 f_1 g(\epsilon_+^1\epsilon_+^2 + F(\theta)\epsilon_-^1\epsilon_-^2\cos 2\Phi) \tag{16}$$

where the correlation coefficient, $F(\theta)$, is related to the mean value of α, through relation*

$$F(\theta) = \int_0^1 \alpha p_\theta(\alpha)\,d\alpha = \bar{\alpha} \tag{17}$$

It is also not obvious how assumption (c) can be modified when nonideal polarizers are considered. It is not strictly correct that whatever may happen to either one of the photons following the process described in (b′) will have no effect on the other, distant photon. This becomes clear when we realize that if ν_1 passes through two crossed polarizers (which is possible—although rare—in cases of real, imperfect polarizers) it will be found in a totally unpolarized state, and according to the usual approach the same must be true for ν_2. From a realistic standpoint, this raises some questions, which will not be discussed here and whose clarification would call for new experiments.

8. NONLOCALITY AND RELATIVITY

From the standpoint of special relativity, quantum mechanics raises some problems. As stressed by Dirac, nonlocality is against the spirit of relativity.[16] Even if we do not adhere to realism, the usual view presents a conceptual

*Naturally, the present derivation also applies to experiments using two-channel polarizers. Instead of ϵ_M and ϵ_m, we would use transmission ($T^\parallel, T^\perp$) and reflection ($R^\parallel, R^\perp$) coefficients. The theoretical determination of $F(\theta)$ falls outside the scope of the present approach, which is not intended to substitute quantum mechanics.

difficulty. To show this, I will again consider the experiment represented in Figure 1 but with all four polarizers. In the laboratory frame, ν is detected before ν'. Let the detections of ν and ν' be two events separated by a space-like interval. We can then consider another inertial frame, moving in the same direction as ν' and such that in it ν' is detected before ν. For an observer in the first frame, measuring the state of ν forces ν' into a definite polarization state. Hence, ν' impinges on pol. II in a definite polarization state. Since the Lorentz transformation connecting the two frames does not change the polarization state of ν', the observer in the laboratory frame is able to infer that the same conclusion is valid for an observer in the moving frame. But for the second observer, it is measuring the state of ν' that forces ν into a definite polarization state. Hence, he concludes that ν' impinges on pol. II in an unpolarized state, since to him, when ν' impinges on pol. II the system is still in an entangled state and therefore ν' is in no definite polarization state. Although this is only a conceptual contradiction, with no observable consequences, it *is* a contradiction. The point is that measuring the state of one of the photons of a correlated pair does not inform us as to the state the other photon was in before the measurement; it *forces* the other photon into a definite state. Even if we adhere to the usual interpretation of quantum mechanics, this seems difficult to understand if we do not accept some FTL interaction. If we do accept this notion, however, it is important to know *when* the interaction is triggered. This is indeed crucial and, as has been shown in this chapter, may have far-reaching consequences. It seems that any attempt to clarify the foundations of quantum mechanics will necessarily have ramifications on the special theory of relativity. For example, it seems that a privileged frame, where one of the photons is *really* detected before the other, would have to be introduced. Physicists, however, react strongly to this idea. According to Bell,(17) this possible frame might be an undetectable one. However, if we find a privileged frame difficult to swallow, a privileged, although undetectable frame is even harder to digest. Naturally, if we admit the existence of a privileged frame, the very reason for assuming a principle of relativity is eroded. There is no *a priori* reason why this frame should remain undetected. Therefore, while risking being regarded with suspicion by some of my colleagues, I will advocate Poincaré's viewpoint, according to which the relativity principle is an experimental fact and as such susceptible to endless revision.(18) Perhaps we eventually discover that the Lorentzian point of view is more correct than the Einsteinian.(19)

Some of my predictions seem to conflict with the special theory of relativity. For instance, the experiment proposed in Section 5 considered two situations. In the first (second), ν impinges on pol. I before (after) ν' reaches pol. II. Since the arrivals of each photon at each polarizer can be events separated by a space-like interval, ascertaining which event occurred first will depend on the frame used to describe the experiment. Quantum mechanical formalism, as usually interpreted, predicts the same probabilities for both situations [formula (6)], in agreement with the special theory of relativity. My approach, on the other hand, predicts two different probabilities [formulas (5) and (10)]. As also pointed out in Section 6, in a realistic approach, FTL communication cannot be discarded *a priori*.

Nevertheless, my approach is not necessarily incompatible with special relativity. In the experiment in Section 6, if the splitting of ν' at BS 1 disentangles the two-photon state, or if the splitting of ν at pol. I changes the state of the empty waves on the interferometer, no superluminal communication is possible. Moreover, if the privileged frame in which the probabilities are to be calculated turns out to be the frame in which the source is at rest, there would not, strictly speaking, be just one privileged frame but many—one for each source. Thus, to correctly calculate the probabilities, we would have to know the velocity of the source. This is a perfectly relativistic condition.

In my opinion, quantum nonlocality raises problems that cannot simply be ignored. Thus, it is important to adopt a coherent approach with consequences that can be experimentally investigated.

9. EXTENDING THE APPROACH TO THE CASE OF MOMENTUM CORRELATED PHOTONS

Nonlocality has also been verified in the case of momentum correlated photons.(20) I will now briefly outline how to extend my approach to make it testable here as well, based on an experiment that has recently been proposed(21) and that appears in an expanded version in Figure 6.(22) Phase shifters Φ_1 and Φ_2; beam splitters H_1 and H_2; mirrors M_A, M_B, M_C, and M_D; and detectors U_1, L_1, U_2, and L_2 are used in the original experiment. The quantum mechanical probabilities of coincident detections will be

$$p(U_1, U_2|\Phi_1, \Phi_2) = p(L_1, L_2|\Phi_1, \Phi_2) = \tfrac{1}{4}\eta^2[1 + \cos(\Phi_2 - \Phi_1 + \theta)] \qquad (18a)$$

and

$$p(U_1, L_2|\Phi_1, \Phi_2) = p(L_1, U_2|\Phi_1, \Phi_2) = \tfrac{1}{4}\eta^2[1 - \cos(\Phi_2 - \Phi_1 + \theta)] \qquad (18b)$$

where $p(U_1, U_2|\Phi_1, \Phi_2)$ is the probability of detecting photon 1 at U_1 and photon 2 at U_2, Φ_1 and Φ_2 are the phase shifts, θ depends on the detailed placement of the mirrors and beam splitters, and η is the quantum efficiency of the detectors. A Bell's inequality can be written and (18a,b) used to show that it is violated by quantum mechanics, characterizing a nonclassical correlation. As before, I will assume a FTL interaction. In this case, a measurement to determine the path followed by one of the photons modifies the beams associated to the other photon, establishing a definite phase different between them. To exemplify, I will consider the ideal situation in which $\Phi_2 - \Phi_1 + \theta = 0$. For instance, if photon 2 is detected at U_2, beams A and D acquire a phase difference such that no beam will

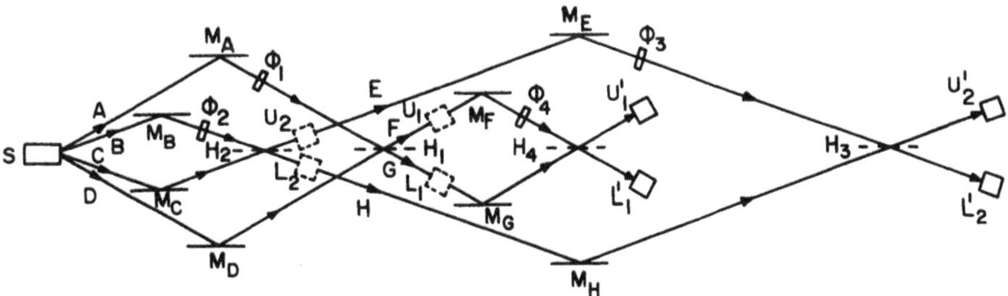

FIGURE 6. A source (S) emits two correlated photons, one into beams A and D, and one into B and C. $\Phi_1, \ldots; H_1, \ldots; M_A, \ldots$; and $U_1, \ldots$ and $L_1, \ldots$ are phase shifters, beam splitters, mirrors, and detectors, respectively. $U_1, \ldots$ and $L_1, \ldots$ indicate the positions of the detectors in the original experiment proposed by Horne, Shimony, and Zeilinger.

follow the path leading to L_1, in agreement with (18). According to my approach, this should be taken as evidence that no empty wave will follow this path. Now, assuming (Section 3) that no detection can trigger the FTL interaction, it has to be triggered by the splitting and recombination of the beams at the beam splitter.

The experiment in Figure 6 is similar to that in Figure 4, Section 4. I will also assume in the present case that the triggering of the action disentangles the two-photon state (as in the former case, there is no empty wave to recombine with the photonic wave). Therefore, using reasoning similar to that in Section 4, we are able to infer that the probabilities of coincident detections do not depend on Φ_3 and Φ_4, in a realistic approach, while according to the usual approach they do. In other words, according to the usual quantum mechanical approach, photon 1 follows both paths F and G at the same time, so to speak, but according to the realistic approach it will follow either one or the other, and this leads to different predictions.

REFERENCES

1. J. F. CLAUSER and A. SHIMONY, *Rep. Prog. Phys.* **41**, 1881 (1978).
2. A. ASPECT, P. GRANGIER, and G. ROGER, *Phys. Rev. Lett.* **49**, 91 (1982).
3. W. PERRIE, A. J. DUNCAN, H. J. BEYER, and H. KLEINPOPPEN, *Phys. Rev. Lett.* **54**, 1790 (1985).
4. J. S. BELL, *Physics* **1**, 195 (1964).
5. F. SELLERI, in: *Quantum Mechanics versus Local Realism: The Einstein–Podolsky–Rosen Paradox* (F. SELLERI, ed.), Plenum Press, New York (1988) and references therein; T. W. MARSHALL and E. SANTOS, *Phys. Rev. A* **39**, 6271 (1989).
6. P. GRANGIER, G. ROGER, and A. ASPECT, *Europhys. Lett.* **1**, 173 (1986).
7. L. C. B. RYFF, *Phys. Lett. A* **136**, 13 (1989).
8. L. C. B. RYFF, Paper presented at the Int. Conf. on the Conceptual Foundations of Quantum Theory, 28 December 1989–2 January 1990, New Delhi, India.
9. L. C. B. RYFF, *Found. Phys.* **20**, 1061 (1990).

10. F. SELLERI, in: *The Wave Particle Dualism* (S. DINER, D. FARGUE, G. LOCHAK, and F. SELLERI, eds.), Reidel, Dordrecht (1984).
11. A. ASPECT, J. DALIBARD, and G. ROGER, *Phys. Rev. Lett.* **49**, 1804 (1982).
12. D. BOHM, B. J. HILEY, and P. N. KALOYEROU, *Phys. Rep.* **144**(6), 323, 349 (1987); J. P. VIGIER, *Lett. Nuovo Cimento* **24**, 258, 265 (1979).
13. L. R. KASDAY, in: *Foundations of Quantum Mechanics* (B. D'ESPAGNAT, ed.), Academic Press, New York (1971).
14. A. SHIMONY, Proc. Int. Symp. Foundations of Quantum Mechanics, Tokyo, 1983.
15. A. SHIMONY, in: *Foundations of Quantum Mechanics* (B. D'ESPAGNAT, ed.), p. 182, Academic Press, New York (1971).
16. F. SELLERI and G. TAROZZI, *Riv. Nuovo Cimento* **4**(2), 1–53 (1981).
17. J. S. BELL, in: *Speakable and Unspeakable in Quantum Mechanics*, Cambridge University Press, Cambridge (1989), specifically papers 15, 16, 18, and 19.
18. M. A. TONNELAT, *Histoire du Principe de Relativité*, p. 125, Flammarion (1971).
19. A. K. MACIEL and J. TIOMNO, *Phys. Rev. Lett.* **55**, 143 (1985).
20. R. GHOSH and L. MANDEL, *Phys. Rev. Lett.* **59**, 1903 (1987); C. K. HONG, Z. Y. OU, and L. MANDEL, *Phys. Rev. Lett.* **59**, 2044 (1987); Z. Y. OU and L. MANDEL, *Phys. Rev. Lett.* **61**, 50, 54 (1988).
21. M. A. HORNE, A. SHIMONY, and A. ZEILINGER, *Phys. Rev. Lett.* **62**, 2209 (1989).
22. The experiment as originally proposed has been performed by: J. G. RARITY and P. R. TAPSTER, *Phys. Rev. Lett.* **64**, 2495 (1990).

CHAPTER 15

WIND EFFECT OF EMPTY QUANTUM WAVES IN A PFLEEGOR–MANDEL-TYPE EXPERIMENT FOR ELECTRONS

MICHAEL SCHMIDT

1. EXPERIMENTS SHOWING THE INTERFERENCE OF INDEPENDENT PHOTON BEAMS

Interference effects produced by the superposition of the light beams from two independent single-mode lasers take place even under conditions in which the light intensities are so low that, with high probability, one photon is absorbed before the next one is emitted by one or the other source. This was first demonstrated by Pfleegor and Mandel.[1] Their experimental setup is shown in Figure 1. From two independent, single-mode He–Ne lasers, light beams with aligned polarization are passed through two attenuators and superimposed at the interference detector R. A part of the beam is split off at the unsilvered glass plates M_1 and M_2 and passed through a pinhole to photomultiplier C. This photo tube functions to register the difference frequency of the two lasers in the form of a beat note. The interference detector R is only activated when the beat frequency falls below 50 kc/s. The interference could only be observed in time intervals on the order or less than the frequency spread of the light, in practice about 20 μs for the used lasers. Since the average number of photons registered per trial was only about 10, correlation techniques were required to demonstrate the interference.

The interference pattern was received on a stack of thin glass plates, each of which had a thickness corresponding to about half a fringe width. The plates were so arranged that light falling on the 1st, 3rd, 5th, etc. plate was reaching

MICHAEL SCHMIDT • Dipartimento di Fisica, Università di Bari, I-70126 Bari, Italy.

Wave–Particle Duality, edited by Franco Selleri. Plenum Press, New York, 1992.

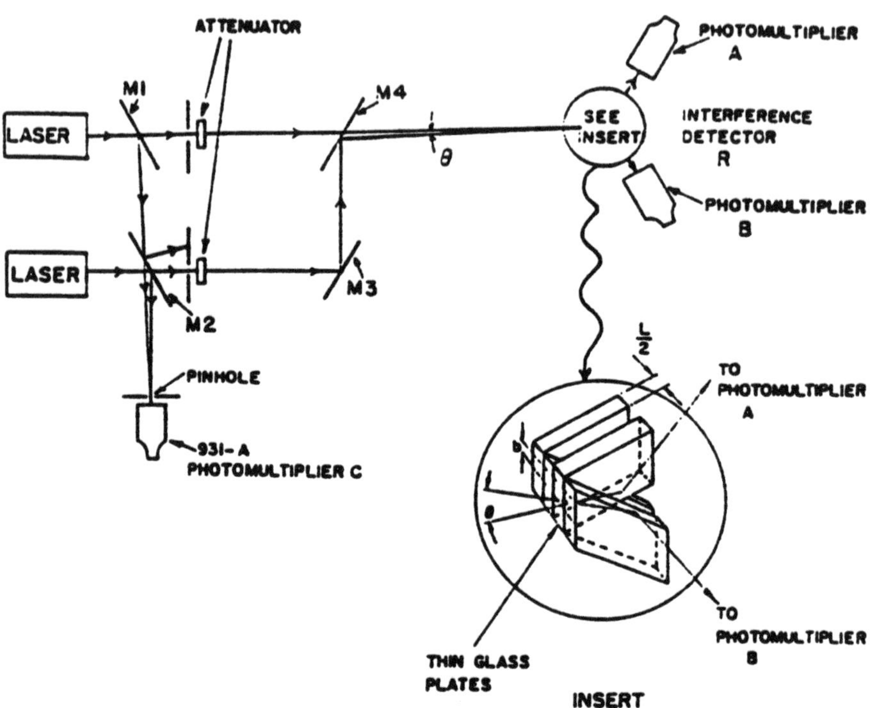

FIGURE 1. The experimental setup of Pfleegor and Mandel. M_1 and M_2 are mirrors; M_3 and M_4 are half-silvered glass plates and θ is the angle between the two light beams. The insert gives details of how the interference is detected.

photo tube A only, and light falling on the even-numbered plates (2, 4, 6, etc.) fed photomultiplier B only. If the half fringe spacing coincides with the plate thickness, the number of counts registered by one of the photomultipliers should increase, while the counting rate at the other detector should decrease. So a negative correlation between the number of counts of channels A and B is to be expected if there is interference between the two independent photon beams. The rather low statistical accuracy of the experiment allowed the existence of the effect to be proven with only four measured values of the correlation coefficient 2–3 standard deviations away from the no-interference line. Therefore, Pfleegor and Mandel(2) repeated the experiment in a modified form to achieve much higher accuracy. This time the correlation coefficient was found about 10 times to be 5–10 standard deviations away from the no-interference line. The results are shown in Figure 2. The existence of low-intensity interference was so established beyond any doubt.

Other authors have reported the interference of independent photon beams. Radloff,(3) for example, produced independent confirmation by recording many times the interference pattern (produced again by two independent single-mode

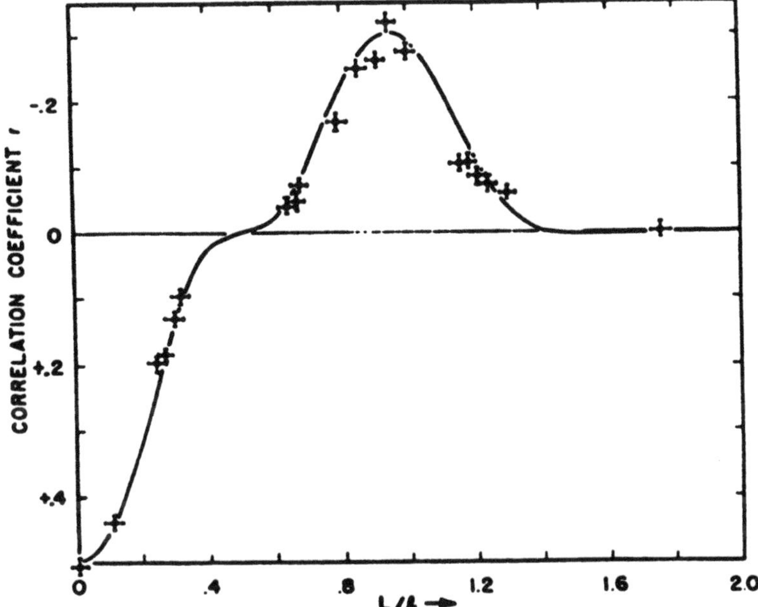

FIGURE 2. The results of Pfleegor and Mandel (crosses) from their second experiment,[2] where the correlation coefficient r is plotted versus L/l (L is the plate thickness and l the fringe halfspacing). There must be a significant negative correlation when L/l is near unity. The theoretical prediction (solid line) for interference fringes is in good agreement with the obtained experimental results.

He-Ne lasers) when it was standing for a short time at the same position on the recorder. Vain'shtein *et al.*[4] repeated the experiment by using a different correlation method analogous to the classical Hanbury–Brown–Twiss method. New methods and a different optical system of the interferometer not only confirmed the occurrence of the interference found by Pfleegor and Mandel, but also enabled the Russian group to determine directly the interference pattern at much lower intensities.

It seemed that with these observations a famous statement of Dirac[5] that ". . . each photon interferes only with itself. Interference between different photons never occurs" was disproved. However, Pfleegor and Mandel were also able:

1. To reproduce their experimental results with a theoretical calculation broadly deduced from that quantum field theory that Dirac had developed coherently with his general point of view;
2. To interpret the low-intensity interference effect by stressing that the localization of a photon at a detector necessarily makes it intrinsically uncertain from which of the two lasers it came. Thus, if the photon wave function represents only our knowledge of the photon behavior, it is

natural to add two wave functions corresponding to the arrival of each detected photon from the two lasers. The squared modulus of this sum would then contain the observed interference effect.

Andrade e Silva and de Broglie[6] gave another interpretation of the Pfleegor and Mandel experiment. For them a photon is a dual structure: particle *and* wave. By particle they mean a very small object that is constantly localized in space. The wave is nothing but an undulatory phenomenon propagating in ordinary three-dimensional space. Therefore, the de Broglie photon, in agreement with Einstein's picture,[7] is a very small region of high concentration of energy which is embodied in a wave in which it constitutes some kind of moving singularity.

With these ideas another explanation of Pfleegor and Mandel's observations is possible. The two lasers continuously emit two undulatory beams, which intersect and interfere with one another in the region R, there giving rise to interference maxima. If one of the two lasers emits a photon, then it will be guided by the wave to reach with larger probability those parts of R where the total wave amplitude is larger. If many photons are falling on a screen S inserted into the region R, the interference pattern becomes visible.

This means that a particle coming from one laser could interact not only with the wave with which it was born, but with a different wave arising from a different laser as well. The particle is thus guided by the sum of the two electromagnetic waves:

$$A_\mu^{(1)}(x,y,z,t) + A_\mu^{(2)}(x,y,z,t)$$

where $A_\mu^{(i)}$ is the four-potential describing the wave emitted from the ith laser ($i = 1,2$) and x,y,z are the coordinates of the particle at time t. Such a conclusion is both absolutely necessary and very natural in the Einstein-de Broglie picture of quantum phenomena, if the Pfleegor–Mandel experiment has to be explained. It is, however, in contradiction with the standard quantum-mechanical description of wave functions of different quantum objects as belonging to different vector spaces and thus are unable to exert any mutual influence.

2. FURTHER DEVELOPMENTS OF THE IDEA OF EMPTY WAVES

The Einstein–de Broglie idea of the reality of particle *and* wave not only holds for photons, but also for all other quantum objects (e.g., electrons, neutrons). To the particle are "attached" the fundamental physical attributes of energy, momentum, and mass, while the wave probably does not carry them at all and for this reason is called "empty," when one is dealing with a wave packet not containing its particle.

Selleri[8] developed from this the idea to test the empty wave concept in stimulated emission. Croca[9] was the first who attached this idea to a testable experimental setup for other than photonic particles, neutrons. Croca's ideas were developed by Croca, Garuccio, and Selleri,[10] who formulated the basic assumption as follows:

> *A neutronic particle propagating in a region of space where several neutronic wave packets of the same frequency are simultaneously present interacts locally with their sum*
>
> $$\Phi = \Phi_1 + \Phi_2 + \cdots + \Phi_n \tag{1}$$
>
> *through the quantum potential*
>
> $$Q = -\frac{\hbar^2 \nabla^2 |\Phi|}{2m|\Phi|} \tag{2}$$
>
> *even when the waves come from different sources. This interaction has the effect, over the statistical ensemble, that the density of the particles becomes proportional to* $|\phi|^2$.

This I call the CGS (Croca–Garuccio–Selleri) assumption.

Schmidt and Selleri[11] have shown that the last part of this assumption is not true in general. They considered the triple-slit experiment of Figure 3 where a fourth slit is constantly closed by the particle detector D_2, another detector D_1 being used for scanning the frequency of arrivals on the "second screen." Two incoherent sources Σ_1 and Σ_2 emit quantum objects of the same nature (electrons, neutrons, etc.) and of the same energy, so that the quantum waves emitted by the two sources have the same frequency. The sources are assumed to have a low intensity of emission, i.e., quantum objects are emitted one at a time at widely separated time intervals. Emission from the upper source Σ_1 can hit on the upper slit only: From this slit a cylindrical wave ϕ_1 will be emitted to the right, sometimes together with a particle. Only cases in which this particle is observed by detector D_1 will interest us, and these only in coincidence with a second particle detection in D_2. It was shown in Ref. 11 that these D_1D_2 coincidences are coupled very often to the emergence of two empty waves ϕ_2 and ϕ_3 from the second and third slit of Figure 3, respectively, if the intensity of Σ_2 is taken to be much smaller (e.g., 100 times smaller) than that of Σ_1 and if a suitable shutter is set in front of Σ_2. For details concerning the experimental apparatus I refer the reader to Ref. 13. What matters here is that in the subset of events that is of interest a particle emerges from the first slit of Fig. 3 in conjunction with its associated wave ϕ_1, while very often only two empty waves ϕ_2 and ϕ_3 emerge from the other open slits.

The particle distribution at the second screen can be calculated in three different ways:

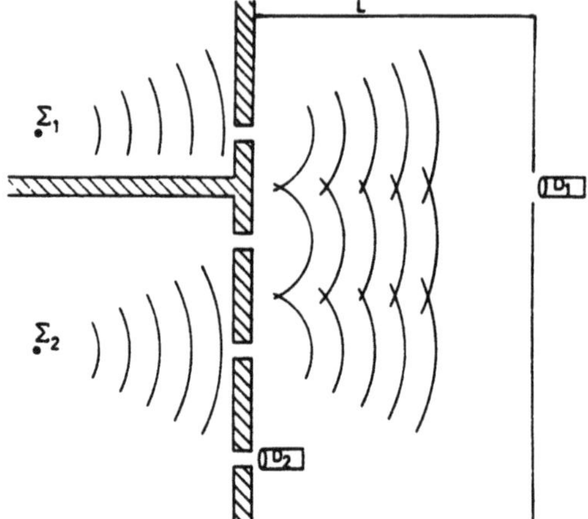

FIGURE 3. Triple-slit experiment for the detection of empty waves. The two incoherent sources Σ_1 and Σ_2 emit quantum objects of the same nature. Since only coincident events of the detectors D_1 and D_2 are registered, and given the very different intensity of the two sources, very often empty waves are coming from the two lower slits while a wave plus a particle are coming from the upper slit.

1. The first one is merely the probabilistic Born interpretation of the wave function

$$\rho_1 = |\phi_1|^2 \tag{3}$$

   In this approach the waves are only mathematical tools for the calculation of probabilities and have otherwise no objective existence: when a particle is detected in D_2 the waves ϕ_2 and ϕ_3 go instantaneously to zero ("reduction of the wave function"), whence (3).

2. The second is linked to the causal interpretation of quantum mechanics, according to which all waves ϕ_1, ϕ_2, and ϕ_3 are objectively real, but also based on the CGS assumption expressed by Eqs. (1) and (2). It leads to

$$\rho_2 = |\phi_1 + \phi_2 + \phi_3|^2 = |\phi_1|^2 + |\phi_2 + \phi_3|^2 \tag{4}$$

   the second equality being a consequence of the incoherent nature of the two sources used in Figure 3.

3. The third one, ρ_3, is numerically calculated by summing particle arrivals on the second screen by using the guidance formula of de Broglie[11]

$$v(x,y,z,t) = \frac{1}{m}\nabla S(x,y,z,t) \tag{5}$$

   obtained from

$$\phi = \phi_1 + \phi_2 + \phi_3 = R\,e^{iS/\hbar}$$

In most cases $\rho_1 = \rho_2 = \rho_3$, as will be shown in the next section. It was Croca[9] who first made the assumption that there are cases where ρ_1 is not equal to ρ_2. Concrete calculations were made in parallel by Croca, Garuccio, and Lepore[12] for photons, and by Schmidt and Selleri[13] for electrons. The last named authors were also able to show that cases also exist where ρ_2 is not equal to ρ_3. This is shown in Figure 4a and 4b. Figure 4a compares ρ_1 and ρ_2, and Figure 4b ρ_2 and ρ_3. The fact that the particles are not able to fill the whole region of space which would be expected with the calculation of ρ_2, we call *wind effect*. It seems that particle trajectories are "blown away" from a part of this region. This is a new feature for particle trajectories, consistent with the pictures of Dewdney *et al.*[14] and of the next section where even in the *normal* double-slit experiment the particle trajectories are not allowed to cross each other. A description of the experimental setup for Figure 3 is given in Ref. 13, while a detailed description of the numerical method is contained in Ref. 15. The difference between ρ_1 and ρ_3 will be used to show that empty waves can be detected in a Pfleegor–Mandel-type experiment for electrons. The difference from the original Pfleegor–Mandel experiment is not only the use of electrons instead of photons, but also the use of two incoherent sources.

3. THE NORMAL DOUBLE-SLIT EXPERIMENT

The normal double slit experiment was designed by Thomas Young and is found in nearly every textbook of quantum mechanics. The experiment was first realized in 1959 by Möllenstedt and Jönsson[16] for electrons, a fact that sometimes seems to be unknown.[17] A first attempt at a causal interpretation was made by Dewdney *et al.*[14] They used a new technique of drawing particle trajectories using the guidance formula of de Broglie. I have repeated their calculations (see Figure 5a) and my results are in good agreement with those of Ref. 14. Furthermore, the distribution at the second screen agrees with that to be expected from the Copenhagen interpretation of quantum mechanics. Instead the authors of Ref. 14 did not show explicitly the distribution ρ at the screen.

I calculated the particle density at the screen in the three different ways explained in the previous section. The results are shown in Figure 5b. In this case there is no different between ρ_1, ρ_2 (solid line), and ρ_3 (asterisks), where ρ_3 is obtained by the summation over 8000 particle arrivals at the screen. For the calculations of ρ_3 a Multi-Channel-Analyzer-Simulator was constructed with a channel width of 0.00002 cm. Particle trajectory arrivals were summed in the channels and the intensity was plotted. The starting point of each trajectory behind the first slit was calculated with a random number generator giving Gaussian distributed random numbers. For details see Ref. 15. The experimental input parameters have been taken from Ref. 14.

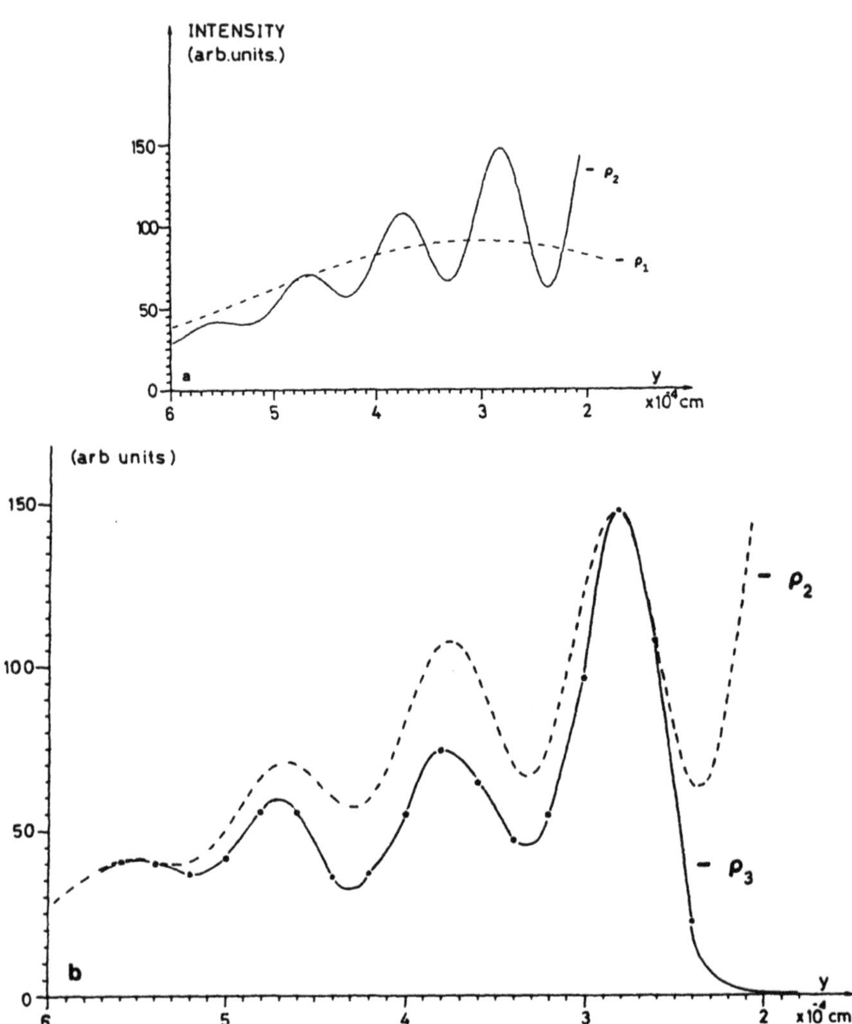

FIGURE 4. For the triple-slit experiment of Figure 3 a comparison is made of differently calculated densities of particle arrivals at the second screen (ρ_1, ρ_2, and ρ_3). (a) The difference between density ρ_1, calculated with the Copenhagen interpretation (dashed line), and the density of ρ_2 derived from the CGS assumption (solid line) is shown. (b) ρ_2 (dashed line) is compared with a very large number of particle arrivals which are obtained with the guidance formula and thus give rise to the interpolated solid line ρ_3.

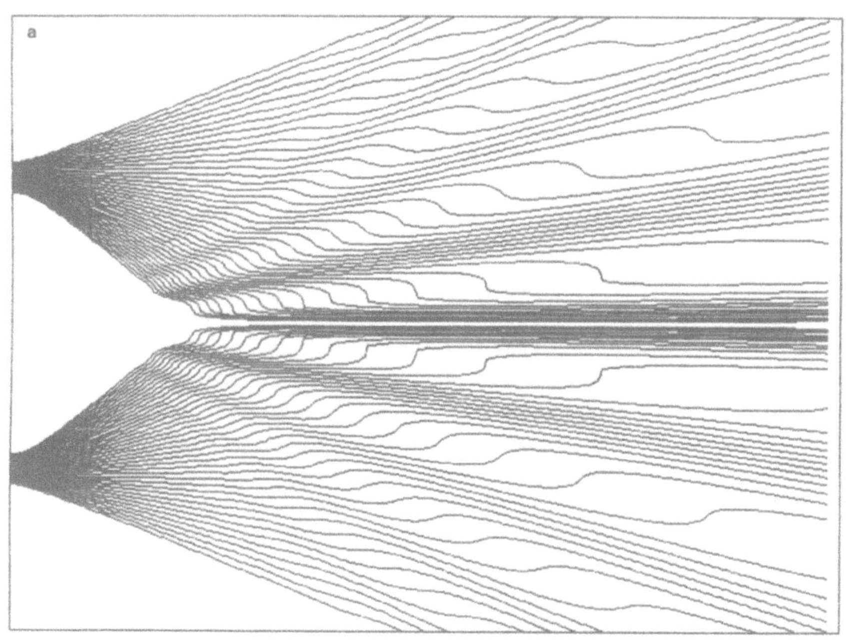

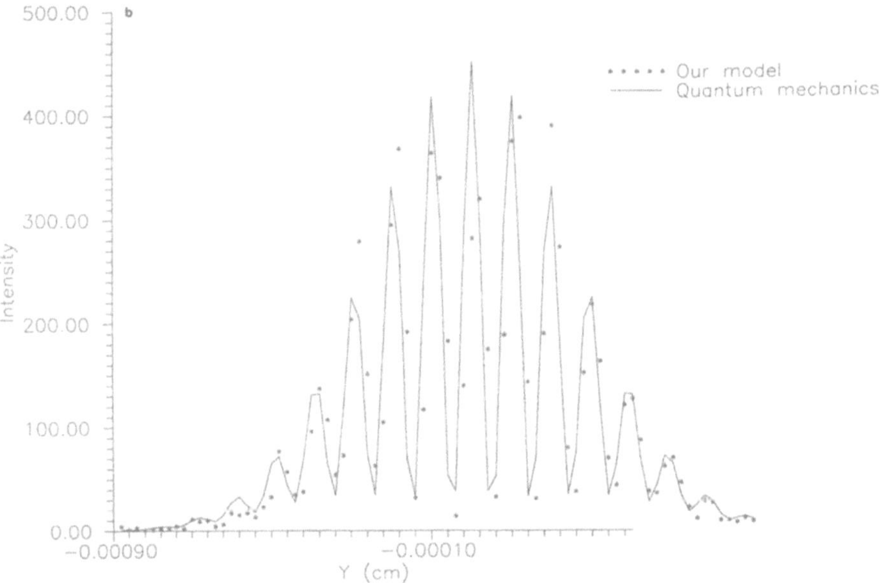

FIGURE 5. (a) The particle trajectories for a double-slit experiment with a single source, showing a very good agreement with those calculated in Ref. 14. (b) The particle distribution of arrivals at the screen when a large number of trajectories are calculated for a double-slit experiment with a single source. This gives the asterisks of the present figure. A calculation of the particle density according to the Copenhagen interpretation is also shown (solid line). The agreement is good, within statistical fluctuations.

4. DOUBLE-SLIT EXPERIMENT WITH TWO INCOHERENT SOURCES

For more than 60 years the double-slit problem has been considered to be completely solved. The experimental results obtained by Möllenstedt and Jönsson[(16)] for electrons, and by Zeilinger *et al.*[(18)] for neutrons have produced striking agreements with the quantum theoretical predictions. At the same time the causal interpretation of quantum mechanics has constantly led to predictions identical to those of the Copenhagen interpretation. From this, many people got the idea that the causal interpretation is at most of philosophical interest and cannot give any new prediction at the empirical level. *It is the aim of the present chapter to show that the situation has evolved in a very interesting way: after the results obtained in Refs. 8–10, 12, 13, one can now discriminate the causal from the Copenhagen "interpretation" of quantum mechanics.*

Starting with the results of Ref. 13 a double-slit experiment was designed (see Figure 6) as follows: Two sources Σ_1 and Σ_2 should emit the same kind of particles, which in our picture are always accompanied by a wave. We have chosen these quantum objects to be electrons because we wanted to compare our results with those of Ref. 14. As one can see, a particle in conjunction with its associated wave coming from the lower source Σ_2 can go through two slits. One is permanently closed by a particle detector D_2. We are only interested in coincidence events where particles are detected in D_1 and D_2, that means: from the lower slit only empty waves are emitted while from the upper slit a particle plus a wave are arriving at the detector. The two sources emit wave packets with coherence length l

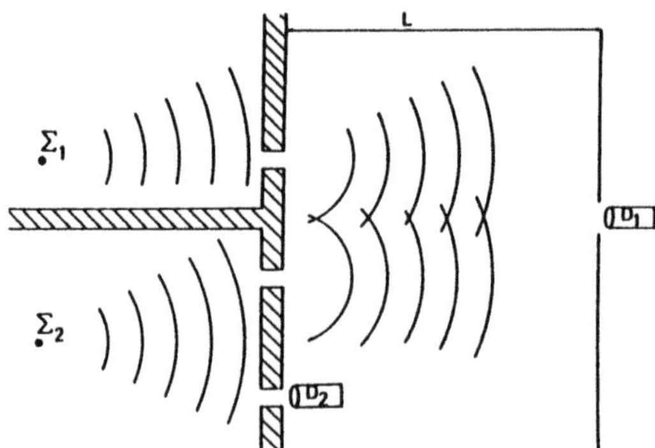

FIGURE 6. A modified Pfleegor–Mandel experiment, where two waves of two different sources should interact. The difference with respect to the original Pfleegor and Mandel experiment is that the two sources Σ_1 and Σ_2 are not in phase and since only D_1D_2 coincidences are registered the empty wave concept can be tested. Given the large difference in the intensities of the two sources, and according to the causal interpretation, a wave with its accompanying particle should emerge from the upper slit while an empty wave comes from the lower slit in the very large majority of the considered events.

equal to their physical length. But l must be known from previous measurements, as well as the group velocity $\mathbf{u}$. In such conditions, every wave packet will take a time $\Delta t = l/u$ to cross any given point. Precisely this Δt defines the coincidence time. If D_2 fires at time t_2, then D_1 must fire at time $L/u + \Delta t/2$ later, if the double detection is to be counted as a true coincidence from electronics (L is the distance between the two screens). This definition of "coincidence" guarantees that the waves accompanying the two particles detected by D_1 and D_2 overlap in the region in front of the second screen.

There is always a nonzero probability that two particles emitted by Σ_2 give rise to D_1–D_2 coincidence. In order to make sure that the particles detected by D_1 come almost always from Σ_1 it is convenient that Σ_1 be more intense than Σ^2, let us say by a factor of 100. This difference must be obtained without any change of the wave accompanying a particle emitted by Σ_2: a suitable toothed wheel can do the job, since it absorbs completely some quantum objects while it lets through unchanged some other ones.

In this chapter I take into account that there is a difference Δt_0 in the starting time t_0 of the two wave packets behind the first screen. Because the starting point x_0 of the wave packet satisfies $x_0 = v_x t_0$, the packets also have different starting points according to their x-coordinate even if v_x were the same for all of them. We also discuss the effect on the particle distribution at the screen when the wave packets have different frequencies.

5. THREE CALCULATIONS OF PARTICLE DENSITY COMPARED

The Copenhagen Interpretation

The particle density ρ at the screen for the double-slit experiment with two incoherent sources is

$$\rho_1 = |\phi_1|^2 \tag{7}$$

where ϕ_1 is the wave function associated with the electron coming from the upper slit in Figure 6. The wave function ϕ_2 of the empty wave coming from the lower slit is assumed to be 0 because of the "reduction of the wave function" postulate which should be valid if the detector D_2 has registered a particle. We have used "Gaussian slits" for our calculations. That means that the particle distribution behind the first screen should be Gaussian. These "Gaussian slits" were first introduced by Feynman.[19] If one starts with a Gaussian wave packet, a time evolution gives only a spreading of this wave packet. At the screen we should expect a Gaussian distribution to build up when many particles arrive on the second screen.

The CGS Assumption

According to the causal interpretation the two waves ϕ_1 and ϕ_2 coming from the lower and the upper slit of Figure 6 are real, even when a particle is detected in D_2. Therefore, the CGS assumption of Section 2 gives

$$\rho_2 = |\phi_1 + \phi_2|^2 = |\phi_1|^2 + |\phi_2|^2 \tag{8}$$

The lack of difference between the squared sum and the sum of the squared wave functions is due to the incoherence of the two sources. Hence, the interference term vanishes. Since we are only interested in particles coming from the upper slit in Figure 6

$$\rho_1 = \rho_2 \tag{9}$$

The Calculation of ρ_3 with the "Guidance Formula" of de Broglie

The total de Broglie wave for the double-slit experiment in every point of space **r** and for all time t can be written

$$\Phi = \sum_{i=1}^{2} R_i \exp(iS_i/\hbar) = R \exp(iS/\hbar) \tag{10}$$

where we wrote

$$\Phi_i = R_i \exp(iS_i/\hbar) \tag{11}$$

($i = 1,2$). The total phase S can be obtained from (10)

$$S = \hbar \operatorname{arctg}\left[\frac{\sum_{i=1}^{2} R_i \sin(S_i/\hbar)}{\sum_{i=1}^{2} R_i \cos(S_i/\hbar)}\right] \tag{12}$$

In the de Broglie theory the particle velocity is given by the gradient of the phase

$$\mathbf{v} = \frac{\nabla(S/\hbar)}{m} \tag{13}$$

From (12) and (13) one can easily obtain

$$v_x = \frac{N_x}{D},\ v_y = \frac{N_y}{D} \tag{14}$$

where

$$D = m \sum_{i,j=1}^{2} R_i R_j \cos[(S_i - S_j)/\hbar] \tag{15}$$

and

$$N_x = -\sum_{i,j=1}^{2} R_i \frac{\partial R_i}{\partial x} \sin[(S_i - S_j)/\hbar] + 1/\hbar \sum_{i,j=1}^{2} R_i R_j \frac{\partial S_i}{\partial x} \cos[(S_i - S_j)/\hbar] \tag{16}$$

with a strictly similar expression for N_y.

These general results can be specialized for Gaussian wave packets. With Belinfante's[20] notation one has

$$\Phi_i(\mathbf{x},t) = (2\pi s_t^2)^{-3/4} \exp[i\mathbf{k}_i(\mathbf{r}_i - 0.5\mathbf{u}_i t) - (\mathbf{r}_i - \mathbf{u}_i t)^2/4\sigma_0 s_t] \tag{17}$$

where σ_0 is the $t = 0$ width of the packet and

$$s_t = \sigma_0[1 + i(\hbar t/2m\sigma_0^2] \tag{18}$$

$$\mathbf{k}_i = m\mathbf{u}_i/\hbar \tag{19}$$

$\mathbf{u}_i$ being the particle velocity on the slit, that can be treated as an initial condition. Furthermore,

$$\mathbf{r}_i = \mathbf{r} - \mathbf{r}_{0i} \tag{20}$$

where $\mathbf{r}$ is the considered point of space and $\mathbf{r}_{0i}$ is the midpoint of the ith Gaussian slit on the screen in Figure 6.

It is easy to identify the ingredients of the local velocity (14) in (17)–(20). One has

$$R_i = (2\pi\sigma_t^2)^{-3/4} \exp[-(\mathbf{r}_i - \mathbf{u}_i t)^2/4\sigma_t^2] \tag{21}$$

$$S_i = \hbar[-3\varphi + \mathbf{k}_i(\mathbf{r}_i - 0.5\mathbf{u}_i t) + (\mathbf{r}_i - \mathbf{u}_i t)^2 \hbar t/8m\sigma_0^2\sigma_t^2] + \xi_i \tag{22}$$

where ξ_i is a random phase that will be taken different from zero only for one of the wave packets.

The new symbols in (21) and (22) are given by

$$\tan 2\varphi = \hbar t/2m\sigma_0^2 \tag{23}$$

and

$$\sigma_t = s_t \exp(-2i\varphi) \tag{24}$$

so that $\sigma_t = |s_t|$.

From (21) and (22) one can next obtain

$$\frac{\partial R_i}{\partial c} = -R_i(x - u_{ix}t - x_{0i})/2\sigma_t^2 \tag{25}$$

$$\frac{\partial S_i}{\partial x_i} = mu_{ix} + \hbar^2 t(x - u_{ix}t - x_{0i})/4m\sigma_0^2\sigma_t^2 \tag{26}$$

$$\frac{\partial R_i}{\partial y} = -R_i(y - y_{0i} - u_{iy}t)/2\sigma_t^2 \tag{27}$$

$$\frac{\partial S_i}{\partial y} = mu_{iy} + \hbar^2 t(y - y_{0i} - u_{iy}t)/4m\sigma_0^2\sigma_t^2 \tag{28}$$

The derivatives (25)–(28) can be introduced in N_x [see (16)] and N_y. The local particle velocity can be calculated through (14), also by using (21) and (22). From the velocity it is only a trivial work to find the trajectory. We will not insist on this here. By summing over a lot of arrivals of particle trajectories at the second screen the density ρ_3 is obtained.

6. NUMERICAL RESULTS OBTAINED WITH THE "GUIDANCE FORMULA"

The numerical values of the fixed parameters in my calculation were:

$$\begin{aligned}
\sigma_0 &= 0.65 \times 10^{-5}\ \text{cm} \\
y_{01} &= 1 \times 10^{-4}\ \text{cm} \\
y_{02} &= -1 \times 10^{-4}\ \text{cm} \\
\mathbf{u}_1 &= (1.3 \times 10^{10}\ \text{cm/s},\ 1.5 \times 10^{4}\ \text{cm/s}) \\
\mathbf{u}_2 &= (1.3 \times 10^{10}\ \text{cm/s},\ -1.5 \times 10^{4}\ \text{cm/s})
\end{aligned}$$

Gaussian wave packets are assumed to arise from the first screen where the slits are located. Actually the slits are only a fiction in such a calculation, the true situation being that of different Gaussian packets, with little initial overlapping, that propagate symmetrically toward the second screen, spread and thus increase their overlapping, and fully interfere in the region containing the second screen. In the ideal situation there are two sources totally in phase, both having the same energy and they emit their particles at the same time t_0, and there is no difference with respect to the situation with only one source in the de Broglie approach. In this case one gets the same picture for the particle trajectories as shown in Figure 5a. As a first test I changed the phase ξ between the two wave packets. The results are shown in Figures 7 to 9. If the phase shift is 0 and 2π the trajectories are not

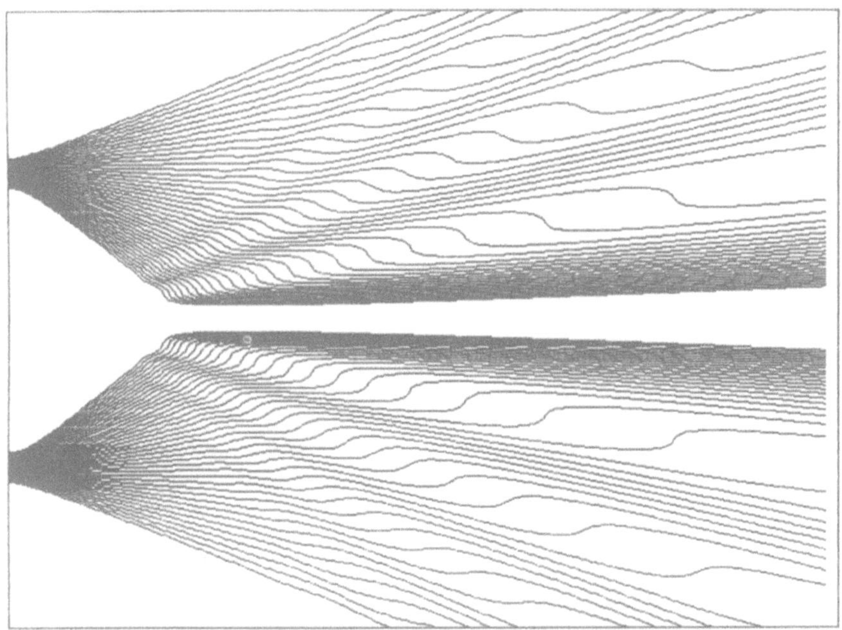

FIGURE 7. Particle trajectories for a double-slit experiment with two sources. While the energy of the particles and the starting time t_0 of the two wave packets from the slits are identical, their phase difference is kept constant and equal to π.

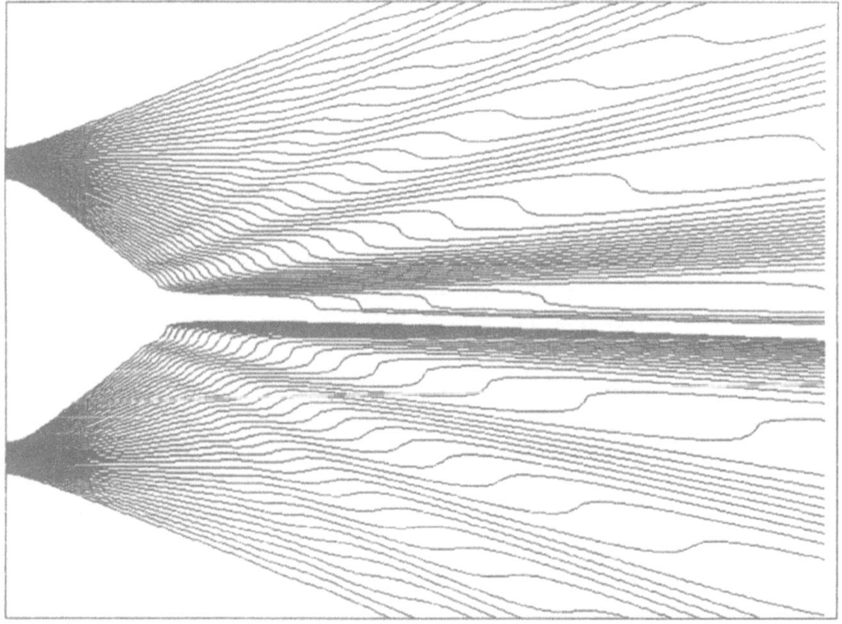

FIGURE 8. Particle trajectories for a double-slit experiment with two sources where the phase difference is $0.75\ \pi$. Energy and t_0 as in Figure 7.

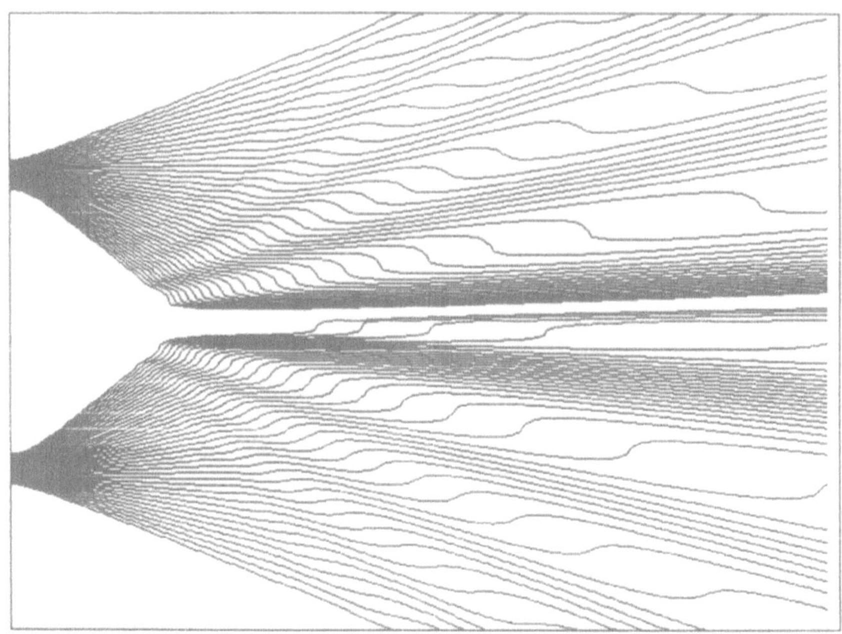

FIGURE 9. Trajectories for electrons where the phase difference of the two wave packets is 1.25 π. Energy and t_0 as in Figure 7.

changed (no difference from Figure 5a). The biggest fringe shift is obtained if $\xi = \pi$ (see Figure 7). The fringe shift decreases in perfect symmetry with the distance to π (see Figures 8 and 9).

Second, I looked at the particle trajectory behavior if the particles were emitted with different energies which is equivalent to a frequency change for the waves. If the energy difference grows, the particle trajectories tend to overlap (see Figures 10 and 11). For larger energy differences the wave packets do not "feel" one another any longer and the particle behaves as in a single-slit experiment (Figure 12).

My main interest was to calculate the density at the second screen. The following parameters were changed with random numbers for one of the wave packets, while for the other wave packet these parameters were fixed: the phase, the starting time t_0, and the energy. All of the following figures were calculated in such a way that the wave packets were coming from the lower and the upper slit of Figure 6, but the particle trajectories were started at the upper slit only. In Figure 13 one sees the density at the screen when the phase shift between the waves is randomly changed between 0 and 2π. With our model I always mean a calculation of the density with the guidance formula of de Broglie, and with the quantum mechanical prediction I mean ρ_1, obtained according to the Copenhagen

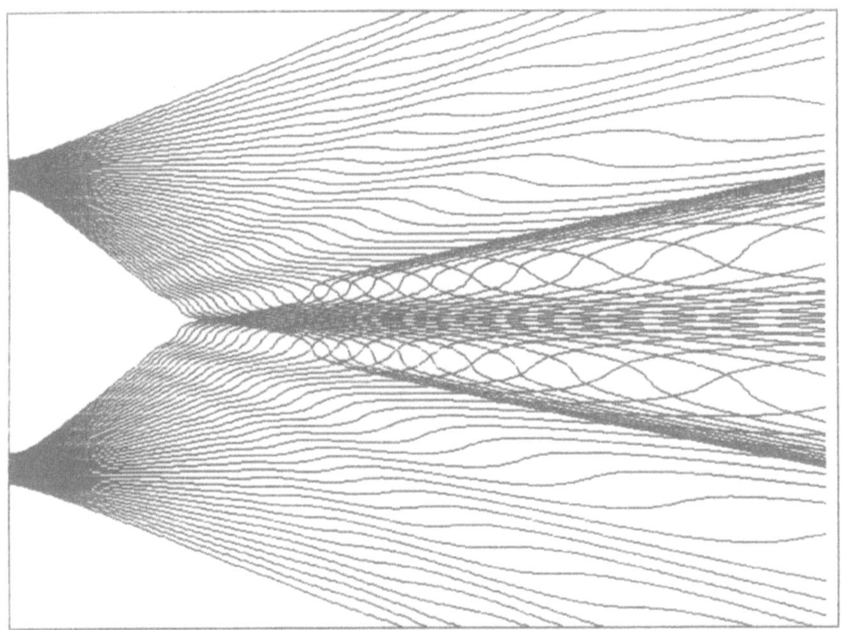

FIGURE 10. Electron trajectories coming from the lower and the upper slit of Figure 6. Here the phase difference and the starting time t_0 are fixed, while the particle energies differ by 1 eV. This energy difference corresponds to momenta different only in the x-direction.

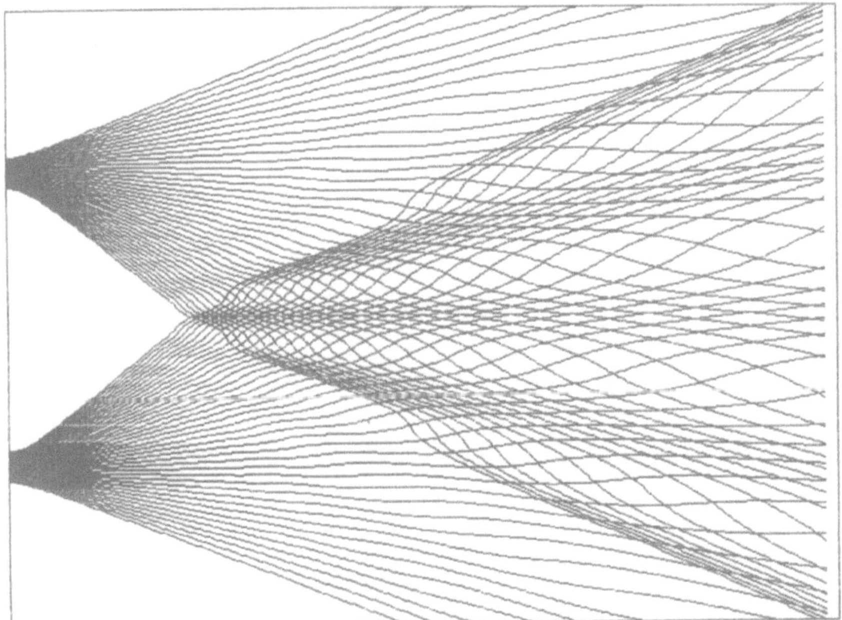

FIGURE 11. Trajectories for electrons where the phase and the starting time are identical but the energies of the two particles differ by about 1.5 eV.

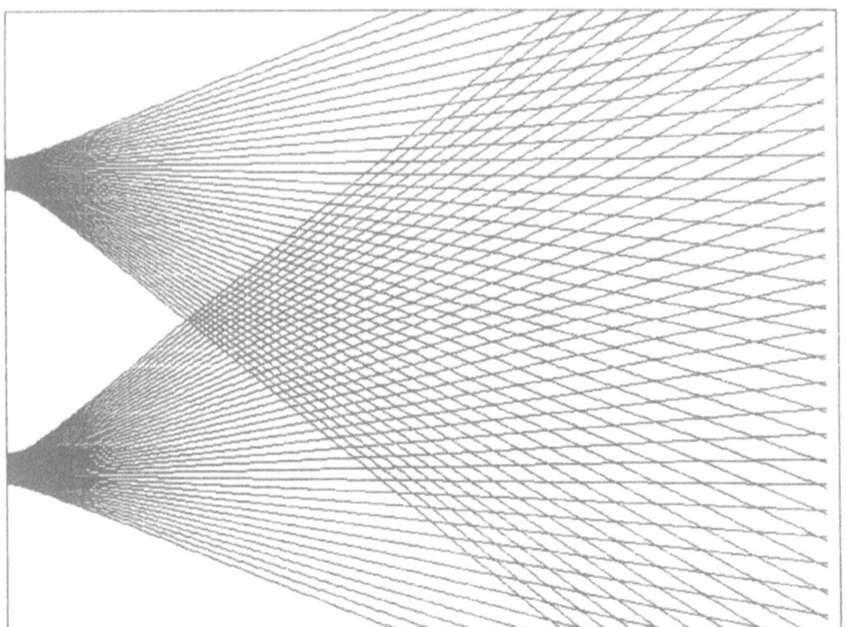

FIGURE 12. When the energy difference of the electrons is 3 eV, the wave packets no longer overlap and the trajectories are only straight lines as they would be in single-slit experiments. This is shown in the present figure.

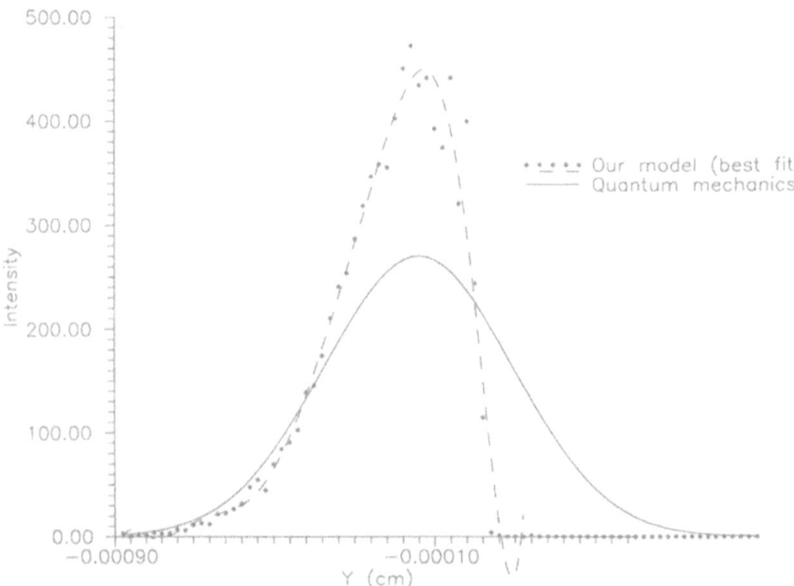

FIGURE 13. Particle distribution at the second screen when the particles start from the upper slit of Figure 6 only, while the waves come from both slits with a random phase difference between 0 and 2π. The asterisks are the obtained data points and the dashed line is the best fit through these points. They are compared with the prediction of the Copenhagen interpretation (solid line).

interpretation. Figure 14 shows the result after varying the energy of one particle in the x-direction randomly between 0 and 4 eV, while the base energy is 46.355 keV. In Figure 15 the starting time t_0 is changed between 0 and 8 fs. This number is so small because after 1 fs the midpoints of the wave packets already have a difference of $2\sigma_0$ in the x-direction, due to the high velocity. In Figures 16 and 17 I altered all of the parameters at the same time with the values given in the figure captions.

> Figure 16 clearly shows that the experimental parameters can be chosen in such a way that a distinction between the causal interpretation and the Copenhagen interpretation (CI) is possible. Figure 17 also shows that if one does not take care of these small differences, one always finds the usual results predicted by CI. In a forthcoming publication I will give an experimental set up where the differences are easier to detect.

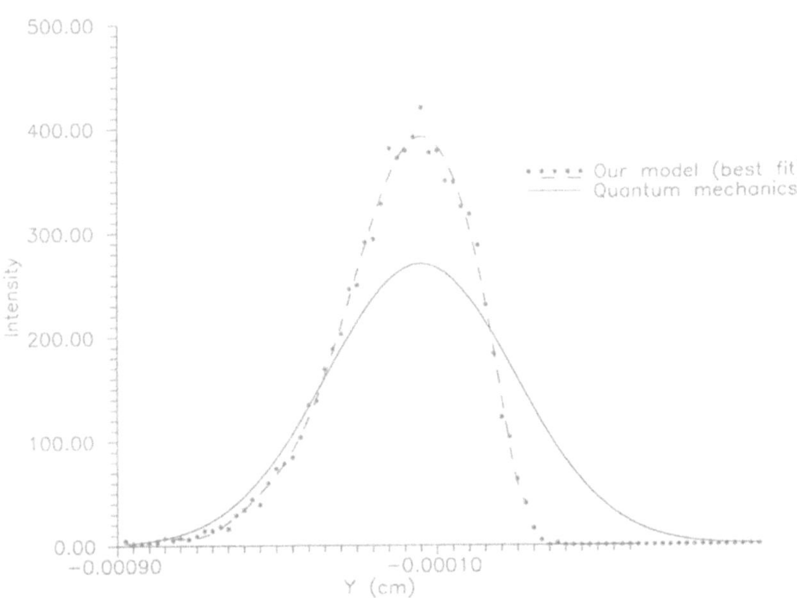

FIGURE 14. Distribution of particle arrivals at the second screen for particles with accompanying waves coming from the first slit, while from the second slit only waves emerge. The asterisks are the result of our calculation obtained with the guidance formula of de Broglie when phases and the frequencies of the waves are identical but the difference in starting time t_0 varies randomly between 0 and 8 fs. A best fit through the data (dashed line) is compared with the results predicted by the Copenhagen interpretation (solid line).

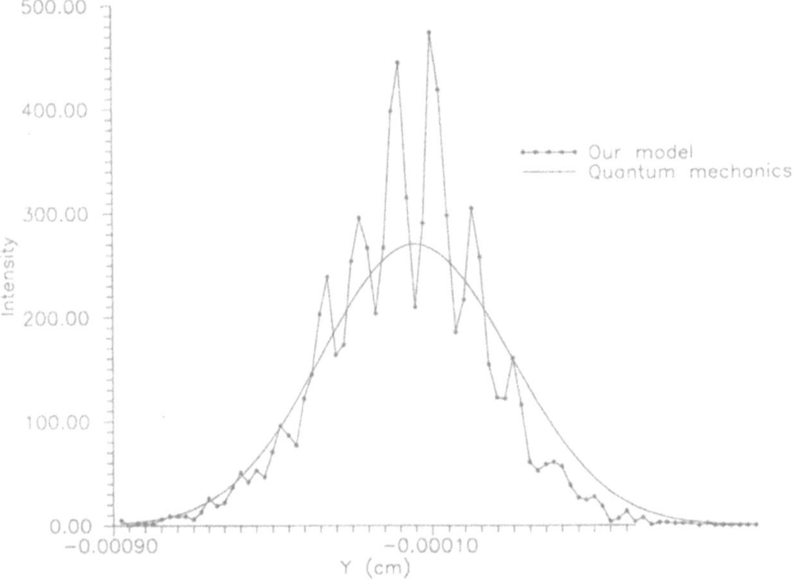

FIGURE 15. Distribution of particle arrivals at the second screen for particles with accompanying waves coming from the first slit, while from the second slit only waves emerge at the same starting time and with the same phase but a different frequency. If the energy difference of the two wave packets varies randomly between 0 and 4 eV the asterisks are obtained with my calculations. A comparison is made with the predictions of the Copenhagen interpretation (solid line).

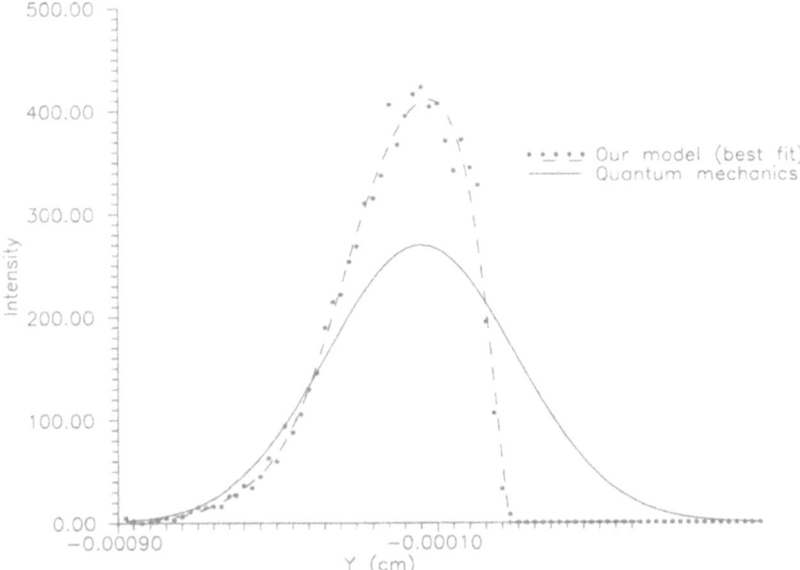

FIGURE 16. A computer calculation of particle arrivals at the second screen for particles with accompanying waves coming from the first slit, while from the second slit only waves emerge. In this figure the two wave packets have a randomly varying energy difference up to 0.5 eV, a randomly varying phase difference between 0 and 2π and a difference in starting times randomly varying between 0 and 2 fs. The difference between the causal interpretation (dashed line) and the Copenhagen interpretation (solid line) is still visible.

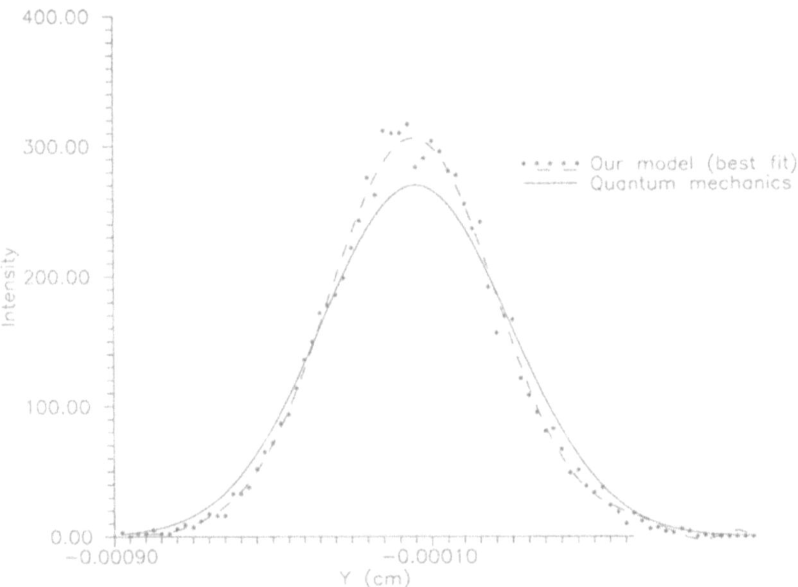

FIGURE 17. A computer calculation of particle arrivals at the second screen for particles with accompanying waves coming from the first slit, while from the second slit only waves emerge. In this figure the two wave packets have a randomly varying energy difference up to 4.5 eV, a randomly varying phase difference between 0 and 2π, and a difference in starting times randomly varying between 0 and 18 fs. Here one sees that the distribution derived with the guidance formula of de Broglie (asterisks for the data and dashed line for the best fit) shows little difference from the distribution which is expected according to the Copenhagen interpretation of quantum mechanics.

ACKNOWLEDGMENTS. I thank Professor F. Selleri for encouragement and valuable criticism concerning the content of this paper. I also thank V. L. Lepore and H. Ziggel for stimulating discussions and the Commission of the European Communities for financial support.

REFERENCES

1. R. L. PFLEEGOR and L. MANDEL, *Phys. Rev.* **159**, 1084 (1967).
2. R. L. PFLEEGOR and L. MANDEL, J. Opt. Soc. Am. **58**, 946 (1968).
3. W. RADLOFF, *Ann. Phys.* **26**, 178 (1971).
4. L. A. VAIN'SHTEIN, V. N. MELEKHIN, S. A. MISHIN, and E. R. PODALYAK, *Sov. Phys. JETP* **54**, 1054 (1981).
5. P. A. M. DIRAC, *The Principles of Quantum Mechanics*, 4th ed., Clarendon Press, Oxford (1958).
6. L. DE BROGLIE and J. ANDRADE E SILVA, *Phys. Rev.* **172**, 1284 (1968).
7. A. EINSTEIN, *Ann. Phys. (Leipzig)* **18**, 639 (1905).

8. F. SELLERI, *Lett. Nuovo Cimento* **1**, 908 (1969); *Found. Phys.* **12**, 1087 (1982).
9. J. R. CROCA, *Found. Phys.* **17**, 971 (1987).
10. J. R. CROCA, A. GARUCCIO, and F. SELLERI, *Found. Phys. Lett.* **1**, 101 (1988). See also: F. SELLERI, *Quantum Paradoxes and Physical Reality*, Kluwer, Dordrecht (1990).
11. L. DE BROGLIE, *Non Linear Wave Mechanics: A Causal Interpretation*, Elsevier, Amsterdam (1960). See also: D. BOHM, *Phys. Rev.* **85**, 166, 180 (1952).
12. J. R. CROCA, A. GARUCCIO, and V. L. LEPORE, *Found. Phys. Lett.* **1**, 101 (1988).
13. M. SCHMIDT and F. SELLERI, *Found. Phys. Lett.* **4**, 1 (1991).
14. C. PHILIPPIDIS, C. DEWDNEY, and B. J. HILEY, *Nuovo Cimento B* **52**, 15 (1979).
15. M. SCHMIDT, Ph.D. thesis, Bremen (1991) (in English).
16. G. MÖLLENSTEDT and C. JÖNSSON, *Z. Phys.* **155**, 472 (1959).
17. G. MÖLLENSTEDT, *Matter Wave Interferometry* (G. BADUREK, H. RAUCH, and A. ZEILINGER, eds.), North-Holland, Amsterdam (1988).
18. A. ZEILINGER, R. GÄHLER, C. G. SHULL, and W. TREIMER, *Symp. Neutron Scattering (Argonne), AIP Conf. Proc.* **89**, 93 (1981).
19. R. P. FEYNMAN and A. R. HIBBS, *Quantum Mechanics and Path Integrals*, McGraw–Hill, New York (1965).
20. F. J. BELINFANTE, *A Survey of Hidden-Variables Theories*, Pergamon, Oxford (1973).

CHAPTER 16

Two-Photon Interference and the Question of Empty Waves

Franco Selleri

1. INTRODUCTION

In this chapter the quantum mechanical probabilities are considered *averages* of other probabilities applying to subensembles of the considered set of quantum systems. Therefore, if p_{qm} is a quantum-mechanically predicted probability for a set S of objects of the same type (e.g., photons, electrons) we will assume that S can be split in a certain number of subsets S_i to each of which a certain probability p_i belongs, such that

$$p_{qm} = \sum_i \frac{N_i}{N} p_i \tag{1}$$

where N_i and N are the populations of S_i and S, respectively.

For a realistically minded physicist there are two good reasons for believing in the previous description:

1. He is led to the conclusion that the experimentally observed violations[(1)] of Bell-type inequalities of the strong type[(2)] are due to the lack of validity in nature of the additional assumptions used in the derivation of such inequalities. He thus considers the following (Clauser–Holt–Shimony–Horne[(3)]) assumption as *false*: "*Given that a pair of photons emerge from two regions of space where two polarizers can be located, the probability*

Franco Selleri • INFN–Sezione di Bari and Dipartimento di Fisica, Università di Bari, I-70126 Bari, Italy.

Wave–Particle Duality, edited by Franco Selleri. Plenum Press, New York, 1992.

of their joint detection by two photomultipliers is independent of the presence and of the orientation of the polarizers."

2. He knows a necessary consequence of local realism in its most general formulation[4]: "*If probabilities are consequences of objectivity real physical properties of the ensembles for which they are predicted, then they must result from averages of probabilities in general different for different subsets composing these ensembles.*"

There are thus both experimental and theoretical reasons for giving a structure to detection probabilities. In quantum theory instead, one describes the detection probability of a photon *always* with a constant numerical factor η, the so-called "quantum efficiency" of the photodetector. Of course, quantum optics had a tremendous success in explaining a very large number of observations, but this can be considered a consequence of the observation of average values in full ensembles. Since the average of a product is in general different from the product of the averages, it is in the case of two-photon observations that one can expect departures from the quantum-theoretical description.

2. DETECTION-PROBABILITY MODEL

When a photon enters a photodetector, consider the quantum efficiency η as an average of different probabilities. For example, assume the detection probability to depend on the photonic wave function ϕ. Now, this is a dangerous assumption, because the transmission through a beam splitter, a polarizer, etc. depends on the wave function. If T and D are the transmission and detection probabilities, respectively, one would have an overall probability

$$T(\phi)\,D(\phi)$$

which in general is different from $T_{qm}\eta$ since a correlation between transmission and detection is in this way established. We need a particular dependence of D on ϕ, such that the well-understood physics of single photons is reproduced. A simple way to achieve this is the following.

Every photon carries a dichotomic variable $\mu = \pm 1$;

$$T(\phi) \text{ does } \textit{not} \text{ depend on } \mu;$$

$$D(\mu, \phi) = \eta + \mu G(\phi) \tag{2}$$

Assuming that the values $\mu = \pm 1$ are equiprobable, one gets

$$\langle T(\phi)\,D(\mu, \phi)\rangle_{\mu} = T(\phi)\eta \tag{3}$$

and the single-photon physics is reproduced if $T(\phi)$ is chosen in conformity to the quantum-mechanical predictions.

For probabilities one must have

$$0 \leqslant \eta \pm G(\phi) \leqslant 1 \tag{4}$$

whence it follows:

$$|G(\phi)| \leqslant \eta; \; |G(\phi)| \leqslant 1 - \eta \tag{5}$$

Assuming that the dependence of G on η is all contained in a positive factor $f(\eta)$ one writes:

$$G(\phi) = f(\eta) F(\phi) \tag{6}$$

with

$$|F(\phi)| \leqslant 1 \tag{7}$$

Given (7), one can satisfy the conditions (5) for all values of $|F|$ only if

$$f(\eta) \leqslant \eta; f(\eta) \leqslant 1 - \eta$$

The simplest function f satisfying both conditions is

$$f^{-1} = \eta^{-1} + (1 - \eta)^{-1}$$

Therefore,

$$f = \eta(1 - \eta)$$

and

$$G(\phi) = \eta(1 - \eta) F(\phi) \tag{8}$$

The simplest choice for F that allows this function to span the whole range of permissible values is

$$F(\phi) = 1 - 2|\psi|^2 \tag{9}$$

where ψ is *proportional* to ϕ. For $|\psi|^2 = 0$, $F = 1$, while for $|\psi|^2 = 1$, $F = -1$. Therefore, (9) fixes the normalization of the wave function: *the maximum value of* $|\psi|^2$ *should be 1*. The dichotomy between the physically real wave ϕ and the probabilistically normalized wave ψ is typical of the de Broglie approach. The difference between ϕ and ψ is, however, reducible to a constant numerical factor.

With reference to Figure 1 we see that the straight line (9) is a possible choice for $F(\psi)$. Line (a) tends to make the effects generated by ψ nearer to zero, and should probably be discarded; line (b) tends instead to magnify the effects of ψ and should be considered as interesting. Line (c) will give the strongest possible effects within the present approach and can be written

$$F_c(y) = 2\,\theta(y) - 1 = \mathrm{Sign}\,(1 - 2|\psi|^2) \tag{10}$$

where $\theta(y)$ is the step function. In order to obtain something similar to curve (b) we can add to $F = y$ a term with three zeros, for $y = -1, 0, +1$. It can be written

$$\alpha(1 + y)y(1 - y)$$

with $\alpha \geqslant 0$, constant. One thus has

$$F_b(y) = y + \alpha y(1 - y^2) \tag{11}$$

From condition (7) one can show that $\alpha \leqslant \frac{1}{2}$. In order to magnify the effects of our model we can take $\alpha = \frac{1}{2}$, so that

$$F_b(y) = \tfrac{3}{2}y - \tfrac{1}{2}y^3 \tag{13}$$

Another possible choice is directly suggested by Figure 1:

$$F_b(y) = \sin\left(\frac{\pi}{2}y\right) \tag{13}$$

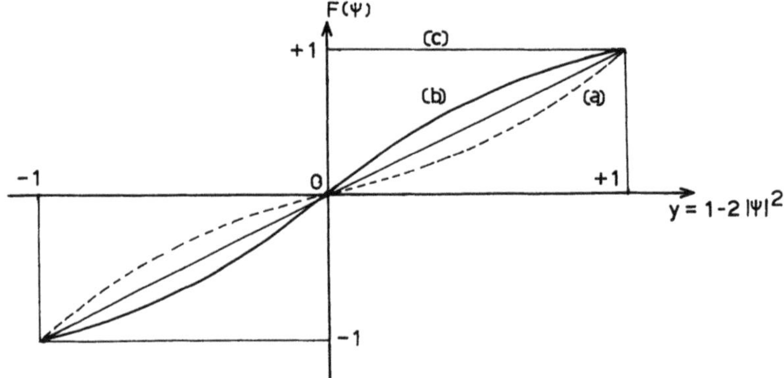

FIGURE 1. Some possible choices of the function $F(\psi)$. The meaning of curves (a), (b), and (c) is discussed in the text.

3. TWO-PHOTON DETECTION

Since the two values of μ are assumed equiprobable, for all measurements performed on independent photons (2) leads to the quantum mechanical probability (3) for all choices of F. The situation changes for two observations performed on correlated photons, if one assumes, for example, that the photons are produced *with the same value of* μ, which, however, will change at random over the values $\mu = \pm 1$ from pair to pair.

Consider the detection probabilities

$$\begin{aligned} D_1 &= [1 + \mu(1 - 2|\psi_1|^2)]\eta_1 \\ D_2 &= [1 + \mu(1 - 2|\psi_2|^2)]\eta_2 \end{aligned} \tag{14}$$

for photon 1 and 2, respectively. In (14) we assumed the same μ for the two photons and took the approximation $\eta_1, \eta_2 \ll 1$, valid in some recent experiments. The μ-averaged product of D_1 and D_2 is given by

$$\langle D_1 D_2 \rangle = [1 + (1 - 2|\psi_1|^2)(1 - 2|\psi_2|^2)]\eta_1\eta_2 \tag{15}$$

In the particular cases where

$$|\psi_1|^2 + |\psi_2|^2 = 1$$

one easily gets the simplified expression:

$$\langle D_1 D_2 \rangle = 4|\psi_1|^2|\psi_2|^2\eta_1\eta_2 \tag{16}$$

In the often-met cases where the transmission probability is given by the squared modulus of the wave function one obtains:

$$\langle T_1 D_1 T_2 D_2 \rangle = 4|\psi_1|^4|\psi_2|^4\eta_1\eta_2 \tag{17}$$

This is a completely new formula, never before met either in de Broglie's or in the standard Copenhagen approach. The fourth power of the wave function has the effect of magnifying all interference phenomena.

4. VIOLATION OF "STRONG" BELL-TYPE INEQUALITIES

With reference to Figure 2, consider two photons (1) and (2) crossing two polarizers and entering two photodetectors. The squared modulus of the probabilistic wave is reduced from unity to

$$|\psi_1|^2 = \cos^2(\lambda - a); \; |\psi_2|^2 = \cos^2(\lambda - b) \tag{18}$$

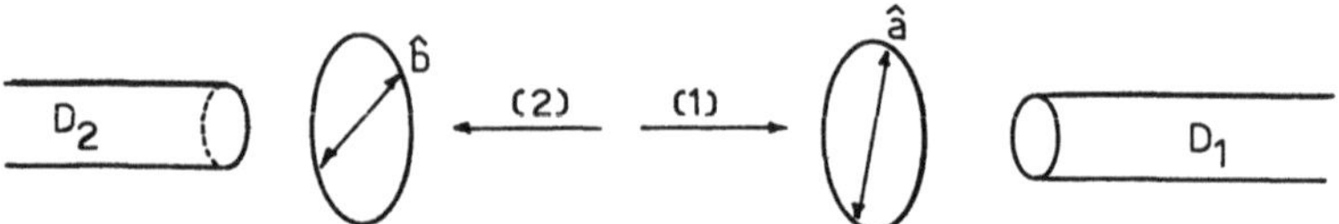

FIGURE 2. Setup of typical experiment on Bell's inequality. Photons (1) and (2) interact with one-way polarizers having axis â and b, respectively, and eventually enter detector D_1 and D_2, respectively.

if the photons had a common linear polarization λ before interacting with the polarizers. For the joint transmission and detection probability we have, in the linear approximation to $F_{1,2}(y)$:

$$P_{12}^{\lambda} = \cos^2(\lambda - a)\cos^2(\lambda - b)\cdot[2 - 2\cos^2(\lambda - a) - 2\cos^2(\lambda - b) + 4\cos^2(\lambda - a)\cos^2(\lambda - b)]\eta_1\eta_2$$

after the average over μ has been performed, and always in the approximation $\eta_1, \eta_2 \ll 1$.

By averaging over λ from 0 to π with constant density we get

$$P_{12} = [2M_{22} - 2M_{24} - 2M_{42} + 4M_{44}]\eta_1\eta_2 \tag{19}$$

where

$$M_{ij} = \langle\cos^i(\lambda - a)\cos^j(\lambda - b)\rangle_\lambda$$

It is a simple matter to show that the following results hold:

$$\begin{aligned} M_{22} &= \tfrac{1}{4}[1 + \tfrac{1}{2}\cos 2(a - b)] \\ M_{24} &= M_{42} = \tfrac{1}{8}[\tfrac{3}{2} + \cos 2(a - b)] \\ M_{44} &= \tfrac{1}{16}[\tfrac{9}{4} + 2\cos 2(a - b) + \tfrac{1}{8}\cos 4(a - b)] \end{aligned}$$

By substituting in (19) one obtains

$$P_{12} = \tfrac{1}{32}[10 + 8\cos 2(a - b) + \cos 4(a - b)]\eta_1\eta_2 \tag{20}$$

This corresponds to a fringe visibility of 73%, while Ou and Mandel[5] obtained 75%.

5. TWO-PHOTON INTERFERENCE ON A SCREEN

In an experiment performed by Ghosh and Mandel[6] the idler and signal photon from a parametric down conversion process hit the same screen (Figure 3).

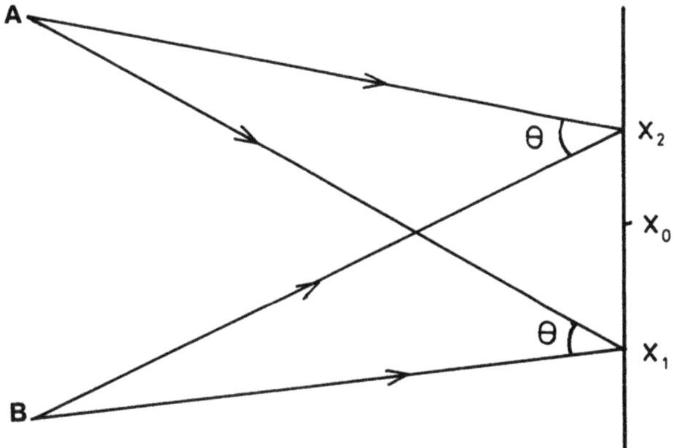

FIGURE 3. Superposition of signal and idler photons from a parametric down conversion process. The points of detection are x_1 and x_2, while x_0 is the symmetry point on the screen.

For this experiment the *total* squared waves in the points x_1 and x_2 contain an interference term between signal and idler wave and are given by

$$\begin{aligned} |\psi_1|^2 &= \tfrac{1}{2}[1 + \cos(\gamma_1 + \Delta)] \\ |\psi_2|^2 &= \tfrac{1}{2}[1 + \cos(\gamma_2 + \Delta)] \end{aligned} \tag{21}$$

where Δ is the *random* relative phase of the signal and idler waves, coming from A and B, respectively, and

$$\begin{aligned} \gamma_1 &= k \sin\theta\, \Delta x_1 = 2\pi\Delta x_1/L \\ \gamma_2 &= k \sin\theta\, \Delta x_2 = 2\pi\Delta x_2/L \end{aligned} \tag{22}$$

where $L = \lambda/\theta$ is the fringe-spacing parameter and

$$\Delta x_1 = x_1 - x_0,\ \Delta x_2 = x_2 - x_0$$

x_0 being the symmetry position on the screen.

In this experiment there are no polarizers or beam splitters and only the double-detection probability should be considered. In the approximation $\eta_1, \eta_2 \ll 1$ one has

$$\langle D_1 D_2 \rangle_\mu = [1 + F(y_1)F(y_2)]\eta_1\eta_2 \tag{23}$$

Notice that from (21) it follows

$$\begin{aligned} y_1 &= 1 - 2|\psi_1|^2 = -\cos(\gamma_1 + \Delta) \\ y_2 &= 1 - 2|\psi_2|^2 = -\cos(\gamma_2 + \Delta) \end{aligned} \tag{24}$$

and therefore the cubic approximation (12) leads to

$$\begin{aligned} F(y_1) &= -\tfrac{3}{2}\cos(\gamma_1 + \Delta) + \tfrac{1}{2}\cos^3(\gamma_1 + \Delta) \\ F(y_2) &= -\tfrac{3}{2}\cos(\gamma_2 + \Delta) + \tfrac{1}{2}\cos^3(\gamma_2 + \Delta) \end{aligned} \tag{25}$$

Since the phase Δ is arbitrary, an average of D_1D_2 must be made over Δ. One thus obtains the double detection probability

$$D_{12} = [1 + \langle F(y_1)F(y_2)\rangle_\Delta]\eta_1\eta_2 \tag{26}$$

where

$$\langle F(y_1)F(y_2)\rangle_\Delta = \tfrac{9}{4}\tilde{M}_{11} - \tfrac{3}{4}\tilde{M}_{13} - \tfrac{3}{4}\tilde{M}_{31} + \tfrac{1}{4}\tilde{M}_{33} \tag{27}$$

with

$$\tilde{M}_{ij} = \langle \cos^i(\gamma_1 + \Delta)\cos^j(\gamma_2 + \Delta)\rangle_\Delta \tag{28}$$

From the easily obtained results

$$\begin{aligned} \tilde{M}_{11} &= \tfrac{1}{2}\cos(\gamma_1 - \gamma_2) \\ \tilde{M}_{13} &= \tilde{M}_{31} = \tfrac{3}{8}\cos(\gamma_1 - \gamma_2) \\ \tilde{M}_{33} &= \tfrac{3}{16}\cos(\gamma_1 - \gamma_2) + \tfrac{1}{8}\cos^3(\gamma_1 - \gamma_2) \end{aligned}$$

one obtains

$$D_{12} = [1 + \tfrac{39}{64}\cos(\gamma_1 - \gamma_2) + \tfrac{1}{32}\cos^3(\gamma_1 - \gamma_2)]\eta_1\eta_2 \tag{29}$$

The predicted fringe-visibility thus turns out to be about 64%, in agreement with the experimental data. If the Ghosh–Mandel geometric factor $[\sin(\pi\Delta x/L)/(\pi\Delta x/L)]^2 \simeq 0.55$ is considered, the visibility is reduced to 35%, still compatible with the rather poor experimental data. From our point of view it is not surprising that the experiment has been done only once and only with three points with large errors: a near 100% visibility of the interference figure was probably impossible to obtain. Our approach could, however, lead to a large visibility, e.g., by adopting curve (c) of Figure 1.

6. TWO-PHOTON INTERFERENCE BEYOND A BEAM SPLITTER

This experiment, performed by Ou and Mandel,[7] is in some respects similar to the previous one (a two-photon interference on a screen is detected), but its essential new feature is the presence of a beam splitter that introduces extra

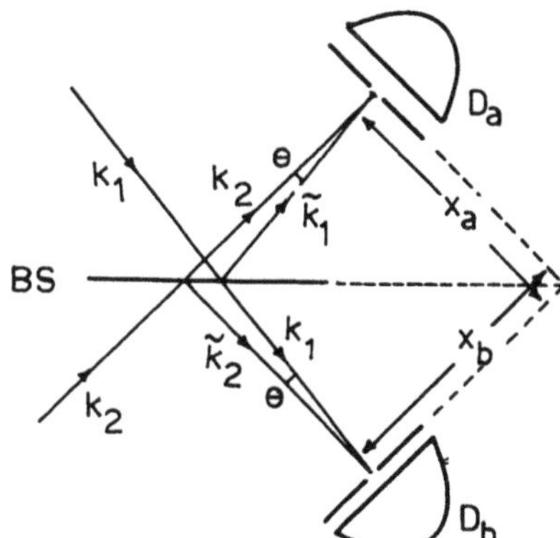

FIGURE 4. Signal and idler photons from a parametric down conversion process interact with the beam splitter BS and are eventually detected in coincidence by photodetectors D_a and D_b.

probability factors and increases the visibility of interference (Figure 4). The squared wave functions are still given by (21), but the double transmission *and* detection probability is now given by

$$D_{12}(x_1, x_2) = |\psi_1|^2 \cdot |\psi_2|^2 \cdot [1 + (1 - 2|\psi_1|^2)(1 - 2|\psi_2|^2)]\eta_1\eta_2 \tag{30}$$

after averaging over the parameter $\mu = \pm 1$ common to the two photons. Substituting (21) in (30) and averaging over Δ gives

$$D_{12}(x_1, x_2) = \tfrac{1}{32}[9 + 8\cos(\gamma_1 - \gamma_2) + 2\cos^2(\gamma_1 - \gamma_2)]\eta_1\eta_2 \tag{31}$$

which has maximum and minimum values in the ratio 19/3, corresponding to a fringe visibility of 73%, in good agreement with observations. Of course, our prediction (30) does not coincide with the quantum theoretical one. It is interesting to notice that near 100% visibility is never observed, in agreement with our predictions. This disagreement with the standard approach is usually explained away by invoking imperfect alignment of the optical apparatus, but it is remarkable that in the dozen or so published experiments a good alignment could never be achieved.

7. THE WANG–ZOU–MANDEL EXPERIMENT

After a suggestion by Croca *et al.*[(8)] for the detection of empty waves, Wang, Zou, and Mandel[(9)] performed an experiment with photon pairs produced in a parametric down conversion process (Figure 5). Their apparatus contained three beam splitters BS_1, BS_2, BS_3 that will all be considered with a 50/50 transmission/reflection probability. In a coincident detection it is only the signal photon that

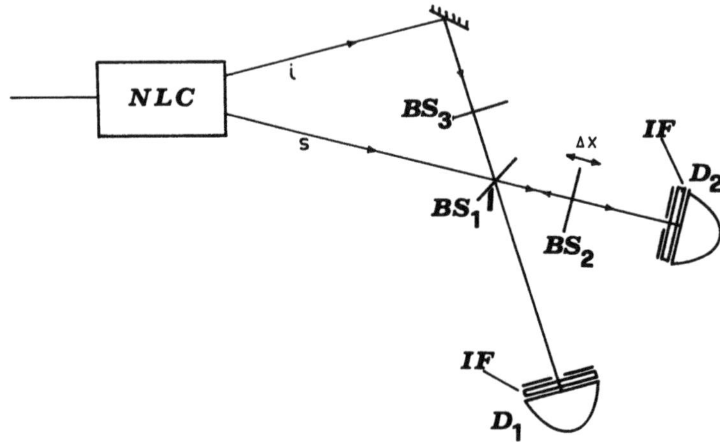

FIGURE 5. Setup of the Wang–Zou–Mandel experiment for the detection of de Broglie waves. Signal (s) and idler (i) photons coming from a nonlinear crystal (NLC) interact with beam splitters BS_1, BS_2, and BS_3 and are eventually detected in coincidence by photomultipliers D_1 and D_2.

can be detected by D_2, and then it is the idler photon that goes to D_1. The experiment was designed to detect a possible influence of the signal *empty* waves reflected by BS_3 and by BS_2 on the probability that the idler photonic particle be transmitted by BS_1 and detected.

Our treatment is very simple for the signal photon: its probability wave is reduced twice in intensity and arrives in the detector with:

$$|\psi_2|^2 = \tfrac{1}{4} \Rightarrow y_2 = 1 - 2|\psi_2|^2 = \tfrac{1}{2} \\ \Rightarrow F_2 = y_2 = \tfrac{1}{2} \tag{32}$$

For the idler photon the wave functions outgoing from BS_3 are given by

$$\psi_3 = \frac{N_3}{\sqrt{2}}\left[e^{i\Delta} - \frac{1}{\sqrt{2}}\right]$$

and (33)

$$\psi_3' = \frac{N_3}{\sqrt{2}}\left[e^{i\Delta} + \frac{1}{\sqrt{2}}\right]$$

where N_3 is a normalizing factor. By imposing the condition

$$|\psi_3|^2 + |\psi_3'|^2 = 1$$

one gets $N_3 = \sqrt{\frac{2}{3}}$. Therefore,

$$\psi_3 = \frac{1}{\sqrt{3}}\left[e^{i\Delta} - \frac{1}{\sqrt{2}}\right] \tag{34}$$

Coming to BS_1 one has

$$\begin{aligned}\psi_1 &= \frac{N_1}{2}\left[e^{i\Delta} - \frac{1}{\sqrt{2}} - \frac{1}{\sqrt{2}}e^{i\theta}\right]\\ \psi_1' &= \frac{N_1}{2}\left[e^{i\Delta} - \frac{1}{\sqrt{2}} + \frac{1}{\sqrt{2}}e^{i\theta}\right]\end{aligned} \tag{35}$$

where θ is the phase shift generated by the displacements of BS_2. The normalization condition $|\psi_1|^2 + |\psi_1'|^2 = 1$ now gives

$$N_1 = \left(1 - \frac{1}{\sqrt{2}}\cos\Delta\right)^{-1/2} \tag{36}$$

Therefore,

$$\psi_1 = \tfrac{1}{2}\left(1 - \frac{1}{\sqrt{2}}\cos\Delta\right)^{-1/2}\left(e^{i\Delta} - \frac{1}{\sqrt{2}} - \frac{1}{\sqrt{2}}e^{i\theta}\right) \tag{37}$$

Obviously, ψ_1 and ψ_1' are normalized in such a way that they give a *conditional probability*. The joint probability for crossing both BS_3 and BS_1 is

$$\begin{aligned}T_{31} &= |\psi_3|^2\cdot|\psi_1|^2\\ &= \frac{\frac{3}{2} - \sqrt{2}\cos\Delta}{12\left(1 - \frac{1}{\sqrt{2}}\cos\Delta\right)}(2 + \cos\theta - \sqrt{2}\cos\Delta - \sqrt{2}\cos(\theta - \Delta))\end{aligned} \tag{38}$$

Averaging over Δ one obtains

$$\langle T_{31}\rangle = [6 + (2 + \sqrt{2})\cos\theta]/24 \tag{39}$$

since

$$J = \frac{1}{2\pi}\int_0^{2\pi} \frac{d\Delta}{1 - \frac{1}{\sqrt{2}}\cos\Delta} = \sqrt{2} \tag{40}$$

For the detection probability one now has

$$y_1 = 1 - 2\,|\psi_1|^2 = \frac{1}{2\left(1 - \frac{1}{\sqrt{2}}\cos\Delta\right)}(\sqrt{2}\cos(\theta - \Delta) - \cos\theta) \tag{41}$$

Therefore, one can obtain

$$\begin{aligned} |\psi_1|^2 y_1 = {} & \frac{1}{4B^2}[4B^2(1 - \cos\theta - 2\cos^2\theta) \\ & + 2B(-4 + \cos\theta + 6\cos^2\theta) + 2 - 3\cos^2\theta] \end{aligned} \tag{42}$$

where

$$B = 1 - \frac{1}{\sqrt{2}}\cos\Delta \tag{43}$$

One also has

$$|\psi_3|^2 = \tfrac{1}{3}(2B - \tfrac{1}{2}) \tag{44}$$

Therefore, the probability that the idler photon crosses BS_3, crosses BS_1, and is detected requires the product of three factors to be calculated:

$$\begin{aligned} |\psi_1\psi_3|^2 y_1 = {} & \frac{1}{12B^2}[8B^3(1 - \cos\theta - 2\cos^2\theta) + 4B^2(-\tfrac{9}{2} + \tfrac{3}{2}\cos\theta + 7\cos^2\theta) \\ & + 2B(4 - \tfrac{1}{2}\cos\theta - 6\cos^2\theta) - \tfrac{1}{2}(2 - 3\cos^2\theta)] \end{aligned} \tag{45}$$

by averaging over Δ, since

$$\langle B\rangle = 1;\ \langle B^{-1}\rangle = \sqrt{2};\ \langle B^{-2}\rangle = 2\sqrt{2} \tag{46}$$

[the second result of course coincides with (40)] one gets after a short calculation:

$$\langle|\psi_1\psi_3|^2 y_1\rangle = \frac{3\sqrt{2} - 5}{6} - \frac{2 + \sqrt{2}}{12}\cos\theta + \frac{4 - 3\sqrt{2}}{4}\cos^2\theta \tag{47}$$

Our final result for the D_1D_2 coincidence probability is then

$$\begin{aligned} & \langle|\psi_1\psi_3|^2(1 + y_1F_2)\rangle = \\ & \tfrac{6}{24} + \cancel{\frac{2 + \sqrt{2}}{24}\cos\theta} + \frac{3\sqrt{2} - 5}{12} - \cancel{\frac{2 + \sqrt{2}}{24}\cos\theta} + \frac{4 - 3\sqrt{2}}{8}\cos^2\theta \end{aligned} \tag{48}$$

Notice the cancellation of the terms linear in $\cos\theta$. There remains

$$\langle|\psi_1\psi_3|^2(1 + y_1F_2)\rangle = \frac{6\sqrt{2} - 4}{24} + \tfrac{1}{2}\left(1 - \frac{3\sqrt{2}}{4}\right)\cos^2\theta \qquad (49)$$

which corresponds to a tolerable fringe visibility of about 8% with twice faster oscillations.

REFERENCES

1. S. J. FREEDMAN and J. F. CLAUSER, *Phys. Rev. Lett.* **28**, 938 (1972); E. S. FRY and R. C. THOMPSON, *Phys. Rev. Lett.* **37**, 465 (1976); A. ASPECT, P. GRANGIER, and G. ROGER, *Phys. Rev. Lett.* **47**, 460 (1981) and **49**, 91 (1982); A. ASPECT, P. DALIBARD, and G. ROGER, *Phys. Rev. Lett.* **49**, 1804 (1982); W. PERRIE, A. J. DUNCAN, H. J. BEYER, and H. KLEINPOPPEN, *Phys. Rev. Lett.* **54**, 1790 (1985).
2. V. L. LEPORE and F. SELLERI, *Found. Phys. Lett.* **3**, 203 (1990).
3. J. F. CLAUSER, M. A. HORNE, A. SHIMONY, and R. A. HOLT, *Phys. Rev. Lett.* **23**, 880 (1969).
4. A. GARUCCIO, V. L. LEPORE, and F. SELLERI, *Found. Phys.* **20**, 1173 (1990).
5. Z. Y. OU and L. MANDEL, *Phys. Rev. Lett.* **61**, 50 (1988).
6. R. GHOSH and L. MANDEL, *Phys. Rev. Lett.* **59**, 1903 (1987).
7. Z. Y. OU and L. MANDEL, *Phys. Rev. Lett.* **62**, 2941 (1989).
8. J. R. CROCA, A. GARUCCIO, V. L. LEPORE, and R. N. MOREIRA, *Found. Phys. Lett.* **3**, 557 (1990).
9. L. J. WANG, X. Y. ZOU, and L. MANDEL, *Phys. Rev. Lett.* **66**, 1111 (1991).

CHAPTER 17

Experiments on the Aharonov–Bohm Effect

Akira Tonomura

1. INTRODUCTION

The Aharonov–Bohm (AB) effect[1] states that electrons can be physically influenced by the magnetic field which the electrons do not touch. This effect is purely quantum-mechanical, since electrons pass through only field-free regions and therefore no force is exerted on them. Although the AB effect had already been tested experimentally[2] just after the prediction, some people questioned[3] even the existence of the AB effect from both theoretical and experimental standpoints. Since the controversy seemed to go on indefinitely, it was apparent that a firm experimental foundation should be established and the questioned ambiguities eliminated by testing whether the phase shift really existed under conditions of no overlap between electrons and magnetic fields. Therefore, a series of new confirmation experiments of the AB effect have been done since 1982.[4] The conclusion was arrived at in 1986[5] that the predicted phase shift was detected even under ideal conditions. Such an experiment could only have been made possible by the recent development of advanced technologies such as electron holography[6] and microlithography. In this chapter, the confirmation experiments of the AB effect are described, especially in relation to wave–particle duality.

2. THE AHARONOV–BOHM EFFECT

The interaction of an electron wave with magnetic fields is described by the Schrödinger equation, in which electromagnetic potentials, rather than fields, are

Akira Tonomura • Advanced Research Laboratory, Hitachi, Ltd., Hatoyama, Saitama 350-03, Japan.

Wave–Particle Duality, edited by Franco Selleri. Plenum Press, New York, 1992.

used. In 1959, Aharonov and Bohm actually solved the Schrödinger equation for the arrangement shown in Figure 1, producing a striking effect. Even if two electron waves travel in field-free regions on both sides of an infinite solenoid, the electron waves are physically influenced to produce a relative phase shift. Aharonov and Bohm attributed this effect to the vector potential surrounding the solenoid, in which the circulation integral does not vanish but is equal to the magnetic flux inside the solenoid. This effect was so far from conventional ways of thinking that various assertions have been made concerning its physical implications. The significance of this effect increased in the late 1970s:[7] The theory of gauge fields was revived as a most probable candidate for the unified theory of all fundamental interactions in nature. In this theory, vector potentials are extended to gauge fields and regarded as a fundamental physical quantity. The AB effect demonstrates the physical reality of gauge fields.

However, in 1978 the existence of this effect was negated by Bocchieri and Loinger.[3] They asserted that previous experimental results were due to the fringing magnetic fields from finite solenoids or whiskers used in these experiments.

On the other hand, Liebowitz[8] presented a classical interpretation of the AB effect: an incident electron passing in a field-free region near the solenoid receives a force producing the AB effect. This is because the field energy inside the solenoid changes due to the overlap of the solenoid's and the electron's magnetic fields, according to his assertion.

In order to clarify these discussions, two experiments were carried out. One demonstrated that an electron interference pattern is formed by the accumulation of just a single-electron interference at one time in the apparatus. The other decisively confirmed the existence of the AB phase shift.

3. FORMATION PROCESS OF ELECTRON INTERFERENCE PATTERN—DEMONSTRATION OF WAVE–PARTICLE DUALITY

An electron interference pattern is always formed in such a way that a single electron exists at one time in the apparatus. Such an experiment is referred to as "impossible, absolutely impossible to explain in any classical way, and has in it the heart of quantum mechanics." Feynman[9] continues, "This experiment has never been done in just this way, since the apparatus would have to be made on an impossibly small scale."

However, such an experiment has now become feasible with the progress of technologies. Electron microscopy allows small-scale magnification, and individual electrons can be observed with a photon-counting detector modified to count electrons. The experiments were actually carried out in a field-emission electron microscope equipped with both an electron biprism[10] and a two-dimensional position-sensitive electron counting system[11] as shown in Figure 2. Electrons

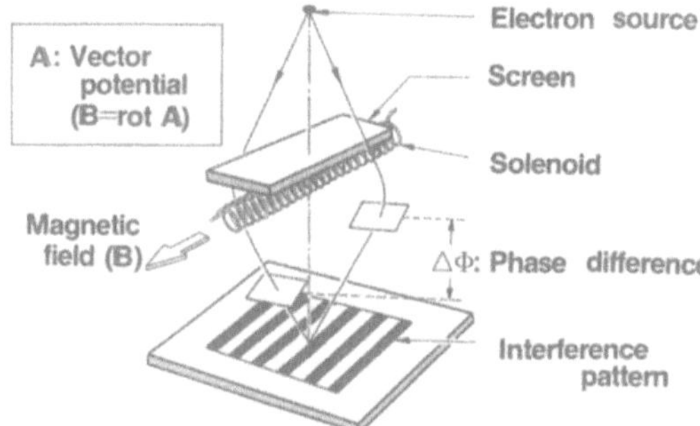

FIGURE 1. Aharonov–Bohm effect.

emitted from a field-emission tip were sent to the biprism. The interference pattern was enlarged by magnifying lenses and recorded by the electron counting system. Electrons could be detected one by one on the TV monitor. When the number of electrons is small, as in Figure 3a, electron distribution seems quite random. However, when the number increases, the interference fringe pattern formed when two electron waves pass through both sides of the biprism is recognizable (Figure 3c–e). Even when the electron arrival rate was as low as 10 electrons/s in the entire field of view, so that at most only a single electron existed at a time, the accumulation of single electrons formed the interference pattern shown in Figure 3e, which occurs when two electron waves pass through both sides of the biprism.

In quantum-mechanical terms, even a single electron can split into two on

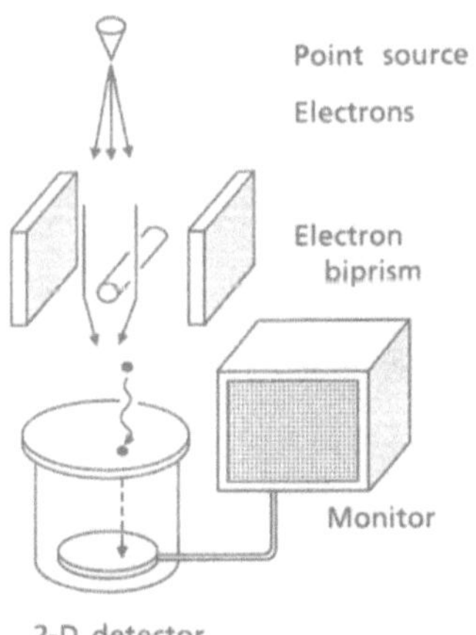

FIGURE 2. Experimental arrangement for single-electron buildup of interference pattern.

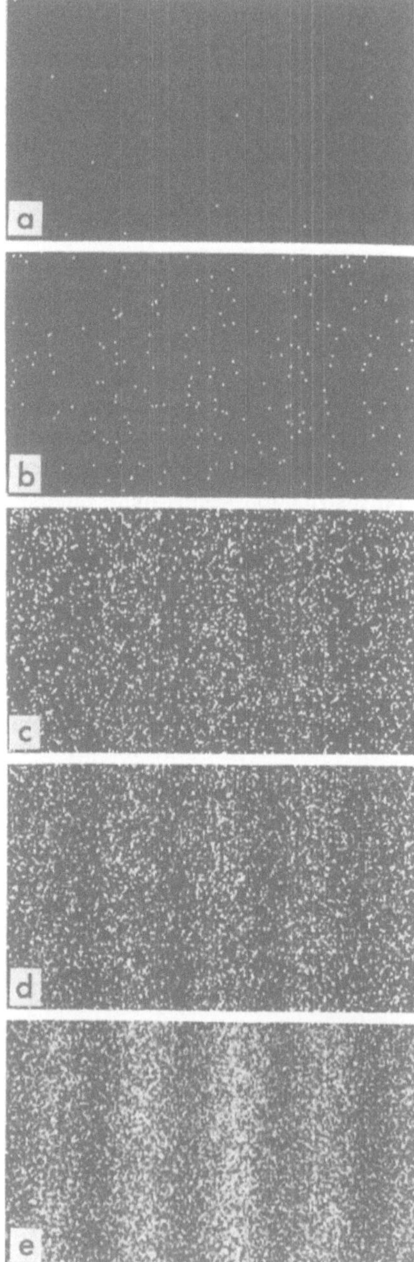

FIGURE 3. Buildup process of electron interference pattern. Number of electrons: (a) 10; (b) 100; (c) 3000; (d) 20,000; (e) 70,000.

both sides of the biprism in the form of a wave function (see Figure 4a). Two partial electrons overlap to interfere on the observation plane and form a probability interference pattern. When detected, the two overlapped partial electrons can be observed as a single electron, never as two. It is interpreted that the measurement makes the extended wave function collapse into a single point instantly.

The mystery of this collapse increases when the experiment shown in Figure 4b is carried out. In this case, two partial electrons having passed through both sides of the biprism filament are thoroughly separated by applying a positive potential to the biprism filament instead of a negative one as in Figure 4a. When measured, a single electron is detected on either side of the biprism filament. Since two partial wave functions must exist on both sides of the filament, the partial wave function on one side of the filament must have jumped to the other side to collapse. This is in spite of the fact that the two partical waves have been completely separated and are traveling in opposite directions. It should be noted here that the coherence length of the electron wave packet in the traveling direction is only 1 μm.

This experiment was designed especially for a demonstration, and therefore the electron frequency was much lower than that of the hologram formation. However, the single-electron accumulation remains the same for any electron interference experiment. The following AB effect experiment was carried out under these single-electron conditions.

4. CONFIRMATION EXPERIMENTS ON THE AB EFFECT

Among a series of experiments, the last considered to be the most conclusive is introduced here. Instead of a straight solenoid, a toroidal ferromagnet was used. An infinite solenoid is experimentally unattainable, but an ideal geometry can be achieved by the finite system of a toroidal magnetic field.[12] Furthermore, the toroidal ferromagnet was covered with a superconducting niobium layer to completely confine the magnetic field.

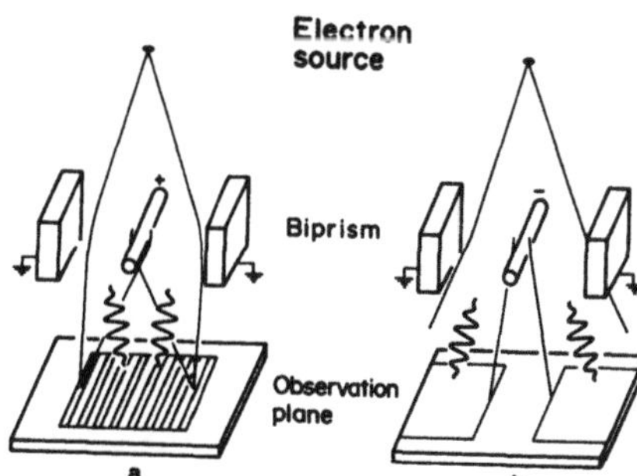

FIGURE 4. Wave packet in the electron biprism. (a) Application of a positive potential to the central filament; (b) application of a negative potential to the central filament.

A sample with such a complicated structure should be smaller than the transverse coherence length of an electron beam, say 10 μm, and therefore, fabricated with the most advanced photolithography techniques. The resultant sample is shown in Figure 5. First, an isolated toroidal sample was fabricated. However, the sample temperature could not be lowered enough, to below the critical temperature T_c (= 9.2 K), to become superconductive. Therefore, the toroidal sample was connected to a niobium plate by a tiny bridge for high thermal conductivity, as seen in the scanning electron micrograph (Figure 5).

The measurement of the relative phase shift was made between two electron waves passing through the inside of the hole and outside the toroid at 5 K. Although measurements were made for many toroidal samples with different magnetic flux values, only two phase shifts were observed, as shown in Figure 6. The phase shift value is either 0 or π. The conclusion is obvious: a relative phase shift of π is produced even when the magnetic field is confined within the superconductor and is shielded from the electron beam. This proves that the AB effect exists.

But why does ΔS take either 0 or π? This quantization of the phase shift provides key evidence for the complete shielding of a magnetic field with the covering superconductor, and is therefore explained in more detail. The mechanism of the phase shift quantization was clarified by the following interference experiment. When the temperature T of the toroidal sample was raised, the

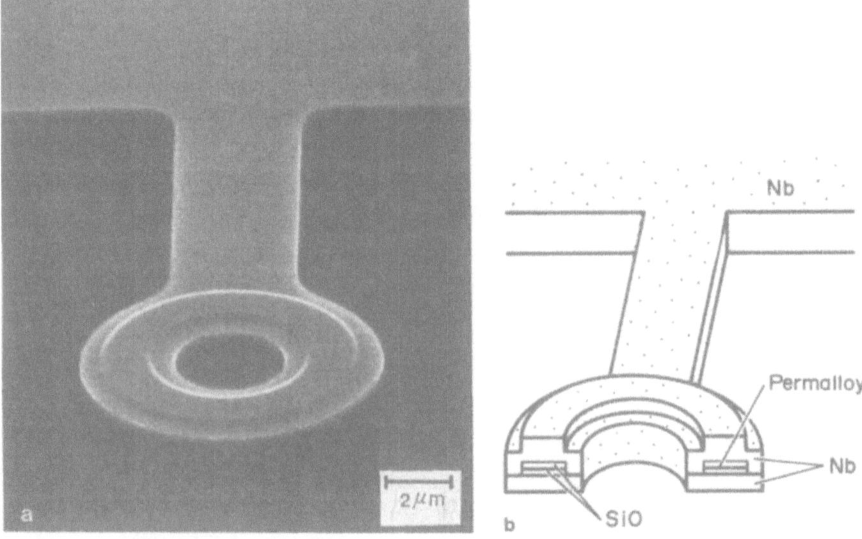

FIGURE 5. Toroidal sample for AB effect experiment. (a) Scanning electron micrograph; (b) schematic.

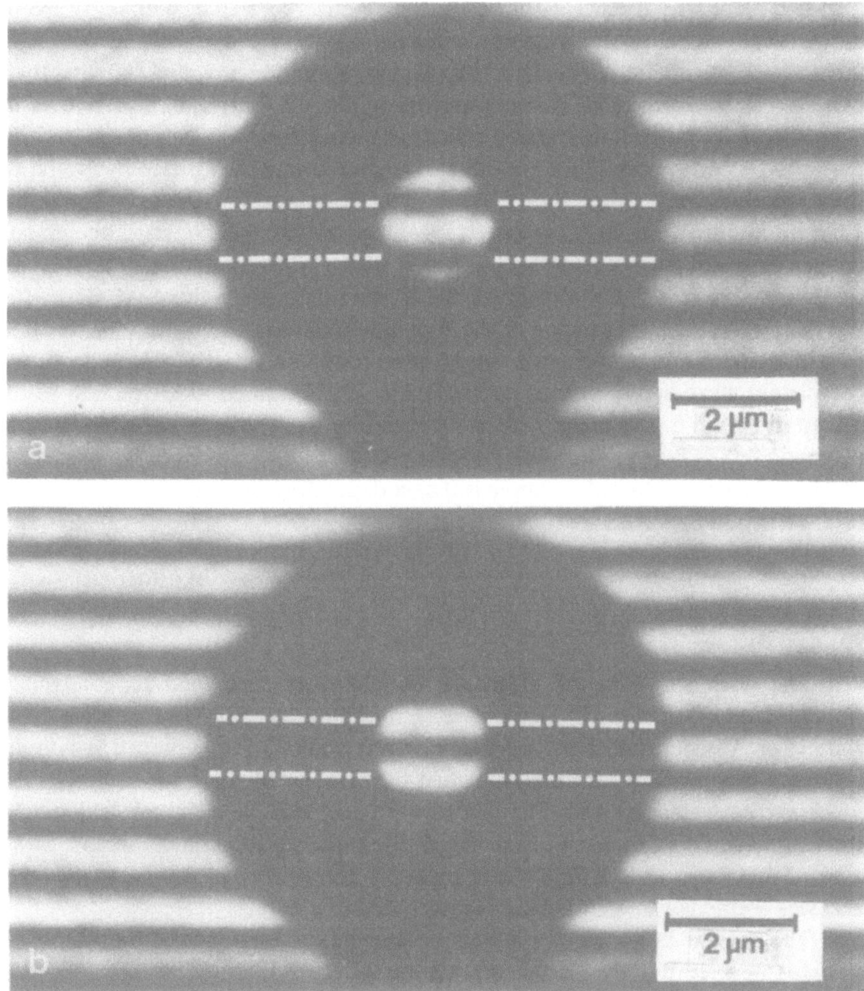

FIGURE 6. Interference patterns of toroidal ferromagnets. (a) Phase shift = 0; (b) phase shift = π.

interferogram changed abruptly when T crossed the superconducting critical temperature T_c: The phase shift value remained either 0 or π, independently of T, as long as a magnetic flux was covered with a superconductor. However, when $T > T_c$, the phase shift value is considered to be determined only by the enclosed magnetic flux Φ and given by $e\Phi/h$. This was also supported by the experimental fact that the phase shift value decreased in proportion to a decrease in magnetization in the Permalloy film, when the sample temperature was further raised to room temperature.

It is clear why the phase shift must have a discrete value, 0 or π. When the temperature of a toroidal sample decreases below T_c, a phenomenon fundamental to superconductivity called "flux quantization" occurs, and can directly be observed in the form of an electron interferogram. At that time, electrons in the superconductor form Cooper pairs, which, in phase, form a coherent wave. When the coherent wave turns once around the magnetic flux, the phase has to be the same, modulo 2π, as the original value. This condition is equivalent to the flux quantization in $h/2e$ units. Conversely speaking, the flux quantization occurs only when a magnetic flux is completely covered with a superconductor. If there is any weak link, the flux is not quantized due to the phase jump there.

Therefore, the occurrence of the flux quantization can assure in the present case that the Nb layer becomes superconducting, and that a magnetic flux is completely surrounded by a superconductor. The obtained experimental results shown in Figure 6 give clear evidence supporting the above explanation. Since a magnetic flux of $h/2e$ produces an electron phase shift of π, the relative phase shift inside and outside a toroidal sample is 0 or π depending on an even or odd number of the trapped flux quanta.

5. CONCLUSIONS

The existence of the AB effect was conclusively confirmed using advanced technologies of electron holography and microlithography. The relative phase shift of π was detected between two electron waves having passed inside and outside a tiny toroidal ferromagnet when the number of the trapped flux quanta inside the covering was odd. The fact that the magnetic field did not leak outside the sample was confirmed by observing the flux quantization phenomenon, which assured that the magnetic field was completely surrounded by the superconductor and was forbidden to leak outside by the Meissner effect.

The present experiments not only confirmed the existence of the AB effect, but also negated the classical interpretation of the AB effect in terms of force concepts (see Ref. 13). The AB effect was confirmed under conditions where there was no chance for two electrons to exist in the experimental apparatus at one time.

REFERENCES

1. Y. AHARONOV and D. BOHM, *Phys. Rev.* **115**, 485–491 (1959); M. PESHKIN and A. TONOMURA, *The Aharonov–Bohm Effect*, Springer-Verlag, Berlin (1989).
2. R. G. CHAMBERS, *Phys. Rev. Lett.* **5**, 3–5 (1966); H. A. FOWLER, L. MARTON, J. A. SIMPSON, and J. A. SUDDETH, *J. APPL. PHYS.* **32**, 1153–1155 (1961); H. BOERSCH, H. HAMISCH, K. GROHMAN, and D. WOHLLEBEN, *Z. Phys.* **165**, 79–93 (1961); G. MÖLLENSTEDT and W. BAYH, *PHYS. Bl.* **18**, 299–305 (1962).
3. P. BOCCHIERI and A. LOINGER, *Nuovo Cimento A* **47**, 475–482 (1978).

4. A. TONOMURA *et al.*, *Phys. Rev. Lett.* **48**, 1443–1446 (1982).
5. A. TONOMURA *et al.*, *Phys. Rev. Lett.* **56**, 792–795 (1986).
6. A. TONOMURA, *Rev. Mod. Phys.* **59**, 637–669 (1987).
7. T. T. WU and C. N. YANG, *Phys. Rev. D* **12**, 3845–3857 (1975).
8. B. LIEBOWITZ, *Nuovo Cimento* **38**, 932–950 (1965).
9. R. P. FEYNMAN, R. B. LEIGHTON, and M. SANDS, in: *The Feynman Lectures on Physics*, Vol. III, pp. 1.1–1.5, Addison–Wesley, Reading, Mass. (1965).
10. G. MÖLLENSTEDT and H. DÜKER, *Z. Phys.* **145**, 377–397 (1956).
11. Y. TSUCHIYA, E. INUZUKA, T. KURONO, and M. HOSODA, *Adv. Electron. Electron Phys.* **64A**, 21–31 (1982).
12. C. G. KUPER, *Phys. Lett.* **A79**, 413–416 (1980).
13. H. ERLICHSON, *Am. J. Phys.* **38**, 162–173 (1970).

INDEX

GPSR Compliance
The European Union's (EU) General Product Safety Regulation (GPSR) is a set of rules that requires consumer products to be safe and our obligations to ensure this.

If you have any concerns about our products, you can contact us on

ProductSafety@springernature.com

In case Publisher is established outside the EU, the EU authorized representative is:

Springer Nature Customer Service Center GmbH
Europaplatz 3
69115 Heidelberg, Germany

www.ingramcontent.com/pod-product-compliance
Ingram Content Group UK Ltd.
Pitfield, Milton Keynes, MK11 3LW, UK
UKHW042019290726
14061UKWH00001BB/73
* 9 7 8 0 3 0 6 4 4 1 6 3 9 *